PROJECT MANAGEMENT

FOR
CONSTRUCTION

David L. Goetsch

PEARSON

Boston   Columbus   Indianapolis   New York   San Francisco   Upper Saddle River
Amsterdam   Cape Town   Dubai   London   Madrid   Milan   Munich   Paris   Montreal   Toronto
Delhi   Mexico City   São Paulo   Sydney   Hong Kong   Seoul   Singapore   Taipei   Tokyo

Editorial Director: Vernon R. Anthony
Senior Acquisitions Editor: Lindsey Gill
Editorial Assistant: Nancy Kesterson
Director of Marketing: David Gesell
Senior Marketing Coordinator: Alicia Wozniak
Marketing Assistant: Les Roberts
Program Manager: Maren L. Beckman
Project Manager: Janet Portisch
Procurement Specialist: Deidra M. Skahill
Art Director: Jayne Conte
Cover Designer: Karen Noferi
Cover Image: Shutterstock
AV Manager, Rights and Permissions: Mike Lackey
Media Director: Leslie Brado
Lead Media Project Manager: April Cleland
Full-Service Project Management: Jogender Taneja, Aptara®, Inc.
Composition: Aptara®, Inc.
Printer/Binder: LSC Communications
Cover Printer: LSC Communications
Text Font: 10/12 ITC Garamond Std

Credits and acknowledgments borrowed from other sources and reproduced, with permission, in this textbook appear on the appropriate page within text.

Library of Congress Cataloging-in-Publication Data is available from the Publisher upon request

6 17

ISBN 10:   0-13-280324-0
ISBN 13: 978-0-13-280324-3

BRIEF CONTENTS

CONTENTS

PREFACE

BACKGROUND

In the fields of construction management, construction technology, construction engineering, civil engineering, and architecture, as well as in other construction-related fields, efficient, effective project management is critical. All construction projects—residential, commercial, industrial, and infrastructure—share common goals: The projects are to be completed on time, within budget, and according to specifications. These goals cannot be achieved without effective project management. Construction projects range from the smallest residential dwelling to the largest skyscraper to roads and bridges, and projects must be well managed if they are to be completed successfully.

This unrelenting demand to complete construction projects on time, within budget, and according to specifications has created a pressing need for specialized education and training for those who manage construction projects. Project managers in construction must know how to manage processes and lead people. The process aspects of project management include: cost estimation, planning/scheduling, procurement, risk management, construction monitoring, and closeout. The people aspects of project management include: leadership, motivation, communication, and efficient/effective management of time, change, diversity, and adversity. Project management has become a specialized field within the broad field of construction, a specialized field requiring specialized instructions in both the process and people aspects of the job.

WHY IS THIS BOOK WRITTEN AND FOR WHOM?

This book is written to fulfill the need for a comprehensive, up-to-date, practical teaching resource that focuses on helping construction and construction-related students become effective project managers. This book is developed in accordance with the specifications contained in *A Guide to the Project Management Body of Knowledge (PMBOK Guide)* maintained by the Project Management Institute (PMI), Pennsylvania. It provides comprehensive coverage of both aspects of project management—process management and leading people—specifically from the perspective of construction projects. Educators and students in such disciplines as construction management, construction technology, construction engineering, civil engineering, architecture, and other construction-related fields will benefit from the material presented herein. The direct, straightforward presentation of material focuses on making the principles of project management practical, understandable, and useful for students. Up-to-date research has been integrated throughout the text along with real-world activities and cases.

ORGANIZATION OF THE BOOK

The text contains 15 chapters organized in two parts. Part One covers all of the process skills needed by construction project managers. Part Two covers all of the people skills needed by construction project managers. The chapters are presented in an order that is compatible with the typical organization of a course in construction project management, and a standard chapter format is maintained throughout the book. In addition to text, photos, and illustrations, each chapter contains a list of chapter topics, summary, key terms and concepts, review questions, and practical application activities. Every other chapter contains a case study of a well-known major construction project that illustrates for students how complex a construction project can be and why effective project management is so critical.

DOWNLOAD INSTRUCTOR RESOURCES FROM THE INSTRUCTOR RESOURCE CENTER

Supplementary teaching and learning materials are provided online. These materials include a PowerPoint presentation covering all chapters in the book, a comprehensive test bank, and an Instructor's Manual. To access supplementary materials online, instructors need to request an instructor access code. Go to www.pearsonhighered.com/irc to register for an instructor access code. Within 48 hours of registering, you will receive a confirming e-mail including an instructor access code. Once you have received your code, locate your text in the online catalog and click on the Instructor Resources button on the left side of the catalog product page. Select a supplement, and a login page will appear. Once you have logged in, you can access instructor material for all Pearson textbooks. If you have any difficulties accessing the site or downloading a supplement, please contact Customer Service at http://247pearsoned.custhelp.com/.

HOW THIS BOOK DIFFERS FROM OTHERS

The approach taken in this book is the result of more than 100 interviews with construction project managers, construction students, and construction professors. Through these interviews the author learned that most textbooks on project management take a generic approach in an attempt to reach the broadest possible market. Consequently, this text focuses solely on construction project management so that all text, illustrations, cases, and activities can be specific to construction and so that concepts can be treated in greater depth than is possible in a generic text.

ABOUT THE AUTHOR

David L. Goetsch is Emeritus Vice President and Professor at Northwest Florida State College. Prior to entering higher education full time, Dr. Goetsch had a career in the private sector that included project management positions in construction, engineering, and manufacturing settings. He served as a project manager in a prestressed concrete firm that designed, engineered, and constructed residential, commercial, industrial, and infrastructure projects ranging from apartment buildings to condominiums to football stadiums to shopping malls to manufacturing facilities to overpasses and bridges. Dr. Goetsch has been selected as Professor of the Year at Northwest Florida State College and the University of West Florida, Florida's Outstanding Technical Instructor of the Year, and was also the recipient of the U.S. Secretary of Education Award for having the Outstanding Technical Program in the United States in 1984 (Region 10).

REVIEWERS

Baabak Ashuri
Georgia Institute of Technology

Suchismita Bhattacharjee
Ball State University

Casey Cline
Boise State University

Denise Gravitt
Western Illinois University

Theodore C. Haupt
Mississippi State University

Francois Jacobs, Ph.D.
California Baptist University

Edward Keeter
Philadelphia University

Daryl L. Orth
Northern Kentucky University

CHAPTER 1

Overview of Construction Project Management

Much of the work of construction firms consists of projects. A single construction firm, depending on its size and core competencies, might be engaged in completing several construction projects at the same time. While working for a prestressed concrete firm, the author once managed three projects at one time: a shopping mall, a bottling plant, and a football stadium. During this period, our firm had more than 20 major projects underway concurrently. At the other end of the spectrum are small firms that take on only one or two projects at a time.

Regardless of the size and complexity of the individual construction firm, the structures they contract to build—from small homes to industrial plants to roads to bridges—are projects that must be planned, budgeted, scheduled, managed, tracked, and completed. Assume that a construction firm wins a contract to renovate and remodel a home for a family that needs more space. The renovation and remodeling would become a project that would have to be planned, budgeted, scheduled, managed, tracked, and completed within a specified timeframe. Assume that a construction firm wins a contract to build a new branch campus for a university. The branch campus would become a project that would have to be planned, budgeted, scheduled, managed, tracked, and completed within a specified timeframe. The same would be true of a construction firm that won a contract to build a new highway including bridges and on/off ramps. Regardless of the size and complexity of the project in question, the basic principles of project management apply to all construction projects.

Construction projects are completed by project teams. Project teams are led by project managers who are responsible for ensuring that their projects are completed on time, within budget, and according to specifications. Consequently, construction students and professionals should be prepared to manage projects and lead project teams. Becoming an effective project manager requires the development of specific process and people skills. Preparing construction students and professionals to be effective project managers is the purpose of this book.

CONSTRUCTION INDUSTRY SECTORS

Construction is a broad term that encompasses several more specialized sectors. As construction has become more complex and more regulated, it has been necessary for construction firms to specialize in a given sector. The specialized sectors for the construction industry are as follows:

- Residential
- Commercial
- Industrial
- Infrastructure

Construction firms that specialize in the residential sector build single-family homes, apartment complexes, and condominiums (see Figures 1.1 and 1.2). More than half of all construction projects in the United States each year are residential projects. Construction firms in the commercial sector build such things as office buildings, retail outlets, restaurants, hotels, motels, colleges, universities, schools, sports arenas, stadiums, hospitals, physician's offices, and health clinics (see Figures 1.3 and 1.4). Construction firms in the industrial sector build such things as manufacturing plants, processing plants, refineries, and mills. Construction firms in the infrastructure sector build roads, bridges, dams, tunnels, sewer systems, and canals (see Figures 1.5 and 1.6).

FIGURE 1.1 2011 House remodeling.

FIGURE 1.2   2011 House remodeling.

FIGURE 1.3   Commercial office building.

FIGURE 1.4 Sports arena.

FIGURE 1.5 Road construction.

FIGURE 1.6 Constructing an overpass ramp.

All of these types of projects must be planned, budgeted, scheduled, managed, tracked, and completed within a specified timeframe. When a construction firm in any of these sectors undertakes a project, its challenge is to complete the project on time, within budget, and according to specifications. This is the principal challenge of project managers in construction. It is a challenge that requires the development of all of the process and people skills explained in this book.

PARTICIPANTS IN CONSTRUCTION PROJECTS

The participants in construction projects make up a diverse group. They include owners and their representatives, architects, engineers, construction professionals (contractors and construction managers), subcontractors, tradespeople, and suppliers (see Figure 1.7). People in all of these categories play a role in construction projects. Consequently, construction project managers must be prepared to interact, collaborate, cooperate, coordinate, and negotiate with people in all of these fields. The roles of the various participants in construction projects can be summarized as follows:

- ***Owners and their representatives.*** An owner is the entity that gets a construction project started. The owner—whether a private individual, private company, public

---

CHECKLIST OF
PARTICIPANTS OF CONSTRUCTION PROJECTS

✓ Owners and their representatives

✓ Architects

✓ Engineers

✓ Construction professionals

✓ Trades People

✓ Suppliers

FIGURE 1.7 Project managers in construction work with diverse teams.

agency, or a group of private individuals—is the entity that has a need for something to be constructed and the money to pay for it (although the money is typically borrowed). The fact that most owners borrow the money to pay for their construction projects does not change the role they play as a participant. The owner is the entity who is ultimately responsible for paying for the project. In some cases, construction firms will work with the owner. In other cases, they will work with someone who has been designated to represent the owner. It is not uncommon for the project's architect to serve as the owner's representative on a construction project.

• *Architects.* Often the first participant hired by the owner is an architect. It is the architect's job to translate the owner's ideas, needs, and requirements into a coherent package that includes architectural drawings and general specifications. The architect's drawings and specifications are, in turn, used by various types of engineers as the basis for developing their plans for the project. The architect often serves as the owner's representative in dealing with the construction firm for the duration of the project. Architectural firms provide more than just the standard design services. They also provide interior design, landscape design, and specification development.

• *Engineers.* Engineers play a critical role in construction projects. Mechanical engineers design and plan the heating, cooling, and ventilation systems for projects. Civil engineers design and plan the site preparation, sewer and water systems, and structural components (e.g., foundation, columns, and beams). Electrical engineers design and plan electrical switchgear, lighting, and electrical conduit systems. Communication and coordination between and among the various types of engineers required on a given construction project, the architect, and the construction firm is essential to the successful completion of a construction project.

• *Construction professionals.* At the center of any construction project are the construction professionals; sometimes referred to as the contractors and subcontractors. Construction professionals are the nucleus of the construction project team. They are responsible for getting construction projects completed on time, within budget, and according to specifications. The construction project manager is a construction professional whose role is much like that of an orchestra conductor. The project manager coordinates the work of all of the other participants to ensure that they operate in a well-coordinated, mutually supportive, systematic way. Without effective project management, a construction project would be like an orchestra in which the various players play what they want when they want to. In such a case, rather than beautiful music the result would be chaos. The same is true of construction projects. In addition to construction project managers, construction firms might also employ construction professionals as estimators, schedulers, and purchasing agents.

- *Tradespeople.* Tradespeople are highly skilled specialists in such fields as electricity, masonry, equipment operation, concrete, flooring installation, elevator installation, insulation installation, roofing, steel erection, plumbing, carpentry, welding, pipe fitting, dry wall hanging, and painting. As construction has become increasingly complex in response to ongoing technological developments, the list of trades has been expanded to include telecommunications, security, and fire protection specialists.
- *Suppliers.* Suppliers play a key role in construction projects. Construction firms must have dependable suppliers that provide the materials and equipment they need to complete construction projects. In dealing with suppliers, cost is an important factor, but dependability is even more important. In addition to providing materials, suppliers are often called on to help architects and engineers decide what materials to specify for a given project. Equipment that is commonly purchased, rented, or leased for construction projects includes cranes, aerial lifts, generators, earthmovers, trenching machines, crushers, cold planers, asphalt pavers, compressors, pumps, and boom trucks.

CONSTRUCTION PROJECT DEFINED

Groups of college students are sometimes required to work together on class projects. When this happens, the group usually meets to: (1) select a group leader to coordinate the activities of individual members and to ensure that all of the work is completed properly and on time and (2) divide the work and assign it to different group members. In this example, the assignment from the professor is a project and the person selected to lead the group is the project manager.

The class project consists of multiple individual assignments that require people, resources, and processes to complete them and that must be carefully coordinated. There is a definite start and end date for the project and specific grading criteria (success criteria). For example, assume that a professor divides his class into five groups and gives each group the assignment of making a presentation to the class on a construction-related topic. The students in each group are to use such resources as computers, books, the Internet, paper, and so on. In developing their presentation, the students apply such processes as research, word processing, graphic imaging, and public speaking. There is a definite date on which the professor made the assignment and a definite date on which each presentation must be made. During the presentation, the professor applies specific success criteria in arriving at a grade.

All projects—whether in college classes or on construction sites—have these same characteristics. Hence, a construction project—regardless of whether it is residential, commercial, industrial, or infrastructure—can be defined as follows:

A construction project is a fully coordinated group of interdependent tasks that are completed by people using resources and processes. Construction projects have definite starting and ending dates, budgets, expectations and specifications (success criteria).

**CHECKLIST OF IMPORTANT
CONSTRUCTION PROJECT CHARACTERISTICS**

✓ Fully coordinated interdependent tasks

✓ People

✓ Processes

✓ Resources

✓ Starting date

✓ Ending date

✓ Success criteria

FIGURE 1.8 All construction projects have these characteristics.

For example, a construction project to build an office complex will consist of several interdependent tasks—site work, footings and foundation, framing, electricity, plumbing, HVAC, masonry, and so on—all of which must be fully coordinated. People performing these interdependent tasks are to use a variety of resources—tools and materials—and processes to complete them. The project will have a starting and ending date, a budget, and definite expectations as set forth in the architectural/engineering drawings and the accompanying specifications.

There are several important concepts in the definition of a project that was earlier presented including the following: fully coordinated interdependent tasks, people, resources, processes, starting date, ending date, and success criteria (see Figure 1.8). Construction project managers should understand all of these concepts and their significance.

Fully Coordinated Interdependent Tasks

Assume that several friends are driving in a car and have a flat tire. The group is in a hurry so it is important to get the tire changed as quickly as possible. To expedite the process, each individual in the group agrees to complete a different task. One individual might get the jack out of the trunk while another retrieves the spare tire. One might loosen the lug nuts while another stands by to take the flat tire off and put the spare tire on. The individual who takes the lug nuts off will stand by prepared to put them back on once the spare tire is in place. The individual who jacked the car up will stand by prepared to jack it down and put the jack and flat tire back in the trunk of the car.

All of the various individual tasks that must be performed in order to change the flat tire are interdependent. This means that one task depends on another for its successful completion, and all of the individual tasks must be done in the proper order for the project to be successfully completed. For example, the flat tire cannot be taken off until the car has been jacked up and the lug nuts have been removed. Then, the car cannot be jacked down until the new tire has been put on and the lug nuts tightened. Construction projects are like this example in that they involve a lot of different but interdependent tasks to be performed, some simultaneously and some in a specific order. Coordinating all of these interdependent tasks and making sure they are performed in the right way and the proper order is the job of the construction project manager.

People, Processes, and Resources

Returning to the example of changing the flat tire, the project—like all projects—required people, processes, and resources. People do the work required to complete the project. In doing the work of the project, people use processes and resources. In the flat tire example, the people riding in the car used such processes as jacking up the car, loosening the lug nuts, removing the flat tire, putting on the spare tire, and tightening the lug nuts. In applying these processes they used resources including a jack, a lug-nut wrench, human strength, a spare tire, human know-how, and time. In addition, the people who did the work of changing the tire were, collectively, a resource. Resources are simply assets that are needed to complete a project. The human resource is typically the most important resource in any construction project.

Starting and Ending Dates

Construction projects begin once a contract has been awarded or shortly thereafter. The contract will contain a definite ending date—a deadline by which all work on the project must be finalized. Ensuring that projects are completed on time is one of the most important responsibilities of project managers. Some contracts received by construction firms contain penalty clauses that are activated if the project is not completed on time and according to specifications. Specifications are the next project component. They specify the success criteria.

Success Criteria

When construction firms receive a contract, it is accompanied by architectural/engineering drawings and specifications. The drawings and specifications show in detail how the project is supposed to turn out—what a successful project will look like. Every kind of project has some type of success criteria—some way of conveying expectations and allowing actual performance to be compared with expected performance. For example, the publisher that received the contract to produce this book needed to know certain things before it could proceed. It needed to know the book's dimensions, what kind of page layout was desired, the type and size of font for regular text and headings, if the book would be produced in color or black and white, page size, what kind of cover was desired (hardback or soft cover), and cover design to name just a few areas of concern. This information was provided in the form of specifications.

One of the challenges facing construction project managers and their teams is to complete projects not just on time and not just within budget, but also according to customer specifications. The drawings, specifications, budget, and contract (with deadlines) contain the success criteria for a construction project. To these must be added one additional set of criteria that apply to all construction projects: applicable local, state, and federal government regulations. In other words, the overall success criteria for construction projects are as follows:

- Stay within budget
- Finish all work on time
- Complete the project according to drawings, specifications, and applicable government codes and regulations

Construction Projects Are Process-Oriented

A Guide to the Project Management Body of Knowledge (PMBOK Guide) is recognized as the authoritative reference for practicing project managers. Published by the Project Management Institute, the *PMBOK Guide* makes the important point that the work of projects is completed through processes.[1] It describes the processes of project management in terms of three components:[2]

- Inputs (documents, plans, designs, specifications)
- Tools and techniques (includes equipment)
- Outputs (residential, commercial, industrial, or infrastructure projects)

The *PMBOK Guide* encompasses more than 40 different processes that are undertaken by people working in five broad process groups as follows:[3]

- ***Initiating.*** The initiating group for a construction project consists of the owner and representatives from the architectural/design firm. In some instances, it will also involve representatives from the construction company—including the project manager—that will build the project.
- ***Planning.*** The planning group is led by the construction firm's project manager. It typically also includes members with expertise in budgeting and scheduling. This group translates the project charter—contract, architectural/engineering drawings, and specifications—into a comprehensive project plan that includes a complete breakdown of the work to be done, a schedule, a budget, tracking mechanisms, and reporting procedures.
- ***Executing.*** The execution group consists of the project manager, purchasing personnel (procurement), the owner or his representative, representatives from the architectural/design firm, representatives from the engineering firms, and representatives from the various subcontractors involved in the project. This group is responsible for getting the project completed on time, within budget, and according to specifications.
- ***Monitoring.*** This group consists of the construction firm's project manager, on-site superintendents, inspectors, quality manager(s), purchasing personnel (procurement monitoring), and accounting personnel for budget monitoring. This group is responsible for monitoring the work of all subcontractors and ensuring that the project stays on schedule.
- ***Closing.*** This group consists of the construction firm's project manager, the owner, representatives of the architectural/design firm. This group is responsible for ensuring that the project has been completed properly and that all aspects of the contract have been satisfied.

In all five of the process groups that are typically involved in construction projects, the project manager is a central figure. The only process group the project manager may not be involved in is the initiation group. This happens when the owner chooses to use the design-bid-build approach for delivering the construction project. With this approach—explained later in this chapter—the construction firm is typically engaged only after the project has been initiated by the owner and the architectural/design firm.

The processes of construction project management require project managers to have a comprehensive knowledge base and skill set that can be divided into nine knowledge areas. In other words, the work performed by project managers who are involved in the five process groups explained above requires specific knowledge and skills that fall into the following categories:[4]

- ***Integration management.*** A construction project is a complex undertaken requiring the work of a long list of entities (e.g., architect/designer, numerous subcontractors, skilled tradespeople). All the work of the various groups and individuals that is necessary to complete a construction project must be orchestrated in such a way that it not just gets done at the right time, but in the right sequence, on time, within budget, and according to specifications. In other words, the work must be fully integrated. Integration is the responsibility of the project manager.
- ***Scope management.*** A project's scope is the entirety of everything that is to be done to complete the project. The scope is based on the contents of the contract, architectural/ engineering drawings, and specifications. Once the scope for a project has been established, it must be managed. Construction projects have a way of going beyond their original scope. This is acceptable as long as the owner, architect/designer, and construction firm agree to modify the contract—typically through the use of change orders—so that the construction firm does not find itself doing work that is outside the contract but not being paid for it. The concept is known as *creep*. It means that after the contract is signed, the owner or architect/designer begins to add requirements that were not part of the original scope for the project. Preventing creep requires careful management of the project scope by the project manager.
- ***Time management.*** Time is always an issue with construction projects. The ability to complete a project on time is one of the major factors that will determine whether a construction company makes a profit or loses money on a given project. Often the contracts for construction projects contain both penalty and incentive clauses. Penalties are assessed for missing deadlines and incentives are awarded for beating them. Once a schedule for a construction project has been established, it is the project manager's responsibility to ensure that it is strictly adhered to. Completing construction projects on time is one of the three basic success criteria that always apply to project management.
- ***Cost management.*** Construction contracts contain one figure that is of supreme interest to the owner and the construction firm: the agreed to price of the project or what is often referred to as the *bottom line*. Construction firms establish the budget for a project based on this bottom line figure. Ensuring that costs do not exceed the budget is an important responsibility of the project manager. Completing construction projects within budget is one of the three basic success criteria that always apply to project management.
- ***Quality management.*** The charter which initiates a construction project consists of at least three components: (1) contract, (2) architectural/engineering drawings, and (3) specifications. Ensuring that the project satisfies the owner's expectations as described in the drawings and specifications is called *quality management*. Quality management is one of the responsibilities of project managers. Completing projects according to specifications is one of the three basic success criteria that always apply to project management.

- *Human resource management.* The work of construction projects is done through various processes, but the processes are operated by people. In construction projects, the human resource is the most important resource. People who are members of project teams differ in their professional backgrounds, personal backgrounds, agendas, ambitions, perspectives, levels of motivation, commitment, and approaches to work. For a construction project team to be effective, its members—people with all of these differences and more—must be molded into a mutually supportive, peak performing team with the common mission of completing the project that brought them together on time, within budget, and according to specifications. This makes leading and managing the human resource one of the most challenging aspects of project management.

- *Communications management.* There are a lot of stakeholders in a construction project. Just a few of these stakeholders include the owner, architect/designer, numerous subcontractors, skilled tradespeople, and various entities within the construction firm. All of these stakeholders need to be fully informed and kept up-to-date concerning expectations, problems, and progress relating to the project. This makes managing the communication process an important responsibility of the project manager.

- *Risk management.* There are risks associated with every construction project. A risk is any factor or circumstance that might prevent the project from being completed on time, within budget, and according to specifications. Common risks associated with construction projects include the weather, material shortages, increases in the cost of important resources after the contract has been signed, accidents and injuries at the job site, natural disasters, and many others. For example, construction firms in Florida must factor the potential for interruptions or destruction from hurricanes into their planning for projects. Construction firms in the Midwest must consider the potential impact of tornados during construction. Firms in California have to factor the potential for earthquakes into their planning. Identifying potential risks and taking the appropriate steps to eliminate or at least mitigate the risks is the responsibility of the project manager.

- *Procurement management.* The ability to efficiently, effectively, and affordably procure the resources needed to complete a construction project is one of the keys to success in project management. Depending on the size and makeup of the construction firm, project managers might procure resources themselves or work with purchasing professionals to procure the necessary resources for their projects. In either case, a well-managed procurement process is an essential part of completing construction projects on time, within budget, and according to specifications.

EVOLUTION OF CONSTRUCTION PROJECTS

There is an evolutionary cycle that applies to construction projects regardless of their size or type from residential to commercial to industrial to infrastructure. The *cradle-to-completion* cycle for construction projects is as follows:

- *Project initiation.* In this phase of the cycle, an owner has an idea for a project, something he or she wants to have built. At this point, the project is just an idea or an expressed need. It might turn out to be a feasible idea or it might just be a dream. That determination is made in the next phase of the cycle: the feasibility analysis.

- ***Feasibility analysis.*** In this phase of the cycle, the owner typically approaches design and/or construction professionals for advice and assistance in determining if his or her idea is feasible. With the help of professionals, the owner completes a feasibility analysis. The feasibility analysis answers such questions as: Can the project be built? Is the project affordable? Will the project be worth what it will cost? Can the project be completed in the desired timeframe? The owner needs to have answers to these questions before he or she can decide whether or not it is feasible to proceed with the project.

- ***Financing.*** In this phase of the cycle, the owner arranges for financing to pay for the project. If financing is available at terms that are acceptable to the owner, the project proceeds to the next phase in the cycle: design.

- ***Project design.*** In this phase of the cycle, the owner works with an architect or designer to develop plans for the project. Design plans for a construction project evolve from conceptual sketches to schematic drawings to detailed architectural and engineering drawings to a comprehensive set of construction documents (i.e., architectural drawings, engineering drawings, specifications, and contract documents).

- ***Request for Proposals or Quotes.*** In this phase of the cycle, the architect/designer develops a request for proposals (RFP) or a request for quotes (RFQ). The RFP or RFQ are made available to construction firms that, in turn, prepare a construction cost estimate, respond with bid packages that demonstrate how they propose to complete the project and at what cost (the construction cost estimate). The architect and owner work together to select the most worthy bid and award that construction firm a contract. This is the traditional design-bid-build delivery system for construction projects. This phase in the cycle is different if the design-build or construction management delivery systems are used. These delivery strategies are explained in the next section.

- ***Construction planning.*** In this phase of the cycle, the construction firm that was awarded the contract for the project finalizes its planning (planning for a construction project actually begins when the firm develops its bid package in response to an RFP or RFQ). This planning includes establishing a project team or teams, developing a work breakdown for the structure, putting all work that must be completed on a schedule, finalizing the budget (converting the construction cost estimate into a budget and performing value engineering), developing plans for mitigating risk, and preparing procurement plans.

- ***Procurement.*** In this phase of the cycle, the construction firm procures the resources it will need to complete the project on time, within budget, and according to specifications. These resources include materials, equipment, tradespeople, subcontractors, and support items (e.g., construction trailer for the on-site project management team, portable sanitation facilities).

- ***Construction.*** In this phase of the cycle, the project is constructed. The project manager tracks progress closely, monitors the quality of performance of all trades and subcontractors, submits change orders when necessary, reports on progress to various stakeholders, and does what is necessary to complete the project on time, within budget, and according to specifications.

- ***Closeout.*** In this phase of the cycle the construction firm completes the project, cooperates with the owner, architect, and engineers in developing the punch list, completes all tasks on the punch list, and turns the project over to the owner.

DELIVERY STRATEGIES FOR CONSTRUCTION PROJECTS

When an owner has a need and decides to move forward with having it constructed, there are a number of different approaches that can be taken. The most common of these are as follows: (1) design-bid-build, (2) design-build, and (3) construction management. Each of these approaches has its advantages and disadvantages, but all of them require effective project management. Consequently, construction students and professionals should be familiar with these commonly used delivery strategies.

Design-Bid-Build Approach

With this approach, the owner contracts with an architect or a designer to develop a set of plans and accompanying specifications. With the *design package* completed, the owner requests bids from construction firms that might be interested in and capable of taking on the project. The lowest bid that appears to meet all of the requirements from the construction firm that appears to be able to complete the project is accepted. At this point—theoretically—the owner has a locked-in price and knows how much the project will cost him. The construction firm that submitted the winning bid then completes the project within the specified budget and contracted deadline. Again, this is theoretically how this approach works.

In reality, the design-bid-build approach to construction projects is only as good as the plans and specifications upon which the bids from construction firms were based. If the architect/designer overlooks some aspect of the project, leaves something out that must be included, or errs in any way the construction firm will submit change orders which can increase the cost of the project. In addition, if the owner changes his mind about some aspect of the project after the contract has been signed, the construction firm will submit change orders to cover the costs associated with the change. Construction firms cannot be asked to absorb the costs of errors, oversights, or changes made by architects, designers, or owners.

Another problem with the design-bid-build approach is that there is no collaboration between design and construction professionals before bids are requested. This can cause problems. Sometimes design personnel lack the practical knowledge that construction professionals possess about the details of how a structure must be built. This lack of detailed knowledge can cause design errors. For example, the author once worked on a project in which a commercial building was to have a mezzanine made of prestressed concrete columns, beams, and floor members.

Once completed it would be accessed by an escalator. Had the building been constructed as designed, there would have been no way to place the floor members for the mezzanine in the building. The crane that was to pick them up off of trucks and lower them over the exterior walls of the building could not extend far enough without hitting the walls. Fortunately, a member of the prestressed concrete company's erection crew—the crane operator—saw the plans and raised the issue so that a different erection scheme could be developed. Collaboration between design and construction professionals during the early phases of a project can prevent problems such as this that, otherwise, will result in costly change orders during the construction phase.

Design-Build Approach

With this approach, the owner works with a firm that has the capabilities to design and build the structure that is needed. Design-build firms employ both design and construction professionals who work together from the outset with the owner. In this way, design-build firms avoid some of the problems associated with the design-bid-build approach that result from the lack of collaboration between design and construction professionals. The design-build approach also has the advantage of giving the owner a single point of contact for the entire project (i.e., the design-build firm's designated representative).

The principal disadvantage of the design-build approach is that it puts all of the owner's eggs in one basket. This is because the designer, the builder, and the individual who is selected as the project manager all work for the same firm. The design-build approach can lead to situations in which the project manager is forced to choose between protecting his company's profit margin and revealing errors or other problems that could cost the company money. This lack of objectivity means that project managers must know that they are working for an ethical firm that will back them up when they are forced to point out errors in the design or less than acceptable performance in the construction phase.

Construction Management Approach

With this approach, the owner typically hires a designer to develop the plans and specifications for the project and a construction management firm to manage the actual construction. The construction management firm works closely with the design firm in the developmental stages of the project to ensure that all concerns have been identified and dealt with. Once a complete set of plans and specifications are available, the construction management firm gives the owner a guaranteed maximum cost price for the construction of the project.

Construction management firms like to fast track projects since they can improve their profit margin by completing projects early. There are typically incentives built into the contract for completing work early. Correspondingly, since the cost given to the owner is fixed, they can lose money by finishing late. The success of the construction management approach depends on effective, ongoing communication between the design firm, construction management firm, and the owner.

Construction management firms provide a relatively small team of management personnel to oversee the project and hire subcontractors to perform the hands-on construction work. For example, the author once participated in a project that used the construction management approach to build a 75,000 square foot student services facility on a college campus. The guaranteed maximum price for this project was $21.9 million and the duration of the project was 26 months. The on-site personnel from the construction management firm for this project consisted of a project manager, assistant project manager, administrative assistant, and two work superintendents. All hands-on construction work was subcontracted but managed by this small team of construction professionals.

CONSTRUCTION CONTRACTS

Minimizing risk is important in construction project management. One factor that can affect the amount of risk associated with a construction project is the type of contract awarded. Since minimizing risk is an important responsibility in managing construction projects, construction students and professionals need to understand the types of contracts that are typically awarded to construction firms and the risks associated with each one. These contracts include: (1) single fixed price, (2) unit price, (3) cost plus, and (4) guaranteed maximum price.

Single Fixed Price Contract

With this type of contract, the construction firm agrees to complete the project in question for a fixed amount. This is the type of contract typically used with the design-bid-build approach to construction projects. The weakness of this type of contract is that it depends on the construction firm arriving at an accurate estimate based on plans and specifications that are themselves accurate and complete. If problems occur during the course of construction, either the owner or the construction firm will be liable for absorbing their costs.

Disagreements over which party is actually liable for the additional costs often result in litigation. The cost of litigation can increase the cost of the project for both parties. This weakness has begun to make the single fixed price contract less popular than it once was. Change orders, mediation, arbitration, and litigation can increase the cost of the single fixed price contract substantially over the course of a project and create bad feelings and ill will between the owner and the construction firm.

Unit Price Contract

With the unit price contract, construction firms are asked to submit quotes for their price on each individual unit of the project. For example, assume that an owner asks for quotes to build an office complex. Construction firms interested in bidding for the job might be asked to submit their prices for the site work, foundation, framing, electrical, plumbing, roofing, and finish work. These separate quotes are then totaled to determine each firm's overall quote. The owner then awards the contract to the lowest bidder, provided the firm in question appears to have the necessary capabilities to complete the project and has submitted realistic figures.

The main advantage of this type of contract is that it allows work to begin on a project even before the design is completed. This can be important when there are factors that require a project to be completed by a certain unchangeable deadline. For example, with a firm quote for the site work and foundation in hand, the owner can move forward with that aspect of the project while the design and specifications for the rest of the project are being finalized. The disadvantage of this type of contract is that the ultimate cost of the project is not known until it is completed.

Cost Plus Contract

With a cost plus contract, the owner and construction firm determine what work needs to be done. The construction firm agrees to do the work for the cost of materials and labor plus a specified fee. The fee is typically a percentage of the cost of the project. The owner reimburses the construction firm for the costs of materials and labor. An important aspect of the cost plus contract involves the owner and construction firm agreeing up front on exactly what materials and labor will be considered reimbursable. This type of contract is used when it is necessary to get started on a project before its full scope is understood so that the project can be fast-tracked.

To succeed, this type of contract requires close collaboration between the owner and the construction firm from the outset of the project through its completion. Trust between the owner and the construction firm must be established, nurtured, and maintained. The principal weakness of the cost plus contract is disagreements between the owner and the construction firm over what is reimbursable. Because this type of contract is used when the scope of the project is not well-defined, there is always the risk that the owner will think the construction firm is asking to be reimbursed for work that was not necessary or could have been done less expensively.

Guaranteed Maximum Price Contract

This type of contract is exactly what the name implies. A maximum price is agreed to between the owner and the contractor and that price may not be exceeded. If the project goes over the guaranteed maximum price, the contractor is liable for the overage.

NEED FOR PROJECT MANAGERS

Construction firms receive contracts from owners that need something built. These contracts become projects that are undertaken by project teams. A project team in a construction firm is like a symphony orchestra: it has a lot of different players—each with a specific instrument and role. Without a conductor to lead, coordinate, and facilitate, the symphony members are more likely to produce noise than music. Like orchestras, project teams in construction firms are often cross-functional in nature. This means they are composed of individuals with various types of expertise.

Project teams in construction firms, like an orchestra, need a conductor who can meld the members into one coherent team and keep them on task and on time. That conductor is the project manager. Without project managers who can get a disparate group of people working together as a well-coordinated, mutually supportive team, construction projects can turn into disjointed undertakings bordering on chaos.

In a competitive business environment, construction firms excel by completing the projects they undertake on time, within budget, and according to specifications. Those that cannot meet these basic success criteria lose business to other firms that can. Ensuring that projects are completed on time, within budget, and according to specifications is the job of the project manager. Without a conductor, orchestras are likely to make more noise than music. Without a project manager, project teams are likely to make more problems than progress.

CONSTRUCTION PROJECT MANAGEMENT SCENARIO 1.1

Which Delivery Approach Should We Use?

Dale Cartwright is a project manager for ABC Construction, Inc. He has worked for several different construction companies over the course of his career and managed construction projects using all three of the most common approaches: design-bid-build, design-build, and construction management. Cartwright even teaches a night course in construction project management at the local university. Consequently, it came as no surprise when he was asked to serve on an ad hoc construction committee for the university. The university plans to build new dormitories to accommodate rapid growth in the student population, and the university president needs to decide how to approach the project. In a committee meeting, the president asked Cartwright for his opinion concerning the delivery approach the university should select for building the new dormitories: design-bid-build, design-build, or construction management. The last time the university had a new building constructed, the design-bid-build approach was used and the cost overruns were enormous. As a result, the president is being encouraged by his board of trustees to use either the design-build or construction management approach this time.

Discussion Questions

In this scenario, Dale Cartwright is being asked for his views on the three most common construction delivery approaches. Put yourself in Cartwright's place. How would you advise the university president in this matter? Which of the three approaches would you recommend and why?

PHASES OF CONSTRUCTION PROJECTS

Regardless of size and complexity, all construction projects go through several phases. These phases include: (1) initiation, (2) design, (3) project planning, (4) procurement, and (5) construction. These five phases are all interrelated and interdependent. Although project managers in construction firms are concerned primarily about the planning and construction phase of projects, they need to know what takes place in all five phases. Depending on the size and type of construction firm, project managers may be involved in all five phases of a construction project. In fact, this is often the case with design-build and construction management firms.

- *Project initiation.* A construction project is initiated when an owner has an idea or recognizes a need. The owner—whether an individual, a private corporation, or a government agency—considers the feasibility of the project and makes a decision to either proceed, delay, or drop the idea completely. If the decision is to proceed, the owner begins the process of arranging financing in the appropriate form (e.g., loan, government appropriation, seeking investors). If the financing is forthcoming, the owner proceeds to the next phase of the process: design.
- *Project design.* When an owner has a feasible idea for a construction project and can secure the necessary funding, the project moves into the design phase. The owner hires an architect/designer to translate his or her idea into a set of plans and specifications. These plans, specifications, and the contract become the project charter. Once a construction firm has a project charter, the project moves into the next phase: planning.
- *Project planning, scheduling, and risk management.* Much of the planning for a construction project actually begins before the construction firm has a contract. This is especially the case when the firm has to submit a bid or a quote in order to win the contract. In preparing a bid package or a quote, the construction firm must develop at least a rough breakdown of the work involved and estimate the cost of doing the work. However, once a contract has been received, the firm's appointed project manager leads the effort to develop a comprehensive plan for completing the project on time, within budget, and according to specifications. That plan will include a detailed breakdown of the work to be done—sometimes called the *work breakdown structure*. It will also contain a schedule for the work, a budget for the project, and a list of the reports the project manager will submit on a regular basis. Some construction firms pull all of these components together in a comprehensive project management plan. Figure 1.9 is the table of contents for the project management plan for a large commercial building. In addition, part of the planning process identifies risks associated with the project in question and develops strategies for minimizing the risks. Risk management strategies

Project Management Plan Commercial Building

TABLE OF CONTENTS

FIGURE 1.9   Contents of a sample project management plan.

can have an effect on any of the construction documents, but most often they are built into the contract.

- **Project procurement.** Construction projects require a variety of resources. These resources are secured through the procurement process—sometimes called the *purchasing process*. Hiring subcontractors is part of the procurement process. Project managers in smaller firms may actually manage the procurement process. In larger firms, project managers work with procurement professionals and specialists to secure the resources they need for their projects.
- **Construction.** Once the owner's ideas have been translated into a design, and the project plan has been developed, and the necessary resources have been procured, the structure in question must be constructed. During the construction phase, the project manager's job involves tracking, monitoring, and reporting. Project managers should never make the mistake of thinking that just because they have a well-developed project plan, the construction phase will automatically proceed according to plan. The work of the various subcontractors must be monitored continually, progress must be tracked carefully, and stakeholders must be kept up-to-date with the latest information about the project on a regular basis. These tasks are the responsibility of the project manager.

INTERNAL VERSUS EXTERNAL PROJECTS

Most projects for construction firms are initiated by a customer—either an owner or an architectural/design firm. Projects that are initiated in this way are external projects because their source comes from outside of the firm. Projects can also be initiated internally. Internal projects are initiatives undertaken by construction firms to enhance some aspect of their operations or their overall competitiveness. In a competitive environment, construction firms sustain themselves by: (1) increasing their business base (more customers and more projects) and (2) reducing the cost of doing business (improving productivity and decreasing costs).

To increase their business base, construction firms must establish a reputation for consistently providing superior value to customers. Superior value is a combination of superior quality, superior cost, and superior service. This is done by consistently completing projects on time, within budget, and according to specifications. Construction firms that provide superior value consistently exceed customer expectations. To reduce costs, construction firms must find ways to enhance productivity and eliminate unnecessary spending. The most competitive construction firms self-initiate internal projects on a continual basis to enhance value and/or reduce costs. They also self-initiate internal projects to develop innovations that give them a competitive advantage in the marketplace, allow them to keep up with the competition, or help them catch up with market leaders that have pulled ahead of them.

The same principles apply when managing internal as well as external projects. Project managers need to be skilled at planning, scheduling, cost and duration estimating, budget control, tracking, reporting, procuring, risk minimization, teambuilding, and leadership. All of the process skills and people skills explained in this book apply to both external and internal projects. Project managers in construction firms should expect to lead both internal and external project teams.

Challenging Construction Project

AMERICA'S INTERSTATE HIGHWAY SYSTEM

Perhaps the most challenging infrastructure-related construction project ever completed in the United States is the Interstate Highway System. The brain child of President Dwight D. Eisenhower, America's Interstate Highway System is known as one of this country's "seven wonders." This system of highways running from coast to coast has been credited with increasing America's gross national product tenfold since it was initiated in 1956.

Construction firms, state governments, and the federal government had to overcome massive obstacles to turn President Eisenhower's dream into reality. Swamps had to be drained, rivers crossed, mountains moved, and gorges spanned. In Louisiana, the construction firm of Boh Brothers Construction had to span the massive Atchafalaya Swamp. The bridge built over the swamp was constructed of precast concrete members that were manufactured at a plant on Lake Pontchartrain and then floated by barge through the many twists and turns of the swamp to where the bridge was being built. The precast members were put in place by cranes mounted on barges. The construction workers had to contend with alligators, mosquitoes, and the unforgiving expanse of the Atchafalaya to complete the span.

In Arizona, the Kiewit Construction Company won the contract to build the stretch of highway that runs through the picturesque Virgin River Gorge, a one-of-a-kind stretch of scenic natural beauty. Building the highway without marring the natural beauty of the Gorge was just one of many challenges Kiewit Construction Company faced. It also had to re-route the Virgin River 12 times so that it could squeeze the highway between the canyon walls of the Gorge.

In Colorado, Straight Creek Constructors—which was a partnership of four construction companies—had to build a tunnel under the Continental Divide. This tunnel is just one of several that had to be blasted through mountains in order for the Interstate Highway System to progress, but it is the longest of them all. More than 70 percent of the tunnel went through rock that could not support itself. Hence, Straight Creek had to build a tunnel support system as it progressed through the mountain. At elevations as high as 11,000 and in some of the country's coldest winters, the construction company had to excavate 524,000 cubic yards of rock. The project required 1,140 people working in three shifts, 24 hours a day six days a week.

Source: Based on Bob Moore Construction, Inc., "Historic Construction Projects." http://www.generalcontractor.com/resources/articles/interstate-highway-system.asp

SPEAKING THE LANGUAGE OF PROJECT MANAGEMENT

One of the most basic skills needed by construction project managers is speaking the language of project management. There are certain terms that are used over and over in project management that constitute the core of the language for this specialized field. In addition to the term *project*, which has already been defined, there are several others that project managers need to understand in order to be conversant in their profession. These terms include the following:

- ***Program.*** Programs are composed of projects. For example, a college degree is sometimes called a *program of study*. The courses that make up the program can be considered projects. In construction, a company might have a commercial building program and a residential building program. All commercial projects undertaken are part of the company's commercial building program. All residential projects undertaken are part of the company's residential building program. Program management is a level higher than project management.

- *Goal.* The goal of a project is its overall purpose (e.g., construct an apartment complex, bank, hospital). The goal of an external project is established by the contract the construction firm receives from the customer. The goal of an internal project is defined by the charter given to the project manager by the firm's higher management team.
- *Objective.* The objective is a major division or deliverable in a construction project. For example, assume that the goal of a certain project is to build a shopping mall. One objective would be to complete the site work. Another would be to pour the footings and foundation. Another would be to frame the structure, and so on.
- *Deliverable.* It is an actual product or service to be completed during the project. A deliverable from an electrical subcontractor would be the electrical system for the building in question. A deliverable from an HVAC supplier might be a new air handler unit. Anything a stakeholder, individual, subcontractor, or supplier agrees to provide as part of the project is that entities deliverable.
- *Scope.* The scope of the project is a comprehensive definition of the project. With construction projects, the scope is defined by the contract, architectural/engineering drawings, and specifications. The scope for a construction project represents the entirety of the work that must be done to complete the project and satisfy the owner's expectations.
- *Tasks and activities.* Objectives consist of tasks, and tasks consist of activities. However, these terms are used somewhat loosely and often interchangeably in construction project management. Some firms refer to the work that must be accomplished as tasks and subtasks while others use the term activities. Which term is used and when is not as important as using them consistently. For example, a residential construction company the author has worked with used the term tasks for the various specific items of work that must be done to accomplish a project objective. Using this approach, assume that one of the objectives in a project plan for building a single-family residence was: *Complete the exterior finish of the home.* This objective could be broken down into the following tasks: (1) install housewrap, (2) install roofing, (3) apply siding, (4) add soffit and fascia, (5) pour driveway and walkway, (6) paint nonsided exterior areas, and (7) landscape the property.
- *Duration.* Duration is the time from beginning to end that is required to complete a task or objective. There can be, and often is, a difference between the actual time a task takes to complete and the scheduled time. The duration is the scheduled time as established during the planning stage of the project. Construction firms often receive incentives for completing the project or components of it in less than their scheduled time (duration).
- *Constraints.* Factors that control, impinge on, inhibit, or restrict the construction company's ability to complete a project on time, within budget, and according to specifications. Common constraints include time, cost/budget, quality expectations, and personnel. Other constraints that sometimes come into play are technology/equipment and facilities. Time, cost/budget, and quality expectations are almost always constraints. Owners typically want the building as soon as they can possibly get it at the lowest price and highest quality. Being able to complete the project within the applicable

constraints is what separates construction firms that excel from those that never exceed mediocrity or even fail. Personnel can become a constraining factor when the organization has insufficient personnel, lacks the personnel with the knowledge and skills required by the project, or has the needed personnel tied up on other pressing projects that require their time and expertise. Technology/equipment can become constraints when they are not readily available, are too expensive, or require skilled operators that are not available.

- **Schedule.** A schedule is a timetable for completing all work in a project. The schedule runs from the starting date of the project until the deadline for completion. All work that must be completed in between these two dates is put on a schedule with targeted completion dates. Some of the work in a construction project must be completed before other work can begin. For example, the site work must be completed before the footings and foundation can be poured. On the other hand, some work can be undertaken simultaneously and be completed in parallel with other tasks. A comprehensive project schedule will reflect not just the beginning and ending date for the overall project, but the beginning date, ending date, and estimated duration for all tasks included in the project.
- **Resource.** A resource is any asset needed to complete a project on time, within budget, and according to specifications. Resources that typically concern construction project managers are time, personnel, money, material, technologies/equipment, and facilities.
- **Processes.** The work of construction is completed by people using processes. A process is a series of standardized steps used over and over to produce a given result. A process is not an end result. Rather, it is a series of events that lead to a predictable end result. Processes commonly used in construction include the initiation, planning, estimating, scheduling, and monitoring processes.

PEOPLE SKILLS IN PROJECT MANAGEMENT

Up to this point, all of the project management concepts and skills explained have been process-oriented. The other side of project management is the people side. To be successful as a construction project manager, one must be effective at: (1) building teams; (2) leading teams; (3) motivating team members; (4) communicating with team members, owners, architects, designers, engineers, and colleagues; (5) managing time; (6) managing change; (7) managing diversity; and (8) managing adversity.

The need to be able to function at a high level of effectiveness in terms of both the process and people sides of the job is what makes excelling as a project manager a challenge. Some construction professionals have excellent process skills but struggle with the people side of the job. At the same time, there are construction professionals who have outstanding people skills but struggle with their process responsibilities. Construction professionals at either end of the process/people continuum will find it difficult to excel as project managers. Only those who commit to developing both the people and the process skills will excel.

The good news is that with effort and commitment both types of skills can be developed. Construction students and professionals can learn to plan, schedule, budget, handle risk, track, report, and procure for projects. At the same time, they can learn to lead, manage, motivate, and communicate with people. Part One of this book is devoted to helping develop the process skills of project management. Part Two is devoted to helping develop the people skills.

CONSTRUCTION PROJECT MANAGEMENT SCENARIO 1.2
I Don't Know If I Want to be a Project Manager

Susan had always been fascinated by construction. Consequently, when the time came for her to declare a college major, Susan decided to look into Construction Management. The program at her university has three tracks: residential, commercial/industrial, and infrastructure. All three tracks emphasize project management. After studying the literature about the program and talking to an advisor, Susan is still unsure of what a project manager does. During her meeting with the advisor, Susan said: "I don't know if I want to be a project manager. I need to know more about the field." The advisor recommended that Susan talk to several students majoring in Construction Management and gave her a list of names.

Discussion Question

In this scenario, Susan is considering a career in construction project management but is not sure if it is what she wants to do. Her academic advisor recommended that she talk with several students majoring in Construction Management. Assume that you are on Susan's list of students to talk to. What would you tell her about what a project manager in construction does and what skills are needed?

VALUE ENGINEERING IN CONSTRUCTION

Value engineering, sometimes referred to as *value analysis*, is a concept in which the owner, designer, and contractor work together to find the best functional balance among the cost, reliability, and performance of the materials and processes used to complete a construction project. The functional question in value engineering is this: *What material or process will satisfy the function in question best when cost, reliability, and performance are all considered?* It is not uncommon for the value engineering process to result in substantial savings in the overall cost of a construction project.

One of the most important benefits of value engineering is that it causes participants in the process to think creatively and to refuse to accept *the-way-we-have-always-done-it* solutions. Done well it can help ensure that the newest best practices in construction processes, the latest technological innovations, and the most up-to-date materials are used in construction projects. However, in order to gain the full benefit of value engineering, the process needs to take place early in the project—in fact, the earlier the better. Ideally, value engineering occurs in the initiation phase of the project so that the design phase can be informed by its findings. If the process is put off until the construction phase, the benefits will be lost due to the cost of making changes to the design and corresponding changes to the construction plan (e.g., construction cost estimate, risk management plan, schedule, and procurement plan).

A value engineering study proceeds in four phases: (1) creative phase, (2) judgment phase, (3) development phase, and (4) presentation phase.[5] These phases involve the following activities:

- ***Creative phase.*** The value engineering team, which should include the construction project manager, reviews the project and brainstorms different ideas. The ideas that seem to have merit to the team are recorded so they can be analyzed more closely.

- ***Judgment phase.*** The value engineering team subjects each idea on the list from the previous phase to careful scrutiny. Upon closer examination, different questions arise like: Is the idea actually workable? Are there any obvious reasons why the idea should be eliminated from the list of potential ideas?
- ***Development phase.*** The value engineering team subjects those ideas that remain on the list to in-depth cost-benefit analysis and life-cycle cost comparisons. In this phase, calculations and sketches are made to document design concepts. Cost summaries are made and technical data are recorded that document the performance capabilities of materials and the viability of new processes and technologies.
- ***Presentation phase.*** The value engineering team presents its findings to the owner and architect explaining its recommendations and providing the accompanying rationale for each.

Rationale for Involving the Project Manager

Ideally, the value engineering process occurs in the initiation stage of a project. Consequently, it is not always possible to involve the construction project manager in the process. However, whenever possible the project manager should be part of the value engineering team to ensure that the contractor's perspective is represented when value decisions are made. For example, when the design-build or construction management delivery systems are used it is possible to involve the project manager much early than when the traditional design-bid-build delivery system is used.

It is not uncommon for designers to develop a plan or make a value decision that looks good on paper and sounds reasonable in an office setting but will not work on the jobsite. Without input from an individual with in-depth construction knowledge and experience—an individual such as a project manager—the perceived benefits of value decisions might be negated by practical, hands-on construction issues the designer either overlooked or was unaware of. This is a common problem in construction projects.

SUMMARY

A construction project is a fully coordinated group of interdependent tasks that are completed by people using resources and processes. Construction projects have definite starting and ending dates and success criteria. The most basic success criteria for all construction projects are to complete projects on time, within budget, and according to specifications. Project managers are needed in construction for the same reason conductors are needed in orchestras. Projects are composed of a number of separate but interdependent tasks all of which must be planned, scheduled, and fully coordinated if they are going to be completed on time, within budget, and according to specifications.

There are four construction sectors: residential, commercial, industrial, and infrastructure. Project managers are needed in all four sectors. Participants in construction projects include owners and their representatives, architects, engineers, construction professionals, tradespeople, and suppliers.

The five process groups that project managers work with on construction projects are initiating, planning, executing, monitoring, and closing. The specific knowledge needed by project managers falls into the following categories: integration, scope, time, cost, quality, human resource, communication, risk, and procurement management. Construction projects are delivered in one of three ways: design-bid-build, design-build, and construction management. Construction projects are typically one of three types: single fixed price, unit price, or cost plus.

Construction projects—large and small—proceed through five phases: initiation, design, project planning, procurement, and construction. Construction project managers may be called upon to manage both internal and external projects. Internal projects are initiated by the construction firm's higher management team to improve productivity and lower costs—thus making the firm more competitive. External projects are initiated by an owner who needs something built and is willing to pay to have it done.

Important concepts in project management include program, goal, objective, deliverable, scope, tasks/activities, duration, constraints, schedule, resource, and processes. People skills needed by construction project managers include building teams, leading teams, motivating team members, communicating with stakeholders, and managing time, change, diversity, and adversity. Value engineering is a process used by owners, designers, and construction professionals to find the best functional balance among cost, reliability, and performance when selecting construction materials and processes.

KEY TERMS AND CONCEPTS

Residential
Commercial
Industrial
Infrastructure
Owners and their representatives
Architects
Engineers
Construction professionals
Tradespeople
Suppliers
Construction project
Fully coordinated interdependent tasks
People, processes, and resources
Starting and ending date
Success criteria
Initiating
Planning
Executing
Monitoring
Closing
Integration management
Scope management
Time management
Cost management
Quality management
Human resource management
Communication management

Risk management
Procurement management
Design-bid-build approach
Design-build approach
Construction management approach
Single fixed price contract
Unit price contract
Cost plus contract
Initiation
Design
Project planning
Procurement
Construction
Internal project
External project
Program
Goal
Objective
Deliverable
Scope
Task and activities
Duration
Constraints
Schedule
Resource
Processes
Value engineering

REVIEW QUESTIONS

1. What are the four sectors of the construction industry?
2. List the participants in a construction project and briefly explain each participant's role.
3. Define the term *construction project*.
4. What are the three success criteria that apply to every construction project?
5. List and describe the five process groups for project management.
6. List and explain the nine knowledge areas for project management.
7. Describe and distinguish among the three delivery approaches for construction projects.
8. List and explain the three most common types of construction contracts.
9. Explain why project managers are needed in construction.
10. List and briefly explain the five phases that all construction projects go through.
11. Distinguish between an internal and an external construction project.
12. What is the difference between a project and a program?
13. Define the following terms as they relate to construction project management: objective, deliverable, scope, task, duration, and constraint.
14. What are the people skills needed by project managers?
15. Define the term *value engineering* and explain the various phases of the value engineering process.

APPLICATION ACTIVITIES

The following activities may be completed by individual students or by students working in a group. Contact a construction firm, and identify an individual who is willing to cooperate in helping complete this project. Ask this individual to identify a project his or her firm has completed. Ask this individual to provide the following information: (a) the work breakdown for the project, (b) the schedule for the project, and (c) the specifications and drawings for the project.

ENDNOTES

1. Project Management Institute, *A Guide to the Project Management Body of Knowledge*, 4th ed. (Newtown Square, Pennsylvania: Project Management Institute, 2008), 412–415.
2. Ibid.
3. Ibid.
4. Ibid.
5. Consulting Engineers of Yukon, "Value Engineering." Retrieved from http://www.cey.ca on February 1, 2012.

CHAPTER 2

Roles and Responsibilities of Construction Project Managers

Chapter One explained what construction project management is. This chapter explains what a construction project manager does. To be effective project managers, construction professionals must develop both process and people skills. This point was made in Chapter One. Process skills allow project managers to participate in or actually carry out the various processes involved in taking a construction project from idea to reality (i.e., initiating, designing, planning, procuring, and constructing). People skills allow project managers to mold individuals into effective project teams and lead them to peak performance in completing construction projects. Both sets of skills are essential for construction students and professionals who want to be effective project managers.

THE CONSTRUCTION PROJECT MANAGER'S FUNCTIONS

When a construction firm receives a contract to build some type of structure, an external project is established and a project manager is appointed. Actually, these two things are sometimes done even before a contract is awarded. When a construction firm decides to undertake an internal project to enhance its competitiveness, a project charter is developed internally and a project manager is appointed. Once a construction professional has been appointed to serve as the project manager for either an external or internal project, that individual will play a well-defined role and have specific responsibilities.

The overall responsibility of a construction project manager is to ensure that projects are completed on time, within budget, and according to specifications. In carrying out this overall responsibility, project managers have a number of specific functions to perform that fall into two broad categories: process functions and people functions. These functions are explained in the next two sections.

PROCESS FUNCTIONS OF CONSTRUCTION PROJECT MANAGERS

When a project has been established and a project manager assigned, a certain prescribed list of responsibilities comes with the assignment. Regardless of whether the project is internal or external, the project manager is responsible for ensuring that the following tasks are

---

CHECKLIST OF THE PROJECT MANAGER'S PROCESS FUNCTIONS

✓ Planning projects

✓ Scheduling projects

✓ Estimating the duration and cost of projects

✓ Procuring project resources

✓ Tracking/monitoring progress and reporting

✓ Handling risk

FIGURE 2.1 Process functions are half of a project manager's job.

completed: developing a plan that includes a comprehensive breakdown of all work to be done, estimating the duration and cost of all work tasks in the project (cost estimation is typically done when the construction company develops its bid or quote for the project), putting all work tasks on a schedule, procuring the necessary resources, tracking/monitoring, reporting progress, and handling risk (see Figure 2.1). These are the key process functions of construction project managers.

Planning Projects

Theoretically, before beginning the planning for a project, project managers must be provided with a project charter, the most important element of which is the contract. However, in reality, construction companies often begin planning in anticipation of receiving a contract. Nonetheless, having a project charter which includes the contract in hand early in the process is important. A project charter can come in several forms. However, with external construction projects it typically consists of a signed contract containing a detailed statement of work, architectural/engineering drawings, the project estimate for budgeting purposes, and specifications. These documents give the construction firm a *charter* to move forward with the project. For internal projects, the charter comes from the construction firm's higher management team.

The various documents that collectively represent the charter for a construction project—regardless of whether the project is external or internal—should answer a number of questions that fall into six categories sometimes referred to as the *Five Ws and One H*: why, who, what, when, where, and how. Project managers must be able to answer these questions before proceeding with a construction project. Questions that should be answered by the project charter include the following (Figure 2.2):

- *Why* is the firm taking on the project? Why is the project important to the firm?
- *Who* will be assigned to the project team?
- *What* specific work tasks must be completed during the course of the project? What resources will be needed to complete the project? What are the budgetary constraints of the project? What types of reports and other project documentation will be required? What are the risks associated with the project? What are the success criteria for the project?

```
┌─────────────────────────────────────────────────────────────────────┐
│                    CHECKLIST OF QUESTIONS THE                         │
│                 PROJECT CHARTER SHOULD ANSWER                         │
│                                                                       │
│      ✓   Why?                                                         │
│                                                                       │
│      ✓   Who?                                                         │
│                                                                       │
│      ✓   What?                                                        │
│                                                                       │
│      ✓   When?                                                        │
│                                                                       │
│      ✓   Where?                                                       │
│                                                                       │
│      ✓   How?                                                         │
│                                                                       │
└─────────────────────────────────────────────────────────────────────┘
```

FIGURE 2.2   A project charter must answer these six questions.

- *When* does the project begin and when must it be completed?
- *Where* is the project located?
- *How* will the work of the project be done? How will the project manager communicate with members of the project team? Higher manager in the construction firm? The owner? The architect/designer? Subcontractors?

Once a project charter has been provided, project planning includes developing a detailed breakdown of all work to be done. This document is called the *Work Breakdown Structure* or *WBS*. The final planning document is a project management plan. Developed for large construction projects, the project management plan is a consolidation of all of the various outputs from the planning process into one comprehensive, fully integrated document. It can be developed in summary form with the more detailed sub-plans attached or referenced, or it can be a detailed compilation of all the sub-plans. Typical contents for a project management plan include at least the following:

- Project scope
- Project requirements
- Project schedule
- Project budget
- Quality management plan
- Human resource plan
- Communications/reporting plan
- Procurement plan
- Risk management plan

Scheduling Projects

Scheduling is an important part of project planning and was introduced in the previous section. However, it is such a specialized planning function that it is explained separately here. Project managers are responsible for either developing comprehensive schedules for all of the work that must be completed in their projects or for working with professional schedulers to develop schedules. There is a great deal of scheduling software available to assist project managers with this critical and often complex responsibility. However, even if using scheduling software, project managers must be able to analyze a work breakdown structure to identify critical paths and dependencies. They must also be able to develop and interpret Gantt charts and PERT charts and various other scheduling tools.

Estimating the Duration and Cost of Projects

Estimating the duration and cost of project activities is typically done during the initiation phase since construction companies must prepare an estimate in order to submit a bid or quote for a project. Consequently, a project manager will have that initial work to use when developing the final cost estimates for the project. Once a work breakdown structure has been developed it can be used as the framework for estimating the duration of each activity in the project. The cost of each activity can then be estimated. This is a critical step for project managers. The more accurate the estimates of duration and cost, the more likely the project will be a success. Inaccurate estimates of cost and duration can cause a construction firm to underestimate what completing the project will cost which, in turn, can eat up the profits the project is supposed to produce.

Estimating duration amounts to estimating the amount of time it will take to complete each individual task required in the project. These estimates, collectively, determine the amount of time that will be required to complete the entire project. Duration estimating is done to a certain level of detail in order to respond to request for proposals (RFP) or request for quotes (RFQ). A common approach is to complete a less detailed estimate in response to an RFP or RFQ and then develop a more detailed estimate when the project is actually established.

Once the work tasks for a project are listed and the duration for each is established, the cost of each activity can be estimated. The overall cost of a given activity is composed of the labor and nonlabor costs associated with it. Labor costs consist of the loaded labor rates for all personnel assigned to the task in question. Loaded labor rates consist of the hourly wages of personnel plus their benefits and any other personnel-related costs tied to individuals. Labor rates also include the cost of outside consultants and subcontractor personnel who will work on the project.

Nonlabor costs include such expenses as temporary facilities that must be built, renovated, or leased (e.g., construction trailers for the job site); travel; hardware and software; equipment; training; teambuilding; and any aspects of the project that are outsourced. Nonlabor costs also include the costs associated with permitting and compliance with local, state, or federal government regulations.

Procuring Project Resources

Project managers are responsible for working with their firm's higher management team, human resources department, and purchasing or procurement department to obtain the various resources needed to complete their projects. Recall from Chapter One that resources are assets needed to complete the project on time, within budget, and according to specifications. Resources for construction projects typically include labor and nonlabor items. Labor needs are people with the knowledge and skills to perform their assigned duties in the project. Outside consultants and subcontractors are resources that must be procured for construction projects. Nonlabor needs are typically materials, equipment/technology, and facilities. For example, it is often necessary to lease or rent large specialized equipment for construction projects.

Labor needs can be met by assigning internal personnel to the project, hiring new personnel, outsourcing some of the labor requirements, or a combination of all these approaches. Nonlabor needs can be met by purchasing materials and purchasing, leasing, or renting equipment and technologies directly from vendors. Equipment and technology needs can also be met by outsourcing them. When facilities are needed they can be built, leased, or acquired through a partnership with another entity. Procurement is a highly specialized field in construction. Consequently, project managers often work hand-in-hand with purchasing professionals to fulfill this aspect of their responsibilities.

Tracking Progress and Reporting

Construction project managers are responsible for monitoring the work, schedule, and budget for a project—a function known as tracking—and for reporting the results periodically to various stakeholders. The schedule and budget for a project are both *living* documents in that they can be and are frequently revised. This is why change management is so important a part of the project manager's responsibilities. Mid-stream changes to the project's contract require change orders to be developed and submitted and the schedule and budget to be revised accordingly. Reports showing the results of continual monitoring are sent to various stakeholders on an agreed upon schedule (e.g., weekly, biweekly, monthly) that is explained in the project management plan.

The progress reports of project managers should contain the following types of information: (1) a description of the work planned for the reporting period as compared with what has actually been accomplished, (2) a planned versus actual statement of the budget for the reporting period, (3) work planned for the next reporting period, (4) problems and concerns as well as how they affect the schedule and budget, (5) problem solutions and adjustments (how problems encountered will be solved and any corresponding adjustments that will be needed in the schedule or budget), and (6) appended material (e.g., charts, graphs, and other graphic material that supports the updates provided in the report).

Handling Risk

With every project undertaken by a construction firm, there are risks. Risk associated with construction projects typically fall into the following categories: money, time, design, quality, site, and weather. In essence, risk is the possibility that things will not go as planned or that factors may surface that might adversely affect the satisfactory completion of the project. To the extent possible, risk should be identified early in the process, preferably even before acceptance of a contract. When identified early, risk can be dealt with in ways that minimize the construction company's exposure, mitigate the risk, or even eliminate it.

Risk-related questions that should be asked about every construction project include at least the following:

- Is the project worth undertaking from a financial perspective (i.e., Is the project going to be more trouble than it is worth?)?
- Have sufficient funds been allocated to the project?
- Can the project be completed within the time allocated?
- Is the design for the project viable?

- Can the project be built as designed?
- Are the materials required for the project actually available at realistic prices?
- Is the equipment required by the project available at realistic prices?
- Are the skilled personnel required by the project available?
- Are there problems with the site that could adversely affect the project?
- Will the weather affect our ability to complete the project on time, within budget, and according to specifications?

No matter how well risk is handled in the early stages before a project is chartered, there will still be some risk associated with most projects. Therefore, project managers must be skilled at risk management. Risk management involves such tasks as: (1) identifying risk associated with a project, (2) evaluating risks to determine the extent of the risk and how it might affect the project, (3) planning appropriate responses to minimize risk or to minimize the firm's exposure should the worst case occur, and (4) implementing the planned responses.

Process skills are essential to effective project management. Without these skills, construction professionals cannot become good project managers. However, even the best process skills are not—by themselves—enough to make an individual an effective project manager. In addition to process skills, construction professionals who want to be project managers must also develop effective people skills.

PEOPLE FUNCTIONS OF PROJECT MANAGERS

In addition to the process functions of project management, there are also people functions. The people functions have to do with getting the best possible performance from all members of project teams while also encouraging them to improve their performance continually. These two factors—peak performance and continual improvement—are at the heart of the people-oriented responsibilities of construction project managers. The people functions are important because peak performance and continual improvement from project team members are essential to completing projects on time, within budget, and according to specifications. No matter how well a project is planned, scheduled, budgeted, monitored, and tracked, the performance of team members is still a major constraint or asset in construction projects. The people functions of project managers are as follows (Figure 2.3):

- *Leadership function.* Project managers must be able to inspire—by their example—team members to make a total commitment to achieving the team's mission of completing the project on time, within budget, and according to specifications. Hence, project managers must be good leaders. Leadership skills are covered in Chapter Eight.
- *Teambuilding function.* Effective project teams do not just happen. They must be built. Teambuilding is the act of molding a disparate group of individuals—all of whom have their own perspectives, motivations, ambitions, and needs—into a well-functioning, mutually supportive team. Consequently, project managers must be good team builders. Teambuilding skills are covered in Chapter Nine.
- *Conflict management function.* Any time people work together in groups there is the potential for conflict. There are so many potentially competing agendas on many construction projects that conflict is almost a certainty. Spirited debates over the best way to do something and other project-related issues should be encouraged in project teams, but spirited debates can quickly become counterproductive conflicts if not handled

---

**CHECKLIST OF THE PROJECT MANAGER'S
PEOPLE FUNCTIONS**

✓ Leadership

✓ Teambuilding

✓ Conflict management

✓ Motivation

✓ Communication

✓ Time management

✓ Change management

✓ Diversity management

✓ Adversity management

FIGURE 2.3   People functions are half of a project manager's job.

properly. When conflict occurs, it can polarize a project team and render it ineffective because time and energy that should be devoted to getting the job done are instead invested in carrying on petty squabbles. Consequently, project managers must be good conflict managers. Conflict management is covered in Chapter Nine.

- **Motivation function.** People come to project teams with different agendas, outlooks, and attitudes. Plus, people have their good days and bad days. Regardless of all this, the project manager's responsibility to ensure peak performance and continual improvement does not change. Consequently, construction project managers must know how to motivate people to give their best effort to the project. Motivation is covered in Chapter Ten.

- **Communication function.** Effective communication is critical in project teams. First, communication is the oil that lubricates the gears of human interaction. People who work in teams need to know what is expected of them, when it is expected, why it is expected, and how they will be evaluated against expectations. They also need to be kept up-to-date on the daily information flow, on changes that affect them, and on anything else that will help them perform at peak levels and continually improve. Further, they need someone in a position of authority—in this case the project manager—who will listen when they have concerns, complaints, ideas, or recommendations. All of this means that project managers must be effective communicators, which includes being good listeners.

- **Time management function.** Time is always an important consideration for construction project managers. When a contract is received for an external project or a charter is provided for an internal project, time will be one of the most prominent of the various success criteria. Remember that completing projects on time is always a success criterion in construction projects. Consequently, time management is an important function for project managers. They must be good at managing their own time and that of their project teams.

- *Change management function.* Change is not only an ever-present probability for construction project managers, but it also is a predictable reality. Of course, project managers cannot predict what changes will occur during the course of a project, but they can predict that there will be changes. Owners will change the project specifications, a change to the project deadline will be negotiated, important team members will be pulled away to work on other projects, or a risk factor will come into play in its worst-case scenario. Further, the maxim that everything takes longer than you think frequently applies in project management. This maxim and its corresponding reality can force adjustments to be made in the project schedule. Because change is a fact of life in construction project teams, change management is an important function for project managers.
- *Diversity management function.* The American workplace is one of the most diverse in the world, and construction project teams and firms reflect this diversity. It is predictable that the members of project teams will have different backgrounds, genders, races, cultures, worldviews, perspectives, levels of education and experience, agendas, and attitudes. Molding people of diverse backgrounds into mutually supportive team members who work together to ensure that the team performs at its best is an important function of the project manager. Diversity in project teams has the potential to strengthen the team or to cause it to devolve into counterproductive factions separated by diversity. Ensuring that diversity is an asset in project teams is the responsibility of the project manager.
- *Adversity management function.* Project teams will go through times of adversity. Like change, adversity is a predictable fact of life for project managers and the members of project teams. A serious accident might occur at the job site. The construction firm may fall upon hard economic times. Changes in project specifications or deadlines may make it difficult to complete the project on time, within budget, and according to specifications. Project teams in the construction industry will encounter adversity. Consequently, leading their teams through adversity is an important function of project managers. The best project managers are those who can give their team members hope and inspiration in even the darkest of times.

CONSTRUCTION PROJECT MANAGEMENT SCENARIO 2.1

I Don't Like Dealing with People

Danny Cutter is a project manager for a mid-sized construction firm. He does an excellent job of planning, scheduling, cost estimating, procuring resources, tracking progress, reporting, and handling risk. In other words, as a project manager, Cutter has strong process skills. In spite of this, the projects he manages always seem to get mired down in problems. His projects are always completed, but never without a steady stream of what seem to be self-inflicted problems. There always seems to be a lot of conflict and confusion among his team members. The firm's personnel and frequently used subcontractors are vocal about not wanting to be assigned to teams led by Cutter.

After the firm finally got through Cutter's latest project, an especially difficult project plagued by problems from the outset, the company's vice president decided she had to get to the bottom of the problem. In a meeting with Cutter, she came straight to the point.

"Danny you are the best planner, scheduler, estimator, and cost controller we have, but your projects never seem to go well. What is the problem?" Cutter was equally frank in his reply. "I can plan projects with my eyes closed and manage the process aspects of projects with my hands tied behind my back. But when it comes to dealing with owners, team members, and higher management it's another story. I don't like dealing with people."

Discussion Questions

In this scenario, Danny Cutter has excellent process skills but lacks people skills. This shortcoming causes problems when he is assigned to lead project teams. Have you ever worked in any setting with someone who was technically competent, but lacked people skills? If so, did this person's shortcoming cause problems? Elaborate on the situation. Unless something changes, will Danny Cutter ever be an effective project manager?

BECOMING A PROJECT MANAGER

Becoming a project manager is often the first major step up the career ladder for professionals in the field of construction. Many top managers in construction firms got their start as project managers. The career continuum unfolds as follows: (1) construction professional; (2) project manager; (3) program manager; (4) senior corporate executive; and (5) chief executive officer (see Figure 2.4). Since this is the logical career progression for professionals in the field of construction, the question that must be answered is this: How does one become a project manager?

According to David Litten of ProjectSmart, the skill set for potential project managers consists of the eight skills:[1]

- ***Learn how to lead and manage.*** Leadership and management are two related but distinct and different concepts. Project managers lead people. They manage budgets, schedules, time, and risk. Leadership involves sharing the vision, inspiring, motivating, and communicating. Managing involves ensuring that work is completed on time, within budget, and according to specifications. Leadership and management overlap and both are important. Consequently, students and professionals who want to be construction project managers must learn to lead and manage.
- ***Learn how to build and lead teams.*** Construction project teams are, by their very nature, cross-functional. This means they are composed of people from varied backgrounds,

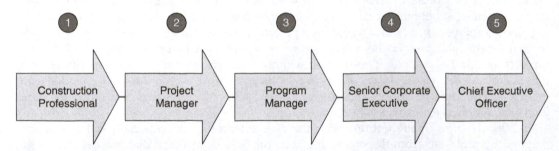

FIGURE 2.4 Career continuum for engineering, quality, and technology professionals.

with different skill sets and different responsibilities to the project. Construction project managers have line authority over only a small number of the members of the project teams they lead. Consequently, they must gain influence with their team members in other ways. Construction professionals who can pull a group of disparate individuals together and mold them into a mutually supportive, well-functioning team are likely to make good project managers.

- ***Learn how to systematically solve problems.*** Construction project managers constantly face both process and people problems. They have to be good at identifying root causes and choosing the best solution alternatives for eliminating the causes. Consequently, students and professionals who want to become construction project managers should practice systematically solving the problems that arise in their studies and jobs. Construction project managers must be good problem solvers.

- ***Learn how to negotiate.*** Project managers must be good negotiators. Negotiating means working with others in mutually supportive ways to reach agreement on an issue of importance. Project managers often negotiate deadlines, schedules, budgets, the use of personnel, and other important factors. In fact, until a project charter is finalized, many of its elements are negotiable. The best case-scenario is when construction project managers are involved from the outset in the initiation stage of projects and when they help develop the firm's response to an RFP or RFQ. In this way, they have a voice in deciding the negotiable elements of the response. Even after a project has been chartered, negotiation skills are still important. Once a project is established, it may be necessary to negotiate with colleagues to use some of their personnel on a given project. It might even be necessary to negotiate with project team members who report to other managers or supervisors when it is necessary to work late or take on more responsibilities in the project. Consequently, students and professionals who want to be construction project managers should concentrate on developing effective negotiating skills.

- ***Learn how to communicate effectively.*** Effective communication is essential for effective teamwork. Communication is the oil that lubricates the gears of human interaction. It is one of those rare things one cannot do too much of. Becoming an effective communicator requires that project managers be good listeners, clear speakers, concise writers, and adept interpreters of nonverbal cues. Students and professionals who want to be construction project managers should concentrate on developing effective communication skills and continually improving them.

- ***Learn how to organize work.*** Time is always of the essence in construction project management. Project managers who cannot manage their own time will not be able to manage that of their teams or their projects. Once a project is implemented, managing the schedule, budget, and risk are critical tasks for the project manager. In order to do these things, project managers must be well-organized. Consequently, students and professionals who want to be construction project managers should practice getting themselves and their work organized.

- ***Learn how to plan effectively.*** The foundation of effective project management is good planning. Project managers must be able to: (1) look at an overall project and break it down into its component parts, (2) determine what types of personnel they will need as project team members and how many, (3) develop a schedule for the work that must be completed, and (4) develop a budget for the project. Planning skills are essential to effective project management. Consequently, students and professionals who want to be construction project managers should concentrate on becoming good planners.

- **Learn how to estimate and budget effectively.** One of the most fundamental questions a construction firm must answer when considering an RFP or RFQ is this: How much will it cost our firm to complete this project? The answer to this question becomes the baseline for adding in profit and contingencies before submitting a response to an RFP or RFQ. The accuracy of the resulting budget is one of the major determining factors of whether the firm submitting the response will make a profit or lose money. Consequently, project managers must be good at estimating and budgeting. This means they must be good at developing, controlling, monitoring, and adjusting budgets. Estimating and budgeting are essential skills for project managers. Consequently, students and professionals who want to become project managers should work on developing their estimating and budgeting skills.

CHARACTERISTICS OF EFFECTIVE PROJECT MANAGERS

Serving as a construction project manager is both challenging and rewarding. Those project managers who best meet the challenges and, as a result, enjoy the most rewards exhibit the following characteristics on a consistent basis:

- **Advanced process skills.** This characteristic may appear obvious since it has already been emphasized, but it is so important that it must be restated. The most effective project managers develop advanced process skills—they make sure that they become thoroughly accomplished at the planning, scheduling, estimating, budgeting, tracking, reporting, procuring, and risk management aspects of project management. Process competence is a strength shared by the most effective project managers.
- **Advanced people skills.** This characteristic, like advanced process skills, may appear obvious, but it should never be taken for granted. Molding a diverse group of people into a high-performing team and leading them in ways that ensure that projects are completed on time, within budget, and according to specifications is never easy. Strong people skills are assets shared by the most effective project managers.
- **Intellectual curiosity.** The most effective project managers are not blindly conforming robots. Rather, they are intellectually curious. They want to know the *why* behind specifications, success criteria, deadlines, and other factors that apply to their projects. By understanding why certain factors apply, they are better prepared to make informed decisions when questions arise or when problems occur.
- **Commitment.** The most effective construction project managers are committed to completing their projects on time, within budget, and according to specifications. They maintain a can-do attitude that can be summarized as follows: *I will do everything necessary—within the limits of legality and ethics—to successfully complete my projects.* Project managers who are committed go beyond just trying or even trying their best. They are determined to get the job done. Consequently, they approach problems, changes, and other inhibitors not as insurmountable barriers, but as roadblocks to go around, over, or through.
- **Vision and insight.** The most effective project managers have both vision and insight. In other words, they understand the big picture and where their project fits into it

(vision). They also understand the details of their project and what must be done to ensure that these details come together in ways that satisfy the vision of their firm and the owner (insight).

- *People orientation.* The most effective project managers understand that even the best planned, best scheduled, best budgeted, and most carefully monitored project is ultimately dependent on people for its success. Owners, team members, the firm's higher management, the customer's representatives, vendors, and subcontractors are all people with their own needs, interests in the project, agendas, and egos. Consequently, effective project managers learn to view people as assets, treat them with respect, and give credit where credit is due. They understand that the way they treat stakeholders will shape how the stakeholders view the project in question.
- *Character.* The most effective project managers are trusted by all stakeholders of their projects (i.e., owners, team members, higher management, architects, engineers, subcontractors, and suppliers). They understand that stakeholders who do not trust them will not cooperate and collaborate in making sure that projects are completed successfully. Consequently, effective project managers make a point of being honest, dependable, and ethical in all of their dealings with stakeholders. Further, they insist that stakeholders reciprocate.

In addition to these characteristics, Duncan Brodie of ProjectSmart offers the following additional characteristics of effective project managers:[2]

- *Focus on solutions.* When problems arise, as they surely will, effective project managers focus on solutions. They do not let themselves become frozen by fear of the consequences of failure, nor do they fall into the trap of paralysis by analysis. Rather, they approach problems systematically, analytically, and with a mind to solving them.
- *Participative and decisive.* Construction project managers lead project teams. Consequently, when decisions must be made, they ask for input from the team members who will have to carry them out. This does not mean they ask for a vote. Rather, it means they get input from those who will be affected by the decision before making it. Once they have collected the input of their team members—the participative aspect of decision making—effective project managers make what they believe to be the best possible decision under the circumstances they face. This is the decisive aspect of their decision making.
- *Focus on the customer.* The best construction project managers have a customer focus. This is important because, ultimately, it is the owner they must please. How would the customer want us to do this? This question is always at the forefront of a good project manager's thoughts. In fact, there will be times when project managers will be forced to play the role of advocate on behalf of the customer. A project that satisfies the customer in all of its aspects is a firm's best marketing tool for generating future work.
- *Focus on win-win outcomes.* The owners who initiate construction projects have their own needs and agendas. Construction firms have their needs and agendas. In fact, all stakeholders in a construction project have their own needs and agendas. The best project managers are those who can keep the needs of all entities affected by problems with the project at the forefront of their thinking and concentrate on finding solutions that allow both parties to benefit. Project managers who think they benefit when their

firm wins and the owner or other key stakeholder lose are shortsighted. Win-lose solutions dampen relationships and make the possibility of future contracts with the same owner unlikely. However, win-win solutions build relationships that result in future contracts.

- *Lead by example.* The best construction project managers expect their team members to "do as I do" rather than "do as I say." Good project managers are good leaders. Good leaders inspire team members to give their best by consistently exemplifying everything they expect of team members. Members of project teams are just like anyone else. They are more likely to follow the leader's example than his words.

- *Get the best from all stakeholders.* Construction project management is a team sport. In order to succeed, project managers must be adept at getting the best from all stakeholders in a project. This means they must take full advantage of the strengths or team members while minimizing their weaknesses. It also means they do what is necessary to pull the individual team members together into a unified whole that performs better than the sum of its individual parts.

PROJECT MANAGEMENT CERTIFICATIONS

Obtaining credentials in a field is one of the prerequisites to climbing the career ladder in that field. Project management is no different. Having a special certification in project management lets construction firms know that individuals have the process and people skills that are essential to effective project management. The Project Management Institute—a leading professional organization for project managers—offers a variety of professional certifications. These include the following:[3]

- *Project Management Professional (PMP)*®. This certification is for individuals who meet certain knowledge and experience requirements. This is the highest level of project management certification.

- *Certified Associate in Project Management (CAPM)*®. This is an entry-level certification with no experience requirement. Professionals in the field of construction who have just been named project managers may wish to test for this certification.

- *Program Management Professional (PgMP)*®. Programs are made up of projects. Therefore, program management is a broader and higher level job than project management. This certification is for experienced program managers who oversee multiple projects that, taken together, are considered a program.

- *PMI Scheduling Professional (PMI-SP)*®. This is focused certification for project management professionals who specialize in the scheduling component of project management. In large firms that handle multiple large projects simultaneously, scheduling is a complex task requiring focused attention. Professionals who work or plan to work primarily as schedulers may wish to test for this certification.

- *PMI Risk Management Professional (PMI-RMP)*®. All project managers are responsible for assessing risk in the projects they manage. However in organizations that handle many large and complex projects simultaneously, risk management is a specialized skill. Professionals who work or plan to work primarily as risk managers may wish to test for this certification.

Project Management Professional Examination

To be certified as a Project Management Professional (PMP), construction project managers must meet specific education and experience requirements and pass an examination. The education and corresponding experience requirements to sit for the PMP examination are as follows:[4]

- High school diploma or an associate degree and 60 months of experience in project management. The 60 months must encompass 7,500 hours of experience.
- Baccalaureate degree with and 36 months of experience in project management. The 36 months must encompass 4,500 hours of project management experience.
- Formal education in project management consisting of a minimum of 35 contact hours. This requirement applies to all candidates regardless of their level of education and length of experience.
- Information about professional certification for project managers is available from the Project Management Institute at www.pmi.org.

CONSTRUCTION PROJECT MANAGEMENT SCENARIO 2.2

You Need to Learn to Look for Win-Win Solutions

Sherry Johns is one frustrated project manager. Before accepting her current position at ABC Construction, Inc., she had been a project manager at a competing firm, Construction Corporation, LLC. ABC Construction had given Johns a substantial salary increase and better benefits to make the jump. Her former colleagues at Construction Corporation think Sherry Johns made a wise career move, but Johns doesn't feel that way. In fact, she is beginning to regret the move and think it was a mistake. Her problem is that the members of ABC Construction's higher management team think differently than their counterparts at Construction Corporation. Her problem is that ABC Construction is structured as a matrix organization. Johns is beginning to think she does not fit in at ABC Construction.

At Construction Corporation, Johns had earned a well-deserved reputation for "looking out for the company" when problems arose in her projects. She had an uncanny ability to find solutions that pinned the blame and the corresponding cost on the owner, architect, engineer, or one of the subcontractors. Since coming to ABC Construction, she has applied this ability quite effectively. However, when owners and other stakeholders have complained to ABC's higher management, they have had a different attitude about her abilities than did her former superiors. In fact, she has even been reprimanded for undermining the trust that existed between an owner and ABC Construction.

In a counseling session with her supervisor recently, Johns was told: "When dealing with stakeholders, you need to learn to look for win-win solutions." Frankly, Johns did not understand what her supervisor meant by this recommendation but she did not want to appear ignorant. Consequently, she did not ask. However, she did intend to find someone who could explain what he meant, and the sooner the better.

Discussion Questions

In this scenario, Sherry Johns is accustomed to finding "one-way" solutions to problems that arise with her projects. She is adept at pinning the blame for problems on any stakeholder other than her construction firm. Her former employer admired this ability but her current employer does not. If Sherry Johns approached you and asked what her supervisor meant by "win-win" solutions, what would you tell her? What would you tell her about why win-win solutions are the best approach when they can be found?

SUMMARY

Construction project managers perform both process and people functions. Process functions include planning, scheduling, estimating cost and duration, procuring, tracking progress, reporting, and handling risk. People functions include leadership, teambuilding, motivation, communication, time management, change management, diversity management, and adversity management.

Project management is a step up the career ladder toward full-time management positions. A common career path for construction professionals is project manager, program manager, senior corporate executive, and chief executive officer. David Litten of ProjectSmart lists the following things that students and professionals who want to be project managers must learn how to do: (1) lead and manage, (2) build and lead teams, (3) systematically solve problems, (4) negotiate, (5) communicate effectively,

(6) organize work, (7) plan effectively, and (8) budget effectively.[5]

Effective project managers have the following characteristics: advanced process skills, advanced people skills, intellectual curiosity, commitment, vision, insight, people orientation, and character. In addition to these characteristics, Duncan Brodie of ProjectSmart adds the following: (1) focus on solutions, (2) participative and decisive, (3) focus on the customer, (4) focus on win-win outcomes, (5) lead by example, and (6) elicit the best from all stakeholders.[6]

Project management certifications are available from the Project Management Institute for construction professionals who meet the education and experience requirements. The various levels of certifications, candidate requirements, and examination information are available from the Project Management Institute at www.pmi.org.

KEY TERMS AND CONCEPTS

Planning projects
Scheduling projects
Estimating the cost and duration of projects
Procuring project resources
Tracking progress and reporting
Handling risk
Leadership function
Teambuilding function
Conflict management function
Motivation function
Communication function
Time management function

Change management function
Diversity management function
Adversity management function
Advanced process skills
Advanced people skills
Intellectual curiosity
Commitment
Vision and insight
People orientation
Character
Project management certifications
Project management professional examination

REVIEW QUESTIONS

1. List and briefly summarize the process functions of project managers.
2. List and briefly summarize the people functions of project managers.
3. When planning a project, what are the six questions that should be asked and answered?
4. List eight skills that make up the skill set for project managers.
5. What are the characteristics of an effective project manager?
6. Explain the various project management certifications that are available to construction professionals.
7. What are the requirements to sit for the Project Management Professional certification examination for an individual who holds a Baccalaureate degree?

APPLICATION ACTIVITIES

The following activities may be completed by individuals or by students working in groups:

1. Identify a project manager in a construction firm who is willing to cooperate and ask this individual the following questions: (a) How did he or she become a project manager? (b) What advice would he or she give a student who wants to become a project manager? and (c) Does this individual recommend becoming certified in project management?

2. Review the process and people functions of project management. Which ones will represent the biggest learning challenge for you? Which ones will come more easily to you? Review the characteristics of an effective project manager. Which of these characteristics do you already have? Which of the ones you will have to develop will come most easily to you? Which of the ones you will have to develop will be the most difficult for you?

ENDNOTES

1. David Litten, "How to Become a Project Manager." Retrieved from http://projectsmart.co.uk/how-to-become-a-project-manager.html on January 14, 2012.
2. Duncan Brodie, "7 Habits of Brilliant Project Managers." Retrieved from http://projectsmart.co.uk/7-habits-of-brilliant-project-managers.html on January 14, 2012.
3. Project Management Institute, "Which Certification is Right for You?" Retrieved from http://www.pmi.org on January 15, 2012.
4. Project Management Institute, "Project Management Professional (PMP)." Retrieved from http://www.pmi.org/Certification/Project-Management-Professional-PMP.aspx on January 20, 2012.
5. David Litten, January 14, 2012.
6. Duncan Brodie, January 14, 2012.

CHAPTER 3

Cost Estimating for Construction Projects

Few, if any, aspects of construction project management are more important than cost estimating. The first thing construction students and professionals need to understand about cost estimating for construction projects is that there is not just one estimate. Rather, there are progressively more detailed and more accurate cost estimates for construction projects. Construction project estimates evolve as the project itself evolves. For example, the owner of a project will want at least a rough cost estimate before deciding to move forward with a project. Such an estimate would necessarily be based on relatively incomplete information. As the project moves through the design phases and the design becomes more and more settled, the cost estimate could become correspondingly more accurate. Consequently, as a construction project moves from initiation to design to construction, the cost estimates for it become increasingly accurate.

Most students of construction project management are concerned primarily with the cost estimates they will help prepare in the future relating to the actual construction costs of a project—the types of estimates used in preparing a bid package in response to a request for proposal (RFP) or request for quote (RFQ). Although this phase of the cost estimation process is the principal focus of this chapter, other phases are covered in sufficient depth that future construction project managers can understand them and even play a contributing role in preparing for these phases. This comprehensive look at cost estimating for construction projects is provided because there will be times when construction project managers are engaged in the estimating process from the very outset of a project as it evolves through the initiation, design, and construction phases. This happens more frequently in larger construction companies than in small companies, but it does happen. Hence, construction students and professionals should be prepared.

APPROACHES TO PREPARING A CONSTRUCTION COST ESTIMATE

As is the case with many key positions in the broad field of construction, there are specialists and generalists who practice the art and science of cost estimating. Larger construction firms often employ specialists who do nothing but prepare cost estimates. Smaller firms rely on project managers to wear several hats, just one of which is cost estimator. Larger and smaller firms occasionally hire consultants to help prepare their estimates. Regardless of whether a construction estimate is prepared by an internal specialist, an outside consultant, or the

---

**CHECKLIST OF
COST ESTIMATION KNOWLEDGE AND SKILLS**

✓ Build a structure mentally.

✓ Work systematically using replicable methodologies.

✓ Document work clearly and logically.

✓ Think critically and investigate to verify data.

✓ Compute labor costs.

✓ Identify cost optimization strategies.

✓ Use extensive hard copy and online reference materials.

✓ Analyze completed estimates.

FIGURE 3.1 Project managers must be able to do these things to be good estimators.

project manager, the same methodologies apply. Further, even if the project manager does not prepare the cost estimate for a construction project, he or she should be involved in the process. Consequently, it is important for construction project managers to understand the basics of construction cost estimating.

GENERIC COST ESTIMATION SKILLS

What must a construction project manager know and know how to do to be an effective cost estimator? What follows is a list of the essentials project managers need to know or know how to do in order to prepare accurate, valid cost estimates for construction projects (Figure 3.1):

- ***Build a structure mentally.*** The scope of a construction project is the sum total of everything that will have to be done to complete it as well as the resources that will be required. Project managers who prepare cost estimates must be able to build the project in their mind without leaving anything out. This is important because the first question a good estimator must answer is not "How much will this project cost?" Before dealing with that important question, an even more pertinent question must be answered: Does our construction firm have the capabilities necessary to even undertake the project? In other words, does the firm have the staff, financial resources, equipment, facilities, and time required? For example, a small construction firm that specializes in single-family residences is not likely to have the wherewithal to take on a large infrastructure or industrial project. To be able to build a project in their mind, project managers must be able to read and understand architectural drawings, engineering drawings, specifications, and contract documents; envision the scope of the project; and match the vision with their firm's capabilities.

- ***Work systematically using replicable methodologies.*** Consistency and uniformity are important factors in construction cost estimating. Consequently, project managers must be able to work systematically using stand methodologies that can be replicated. Estimates for larger projects often require the efforts of several people working simultaneously. If one estimator uses one method while another uses a different method, the consistency of the estimate is undermined. Consistency in the use of estimating methodologies also allows more experienced estimators to check their estimates or, in the event of estimating errors, to quickly find their source.

- **Document work clearly and logically.** It is important to be able to show how any given aspect of a cost estimate was arrived at. When questions arise or problems occur that will require change orders in a project, documentation for the estimate in question becomes critical. Was the problematic estimate based on valid information or invalid information and whose information was it based on (i.e., owner, designer, engineer)? Depending on how they are answered, these questions can save or cost construction firms substantial amounts of money. Proper, comprehensive, logical documentation is often the key to how they are answered. Another instance when proper documentation can come into play is when there are internal disagreements concerning how much cost to assign a given aspect of a project. If estimators disagree, they will typically resolve their disagreement by looking at the historical record. They will find similar situations from past estimates that can be used to guide the current estimate. However, in order to be used in this way the past estimates must be properly, clearly, and logically documented.

- **Think critically and investigate to verify data.** Often an estimate will include bids from subcontractors. One of the worst mistakes an estimator can make is to simply accept a subcontractor's bid at face value. Errors in a subcontractor's bid will become errors in the construction firm's bid. Consequently, project managers must be able to think critically when examining the bids of subcontractors. How accurate have the subcontractor's bids been in the past? Does the bid pass the common-sense test? Is there anything obvious that has been overlooked or left out? Has the bid been purposefully submitted low in hopes of making up the shortages using change orders? Project managers must be able to verify and validate bids submitted by subcontractors before incorporating them into their estimates.

- **Compute labor costs.** Labor costs represent a substantial part of the total costs of any construction project. Consequently, being able to accurately compute them is critical. Project managers must understand and apply the concept of *loaded labor rates*. A loaded labor rate is the hourly wage of a given trade or professional plus other costs borne by the employer, including workers compensation, benefits, and unemployment insurance. It is not uncommon for the loaded labor rate of an individual to be as much as 33 percent higher than his hourly wage. In addition to computing loaded labor rates, project managers must become adept at assigning a specific number of hours to all tasks that must be completed in a project. For example, if the project is a building, they must be able to determine how many hours to allocate for site work, laying the foundation, framing, roofing, electrical work, plumbing, and every other task that must be performed to complete the project.

- **Identify cost optimization strategies.** There are many ways to save money on construction projects. The concept is known as *cost optimization* and project managers who prepare cost estimates must be good at it. Cost optimization involves finding alternative construction methods and materials for meeting the specifications less expensively. This can be especially important when the estimate will be the basis for a construction firm's response to an RFP or RFQ. A construction firm's ability to meet the owner's specified needs less expensively than other firms is one of the keys to success in a competitive environment.

- *Use extensive hard copy and online reference materials.* There are many reference sources available to assist project managers in preparing cost estimates for construction projects (some of these are explained later in this chapter). Project managers must be skilled at identifying valid sources and using them when preparing estimates. Reference materials are particularly valuable for identifying current material and labor costs. Some reference sources use national averages that the project manager must then localize. Project managers will often have their own internal reference materials that pertain to local projects, but when preparing estimates for projects in another location, nationally and regionally normed references are valuable tools.
- *Analyze completed estimates.* Project managers must become skilled at analyzing completed estimates to determine if they pass the tests of common sense and logic. Estimates should be thoroughly analyzed prior to being submitted. Then they should be analyzed again after the project is completed to determine how accurate they were. Discrepancies should be noted and analyzed to determine their source. This information should be recorded and used for improving future estimates. Up front analysis involves comparing the estimate to past projects of a similar type and size, regional averages, and national averages. After the fact analysis involves comparing the estimated costs to actual costs. This type of analysis should also be performed on bids submitted by subcontractors as was explained earlier in the discussion on thinking critically.

CONSTRUCTION PROJECT MANAGEMENT SCENARIO 3.1

His Estimates are Just Well-Informed Guesses

John had years of experience in construction before deciding he wanted to become a construction project manager and try his hand at estimating. He started in the construction business right after graduating from high school. John had worked as a laborer, carpenter, and foreman for more than 10 years when he decided to enroll in college at night to prepare for higher level positions in construction. John wanted to eventually be the CEO of his own construction firm. He enrolled in a Construction Management program at a nearby university.

John is currently studying project management. The class is learning about estimates and John thinks he might have found his niche. He certainly has an excellent grasp of the practical aspects of construction, but his professor is not sure that John should be an estimator. John can envision the details of a construction job, but has trouble seeing the big picture. Further, John likes to take a seat-of-the-pants approach to estimates. He does not like to work systematically, nor does he document his estimates. John's classmates who work in groups with him complain that his estimates appear to be accurate, but they cannot verify how it arrives at his figures. His estimates are not replicable. In fact, they are really just well-informed guesses. A friend and admirer of John in the class wants to help him make the transition from seat-of-the-pants estimator to systematic estimator, but he is not sure what to tell him.

Discussion Questions

In this scenario, John appears to be using his long experience in hands-on construction to make his estimates in class. His estimates are accurate so far, but they are really just informed guesses. Eventually, John will confront an estimate that is outside of his experience. When this happens the quality of his estimates will decline quickly. Assume that you are the friend in this scenario who wants to help John. What should you tell him about the estimating skills he needs to develop and why?

ROLE OF SPECIFICATIONS IN CONSTRUCTION COST ESTIMATES

Guidance for preparing construction cost estimates comes from a variety of sources. One of the most important sources is the collection of construction specifications. The architectural and engineering drawings show the cost estimator what is to be build, but they do not show the specifics concerning the quality of the various phases of construction. For example, the drawings will show windows and doors but not the specific brand, type, or model. Specific quality information is provided in the construction specifications. Consequently, construction cost estimators must have the most up-to-date specifications available if they are to prepare accurate cost estimates.

Construction specifications should be comprehensive and thorough. If done properly they will leave no room for disagreement concerning the expectations of the owner, architect, and engineer—valuable information for the contractor. A typical set of construction specifications for a commercial project is organized as follows:

- Division 0:   Bidding and contract requirements
- Division 1:   General requirements
- Division 2:   Sitework
- Division 3:   Concrete
- Division 4:   Masonry
- Division 5:   Metals
- Division 6:   Wood and plastics
- Division 7:   Thermal and moisture protection
- Division 8:   Doors and windows
- Division 9:   Finishes
- Division 10:   Specialties
- Division 11:   Food service equipment
- Division 12:   Furnishings
- Division 13:   Not used
- Division 14:   Conveying systems
- Division 15:   Mechanical
- Division 16:   Electrical
- Division 17:   Telecommunications
- Division 18:   Audiovisual systems

Each of these divisions is broken down into more specific categories. For example, Division 5—Metals—would include the following categories of specifications at a minimum: (1) structural steel, (2) steel decking, (3) cold-formed metal framing, (4) architectural ornamental handrails and railing, (5) architectural column covers, and (6) miscellaneous metal fabrication. Each of these categories then contains the detailed specifications for its area of concern.

For example, the category "structural steel" would contain detailed specifications for at least the following areas of concern:

- Performance requirements
- Submittals
- Quality assurance
- Delivery, storage, and handling
- Coordination
- Structural steel materials
- Bolts, connectors, and anchors
- Primer
- Fabrication
- Shop connections
- Shop priming
- Galvanizing
- Examination
- Preparation
- Erection
- Field connectors
- Field quality control
- Repairs and protection

Specifications in all of these areas must be taken into account when preparing the portion of the construction cost estimate for structural steel. The actual specifications play a key part in determining the cost. For example, assume that under the heading "Structural Steel Materials" the following specifications appear:

- W shapes: ASTM A 992/A 992M
- Plate and bar: ASTM A 36/A 36M
- Steel pipe: ASTM A 53/A 53M, Type E or S, Grade B

W shapes that meet the criteria to be rated ASTM A 992/A 992M have a different cost than those that are rated to a lower standard. Hence, the construction cost estimator must ensure that not only does he or she determine the correct quantity of W shapes for the project in question, but that he also prices them at the specified rating. This same adherence to detailed specifications must be adhered to throughout the entire project or the construction cost estimate will not be accurate.

TYPES AND LEVELS OF CONSTRUCTION COST ESTIMATES

The two most common types of estimates prepared by construction project managers are unit cost (sometimes called assemblies) and square foot estimates. Square foot estimates multiply standard square foot costs for specific types of projects times the square feet in the project. Unit cost or assemblies estimates are more detailed. With this type of estimate, the project in question is divided into standard elements that correspond to either the products that will be used in construction or the elements of the building itself (i.e., foundation, substructure, superstructure, exterior, roof).

Both types of estimates rely on cost data that is assembled and maintained either by the local firm or by professional third-party companies that are in the business of providing up-to-date construction cost data. The best known of these companies is R.S. Means, Inc. The unit cost or assemblies and square foot estimating methods are explained in this section.

When an owner has an idea or a need for some type of construction, one of the first things he wants to know is how much the project will cost. Because the need for cost estimates begins even before the project's design, there are various levels of construction cost estimates. In each successive level, the estimate should become more informed, more detailed, and more accurate. As is often the case in construction, language can vary. Consequently, the different levels of construction cost estimates sometimes go by different names. What follows are the various levels of construction cost estimates with the most commonly used names for each (Figure 3.2):

- Level 1: Screening or Rough Order of Magnitude Estimates
- Level 2: Preliminary, Conceptual, or Schematic Design Estimates

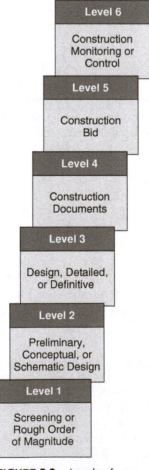

FIGURE 3.2 Levels of construction estimates.

- Level 3: Design, Detailed, or Definitive Estimates
- Level 4: Construction Documents Estimates
- Level 5: Construction Bid Estimates
- Level 6: Construction Monitoring or Control Estimates

The most pertinent of these levels of construction estimates for project managers is Level 5: Construction Bid Estimates. However, since—depending on circumstances—construction managers might be involved in the preparation of estimates at any or all of these levels, they are all explained herein.

Screening or Rough Order of Magnitude Estimates

The screening or rough order of magnitude estimate is used primarily to determine the feasibility of a project. It should provide an owner with sufficient cost information to know if the project is feasible from a financial point of view. If an owner is contemplating a residential construction project—for example, a new home—he or she will need to know approximately how much the home will cost. With this information in hand, the owner can determine if he or she can afford to build the home or if he or she will need to modify his or her dream or even drop the project altogether.

If a business person is considering building a new office complex in which he will lease space to attorneys, physicians, accountants, and other professional personnel, he will need to know the approximate cost. With this information in hand he can determine if the complex is likely to generate sufficient income to make it a good business investment. If a manufacturing firm is considering adding a new product line that will require the construction of a new facility, the firm's executive management team will need to know the approximate cost so that it can be weighed against the amount of new business the new product line is likely to generate.

If the local county commission is considering adding a certain number of miles of new road with accompanying infrastructure elements such as water, sewer, and drainage, it will need to know how much the project will cost. With this information in hand, the county commissioners can make an informed judgment concerning whether to issue bonds, raise taxes, assess road tolls, or defer the project due to cost.

In all of these cases, only very preliminary cost data are used. For example, the approximate cost of the home, office complex, or manufacturing facility would be determined using cost per square foot estimates. Estimators would identify similar projects in similar locations, determine the cost per square foot to build these projects, and apply those costs in preparing rough estimates. The cost of the road or infrastructure project would be determined by applying current cost-per-mile figures to establish a rough estimate. Rough order of magnitude or screening estimates are just that—rough estimates. However, they must be accurate enough to allow those involved to make an informed go/no-go decision for the project.

PREPARING SCREENING/ROUGH ORDER OF MAGNITUDE ESTIMATES.   When an owner takes the first tentative steps toward contracting for a construction project, one of the first questions he needs answered concerns the approximate cost of the project. In order to arrive at an approximate cost so the owner can determine if his idea is feasible, design personnel often use the square foot method for arriving at a rough estimate of the cost. Larger firms often keep their own square foot cost data based on the historical record of their building projects. Square foot data are also available from commercial sources that can located on the Internet. Commercially available cost estimating data typically display national averages that must then be increased or reduced to localize them.

Assume that a restaurant is to be built locally and the design firm contacted by the owner has its own historical data on the mean per square foot costs of other similar projects built in the area. That data would be displayed on a chart such as the one in Figure 3.3. The estimator would refer to the chart of historical data (see Figure 3.3) and find that restaurants in the community are costing $230 per square foot to build. All of the figures in the chart in Figure 3.3, including the $230 per square foot for restaurants, were derived by closely monitoring the actual final costs of other projects that have been built in the community. Final costs of similar projects are divided by their overall square footage. The square footage costs in charts such as the one in Figure 3.3 are averages.

Assume that the owner in this case is interested in building a restaurant that will be 5,800 square feet. By simply multiplying $230 per square foot times 5,800 the design firm could inform the owner that the building will cost him approximately $1.334 million. This, of course, is just a rough estimate, but it is close enough to allow the owner to determine if the project he has in mind is actually feasible.

A problem that sometimes arises when using historical square foot data for a screening or rough order of magnitude estimate is that the data are old. In cases such as this, estimators apply a percentage increase to the square foot cost figures before multiplying them times the

Square Foot Costs				
Building Type	**Median Cost per Square Foot**	**Typical Size Gross Square Foot**	**Typical Range Gross Square Foot**	
Apartments, Low Rise	$114.00	21,000	9,700 –	37,200
Apartments, Mid Rise	$144.00	50,000	32,000 –	100,000
Apartments, High Rise	$164.00	145,000	95,000 –	600,000
Medical Offices	$186.00	6,000	4,000 –	15,000
Banks	$254.00	4,200	2,500 –	7,500
Offices, Low Rise	$161.00	20,000	5,000 –	80,000
Offices, Mid Rise	$161.00	120,000	20,000 –	300,000
Offices, High Rise	$204.00	260,000	120,000 –	800,000
Dept. Stores	$107.00	90,000	44,000 –	122,000
Restaurants	$230.00	4,400	2,800 –	6,000
Retail Stores	$113.00	7,200	4,000 –	17,600
Factories	$102.00	26,400	12,900 –	50,000
Supermarkets	$111.00	44,000	12,000 –	60,000
Garages, Parking	$69.00	163,000	76,400 –	225,300
Hospitals	$305.00	55,000	27,200 –	125,000
Warehouses	$81.00	25,000	8,000 –	72,000
Warehouse & Office	$89.00	25,000	8,000 –	72,000

FIGURE 3.3   Square foot costs such as these are local, regional, or national averages.

Owner's Proposed Project

An owner wants to build a high-rise apartment complex containing 150,000 square feet on the beach in Destin, Florida. The design/construction firm built a similar complex in Fort Walton Beach, Florida, two years ago. The cost of the project in Fort Walton Beach was $24.6 million. The design/construction firm knows that the data for the Fort Walton Beach project are two years old and that building cost in the two cities are different. The company's cost accounting department maintains time and location indices for the most locations in which it builds. The time and location index for Fort Walton Beach is 32. The time and location index for Destin is 88.

$$\frac{\text{FWB building cost}}{\text{FWB index}} = \frac{\text{Destin building cost}}{\text{Destin index}}$$

$$\frac{\$24,600,000}{32} = \frac{\text{Destin building cost}}{88}$$

$$\frac{24,600,000 \times 88}{32} = \$67,650,000$$

Estimated project cost = $67,650,000

FIGURE 3.4 Adjusting square foot cost using time and location indices.

actual square footage of the project in question. For example, assume that the $230 figure used in the restaurant example is two years old. The estimator knows that local construction costs have increased by 6 percent during this two year period. The $230 per square foot figure would be increased by 6 percent ($230 × 1.06 = $243.80) to establish a current per square foot cost of $244.

Another situation that arises when preparing screening or rough order of magnitude estimates is that the square foot data available are for one location but the construction project will be in another location. Further, the square foot data available are old and must be adjusted before being used. This means that two adjustments must be made before preparing the estimate: (1) location cost adjustment and (2) time cost adjustment. Figure 3.4 contains an example of how this type of situation is handled by estimators.

An owner wants to build a high-rise apartment complex on the beach in the resort town of Destin, Florida; a potentially costly project. Consequently, he is very interested in the cost of the project. The design/construction firm he approaches with the idea built a similar project on the beach in another resort community in Florida: Fort Walton Beach. The firm's cost accounting department maintains accurate records of its building costs for different types of projects and charts them for estimators (see Figure 3.3). The firm also maintains time and location indices for the various cities in which it typically builds.

The data available for estimators come from the project in Fort Walton Beach. The data are accurate, but they are two years old. A further complication is that building costs in Fort Walton Beach and Destin are not the same. The firm has its own time and location indices for both towns. Firms that do not maintain this type of data can find it easily through commercial sources on the Internet for most cities in the United States. With the cost and date of the Fort Walton Beach project known, the approximate cost of the Destin project was calculated by applying the time and location indices for each town.

Preliminary, Conceptual, or Schematic Design Estimates

When a firm moves into the schematic phase of the design process, the scope of the proposed construction project becomes to be better defined. As a result, the cost estimate can, in turn, be better defined. This is the nature of construction cost estimates. At each successive level, the estimate becomes better defined and more accurate. In this step, the designer develops schematic drawings, rough renderings or three-dimensional computer models, elevations, sections, and generic specifications. Soil tests, site data, and utility requirements are also pulled together in preliminary form. The data from these various preliminary tools are used to help the owner and designer make decisions about the design that can decrease or increase the cost.

It is not uncommon for Level 2 cost estimates to result in changes to the owner's plans—both major and minor. In order to stay within the owner's budget, it might be necessary to reduce the size of the project, make room and space planning changes, or even lower the owner's expectations. It is often in this stage of estimating the cost of a project that the owner is confronted with the hard realities of his dream. On the other hand, it is not uncommon for the design firm to be able to show the owner how he or she can embellish on his or her dream as a result of more accurate data.

Design, Detailed, or Definitive Estimates

Estimates at this level continue the trend of becoming more defined and more accurate. By now the design firm knows most of what it will need to know to complete the plans for the project. Space and room planning have been firmed up; materials have been selected; schedules for finishes, doors, partitions, and hardware are under development; the engineering design is at least partially complete; equipment layouts have been planned; and specifications have been outlined and the outlines are being filled out with definite information. At this stage in the process, the design firm can begin to the value engineering process in which it seeks ways to meet the owner's expectations and satisfy the specifications less expensively.

COMMONLY USED FORMATS FOR COST ESTIMATING

It is important to take a systematic approach to construction cost estimating. To assist construction professionals in this regard, there are two commonly used models. One is based on the *unit price* approach to construction cost estimating. The other is based on the *assemblies* approach. A widely used model for making unit price cost estimates is the CSI 50-Division MasterFormat® developed by the Construction Specifications Institute or CSI. A widely used model for making assemblies cost estimates is the UNIFORMAT II-12 Division Format. Construction project managers should be familiar with both of these cost-estimating models.

Unit Price Model: Construction Specifications Institute/CSI 50-Division MasterFormat®

The CSI 50-Division format was developed to promote better communication among the technical stakeholders in construction projects (i.e., architects, engineers, specification writers, construction firms, suppliers, and consultants). The format was designed to promote more detailed, more systematic specifications and to reduce errors that result from incomplete information. It is a product-based system. Figure 3.5 shows the 50 divisions

General Requirements Subgroup

- Division 01 – General Requirements

Facility Construction Subgroup

- Division 02 – Existing Conditions
- Division 03 – Concrete
- Division 04 – Masonry
- Division 05 – Metals
- Division 06 – Wood, Plastics, and Composites
- Division 07 – Thermal and Moisture Protection
- Division 08 – Openings
- Division 09 – Finishes
- Division 10 – Specialties
- Division 11 – Equipment
- Division 12 – Furnishings
- Division 13 – Special Construction
- Division 14 – Conveying Equipment
- Division 15 – RESERVED FOR FUTURE EXPANSION
- Division 16 – RESERVED FOR FUTURE EXPANSION

Facility Services Subgroup

- Division 20 – RESERVED FOR FUTURE EXPANSION
- Division 21 – Fire Suppression
- Division 22 – Plumbing
- Division 23 – Heating Ventilating and Air Conditioning
- Division 24 – RESERVED FOR FUTURE EXPANSION
- Division 25 – Integrated Automation
- Division 26 – Electrical
- Division 27 – Communications
- Division 28 – Electronic Safety and Security
- Division 29 – RESERVED FOR FUTURE EXPANSION

Site and Infrastructure Subgroup

- Division 30 – RESERVED FOR FUTURE EXPANSION
- Division 31 – Earthwork
- Division 32 – Exterior Improvements
- Division 33 – Utilities
- Division 34 – Transportation
- Division 35 – Waterways and Marine Construction
- Division 36 – RESERVED FOR FUTURE EXPANSION
- Division 37 – RESERVED FOR FUTURE EXPANSION
- Division 38 – RESERVED FOR FUTURE EXPANSION
- Division 39 – RESERVED FOR FUTURE EXPANSION

Process Equipment Subgroup

- Division 40 – Process Integration
- Division 41 – Material Processing and Handling Equipment
- Division 42 – Process Heating, Cooling, and Drying Equipment
- Division 43 – Process Gas and Liquid Handling, Purification and Storage Equipment
- Division 44 – Pollution Control Equipment
- Division 45 – Industry-Specific Manufacturing Equipment
- Division 46 – Water and Wastewater Equipment
- Division 47 – RESERVED FOR FUTURE EXPANSION
- Division 48 – Electrical Power Generation
- Division 49 – RESERVED FOR FUTURE EXPANSION

FIGURE 3.5 This is the format recommended by the Construction Specifications Institute.

established by the CSI. Estimators use the 50 divisions to organize their estimates of projects. A separate estimate is developed for each division on the list that applies to the project in question. The sum of the various divisions equals the estimate for the overall project.

Systems or Assemblies Model: UNIFORMAT II System Format

The systems or assemblies model groups several different trades into broad assemblies that match the actual systems of a building (e.g., foundation, substructure, superstructure, exterior, roofing). The most widely used model for making assemblies estimates is the UNIFORMAT II Systems model. UNIFORMAT II comes from ASTM E1557: *"Standard Classification for Building Elements and Related Sitework."* Its purpose is to provide a standardized, systematic format for tying together building designs, specifications, and estimates. The UNIFORMAT II model improves communication among stakeholders in construction projects and provides estimators with a standard format that ensures uniformity from project to project. This, in turn, improves the reliability of the data used in making estimates.

Each element in the UNIFORMAT II model is a major functional component that is present in most buildings. Whereas the CSI MasterFormat® is product-based system, UNIFORMAT II is building assembly based. The UNIFORMAT II elements are as follows:

- A: Substructure
- B: Shell
- C: Interiors
- D: Services
- E: Equipment and furnishings
- F: Special construction
- G: Building sitework

The cost data for each of these elements is available in book form and online from R.S. Means under "assemblies cost data." Figures 3.6 and 3.7 show the type of cost data provided by R.S. Means.

Comparison of Unit Price and Assemblies (Systems) Estimating Models

Construction project managers will use both the unit price and the assemblies models for making cost estimates depending on the stage of the project. Consequently, it is important for construction project managers to understand the differences between these two widely used models. R.S. Means describes the difference between the unit price model and the assemblies model (also referred to a "systems" model) of cost estimating as follows: "In a Unit Price estimate, each item is normally included along the guidelines of the 50–Division

B10 Superstructure

B1010 Floor Construction

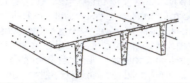

Most widely used for moderate span floors and roofs. At shorter spans, they tend to be competitive with hollow core slabs. They are also used as wall panels.

System Components	QUANTITY	UNIT	COST PER S.F.		
			MAT.	INST.	TOTAL
SYSTEM B1010 235 6700					
PRECAST, DOUBLE "T", 2" TOPPING, 30' SPAN, 30 PSF SUP. LOAD, 18" X 8'					
Double "T" beams, reg. wt, 18" x 8' w, 30' span	1.000	S.F.	10.21	1.66	11.87
Edge forms to 6" high on elevated slab, 4 uses	.050	L.F.	.01	.20	.21
Concrete ready mix, regular weight, 3000 psi	.250	C.F.	1.01		1.01
Place and vibrate concrete, elevated slab less than 6", pumped	.250	C.F.		.37	.37
Finishing floor, monolithic steel trowel finish for finish floor	1.000	S.F.		.82	.82
Curing with sprayed membrane curing compound	.010	C.S.F.	.06	.09	.15
TOTAL			11.29	3.14	14.43

B1010 234 — Precast Double "T" Beams with No Topping

	SPAN (FT.)		SUPERIMPOSED LOAD (P.S.F.)	DBL "T" SIZE D (IN.) W (FT.)	CONCRETE "T" TYPE	TOTAL LOAD (P.S.F.)	COST PER S.F.		
							MAT.	INST.	TOTAL
1500	30		30	18x8	Reg. Wt.	92	10.20	1.66	11.86
1600		RB1010	40	18x8	Reg. Wt.	102	10.35	2.14	12.49
1700		-010	50	18x8	Reg. Wt	112	10.35	2.14	12.49
1800			75	18x8	Reg. Wt.	137	10.40	2.22	12.62
1900			100	18x8	Reg. Wt.	162	10.40	2.22	12.62
2000	40		30	20x8	Reg. Wt.	87	7.80	1.38	9.18
2100			40	20x8	Reg. Wt.	97	7.90	1.65	9.55
2200		RB1010	50	20x8	Reg. Wt.	107	7.90	1.65	9.55
2300		-100	75	20x8	Reg. Wt.	132	7.95	1.78	9.73
2400			100	20x8	Reg. Wt.	157	8.10	2.17	10.27
2500	50		30	24x8	Reg. Wt.	103	8.40	1.24	9.64
2600			40	24x8	Reg. Wt.	113	8.45	1.51	9.96
2700			50	24x8	Reg. Wt.	123	8.50	1.62	10.12
2800			75	24x8	Reg. Wt.	148	8.50	1.64	10.14
2900			100	24x8	Reg. Wt.	173	8.65	2.03	10.68
3000	60		30	24x8	Reg. Wt.	82	8.50	1.64	10.14
3100			40	32x10	Reg. Wt	104	9.90	1.35	11.25
3150			50	32x10	Reg. Wt.	114	9.80	1.17	10.97
3200			75	32x10	Reg. Wt.	139	9.85	1.26	11.11
3250			100	32x10	Reg. Wt	164	9.95	1.58	11.53
3300	70		30	32x10	Reg. Wt.	94	9.85	1.24	11.09
3350			40	32x10	Reg. Wt.	104	9.85	1.26	11.11
3400			50	32x10	Reg. Wt.	114	9.95	1.58	11.53
3450			75	32x10	Reg. Wt.	139	10.05	1.88	11.93
3500			100	32x10	Reg. Wt.	164	10.30	2.51	12.81
3550	80		30	32x10	Reg. Wt	94	9.95	1.57	11.52
3600			40	32x10	Reg. Wt.	104	10.20	2.20	12.40
3900			50	32x10	Reg. Wt.	114	10.30	2.50	12.80

FIGURE 3.6 B10 Superstructure—B1010 Floor Construction.
Courtesy of R.S. Means Company.

B20 Exterior Enclosure

B2010 Exterior Walls

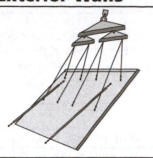

The advantage of tilt up construction is in the low cost of forms and placing of concrete and reinforcing. Tilt up has been used for several types of buildings, including warehouses, stores, offices, and schools. The panels are cast in forms on the ground, or floor slab. Most jobs use 5-1/2" thick solid reinforced concrete panels.

Design Assumptions:
Conc. fc = 3000 psi
Reinf. fy = 60,000

System Components	QUANTITY	UNIT	COST EACH		
			MAT.	INST.	TOTAL
SYSTEM B2010 106 3200					
TILT-UP PANELS, 20'X25', BROOM FINISH, 5-1/2" THICK, 3000 PSI					
Apply liquid bond release agent	500.000	S.F.	10	65	75
Edge forms in place for slab on grade	120.000	L.F.	36	400.80	436.80
Reinforcing in place	.350	Ton	329	376.25	705.25
Footings, form braces, steel	1.000	Set	415.80		415.80
Slab lifting inserts	1.000	Set	54		54
Framing, less than 4" angles	1.000	Set	70.30	788.50	858.80
Concrete ready mix, regular weight, 3000 psi	8.550	C.Y.	931.95		931.95
Place and vibrate concrete for slab on grade, 4" thick, direct chute	8.550	C.Y.		204.86	204.86
Finish floor, monolithic broom finish	500.000	S.F.		370	370
Cure with curing compound, sprayed	500.000	S.F.	30.75	44.50	75.25
Erection crew	.058	Day		649.60	649.60
TOTAL	500.000		1,877.80	2,899.51	4,777.31
COST PER S.F.			3.76	5.80	9.56

B2010 106	Tilt-Up Concrete Panel	COST PER S.F.		
		MAT.	INST.	TOTAL
3200	Tilt up conc panels, broom finish, 5-1/2" thick, 3000 PSI	3.75	5.80	9.55
3250	5000 PSI	3.88	5.70	9.58
3300	6" thick, 3000 PSI	4.13	5.95	10.08
3350	5000 PSI	4.28	5.80	10.08
3400	7-1/2" thick, 3000 PSI	5.25	6.15	11.40
3450	5000 PSI	5.45	6.05	11.50
3500	8" thick, 3000 PSI	5.65	6.30	11.95
3550	5000 PSI	5.90	6.20	12.10
3700	Steel trowel finish, 5-1/2" thick, 3000 PSI	3.75	5.90	9.65
3750	5000 PSI	3.88	5.80	9.68
3800	6" thick, 3000 PSI	4.13	6.05	10.18
3850	5000 PSI	4.28	5.95	10.23
3900	7-1/2" thick, 3000 PSI	5.25	6.25	11.50
3950	5000 PSI	5.45	6.15	11.60
4000	8" thick, 3000 PSI	5.65	6.40	12.05
4050	5000 PSI	5.90	6.30	12.20
4200	Exp. aggregate finish, 5-1/2" thick, 3000 PSI	4.07	5.90	9.97
4250	5000 PSI	4.21	5.80	10.01
4300	5" thick, 3000 PSI	4.46	6.05	10.51
4350	5000 PSI	4.60	5.95	10.55
4400	7-1/2" thick, 3000 PSI	5.60	6.30	11.90
4450	5000 PSI	5.80	6.15	11.95

FIGURE 3.7 B20 Exterior Enclosure—B2010 Exterior Walls.
Courtesy of R.S. Means Company.

MasterFormat of the Construction Specifications Institute. In a Systems estimate, the same items are allocated to one of seven major group elements in the UNIFORMAT II organization. Certain items that were formerly grouped into a single trade breakdown must now be allocated among two or more systems. An example of this difference would be concrete. In a Unit Price estimate, all the concrete items on the job would be priced in the 'concrete' section of the estimate, CSI Division 03. In a Systems estimate, concrete is found in a number of locations. For instance, concrete is used in all of these systems: Division A10, Foundations; Division A20, Basement Construction; Division B10, Superstructure; and Division B20, Exterior Closure."[1]

Construction Document Estimates

In this phase the various construction documents—floor plans, elevations, sections, schedules, engineering drawings, equipment layouts, and specifications—are either complete or at least 90 percent complete. In this step estimators validate and correct previous estimates based on having complete or almost complete information. In this phase the construction documents are checked to: (1) ensure they are ready to be used to guide and evaluate the bids of subcontractors and (2) determine if modifications to the design have affected the scope of the project (a concept known as "design creep" or "scope creep").

Construction Bid Estimates

Although construction project managers might be involved in the development of estimates at any of the levels presented, it is the construction bid estimate that they are most likely to participate in developing. In larger construction firms that maintain a staff of estimators, the project manager will be a participant in developing construction bid estimates. In smaller firms the project manager will prepare the estimate. Construction bid estimates can be developed at this level (Level 5) when the construction documents are complete or they can be developed at Levels 3 or 4 depending on the delivery system that will be used.

If the delivery system is the design-bid-build approach, construction bid estimates will be based on complete construction documents. If the design-build or construction management approaches are used, the construction bid estimate can be prepared earlier (Level 3 or 4). Regardless of when they are prepared, the construction bid estimate is used as the basis of the construction firm's response to the owner's RFP or RFQ. It is also used to evaluate the bids that will be received from subcontractors.

Construction bid estimates typically use the assemblies or unit cost method explained earlier. Whether the CSI's 50-division format or the UNIFORMAT System is used is up to the preferences of the estimator's firm. The final product that results from the construction bid estimate is a comprehensive summary document like the one shown in Figure 3.8. Because it corresponds with the way a building is actually constructed, the Uniformat System has gained in popularity and appears to be the more commonly used system.

PROJECT BUDGET

Job:

Cost Code	Work Type	Quantity	Unit	Direct Labor Cost	Labor Unit Cost	Equipment Cost	Equipment Unit Cost	Material Cost	Overhead Cost
01500.00	Move In	ls		$	$	$	$	$	$
01700.00	Clean Up	ls		$	$	$	$	$	$
			Subtotals	$		$		$	$
32220.10	Excavation, unclassified		cy	$	$	$	$	$	$
32222.10	Excavation, structural		cy	$	$	$	$	$	$
32226.10	Backfill, compacted		cy	$	$	$	$	$	$
31350.00	Piledriving rig, mobilization & demobilization	job	ls	$	$	$	$	$	$
31361.10	Piling, steel, driving		lf	$	$	$	$	$	$
			Subtotals	$		$		$	$
	Concrete								
03150.10	Footing forms, fabricate		sf	$	$	$	$	$	$
03150.20	Abutment forms, prefabricate		sf	$	$	$	$	$	$
03157.10	Footing forms, place		sf	$	$	$	$	$	$
03159.10	Footing forms, strip		sf	$	$	$	$	$	$
03157.20	Abutment forms, place		sf	$	$	$	$	$	$
03159.20	Abutment forms, strip		sf	$	$	$	$	$	$
03157.30	Deck forms, place		sf	$	$	$	$	$	$
03159.30	Deck forms, strip		sf	$	$	$	$	$	$
03251.30	Concrete, deck, saw joints		lf	$	$	$	$	$	$
03311.10	Concrete, footings, place		cy	$	$	$	$	$	$
03311.20	Concrete, abutments, place		cy	$	$	$	$	$	$
03311.30	Concrete, deck, place & screed		sy	$	$	$	$	$	$
03345.30	Concrete, deck, finish		sf	$	$	$	$	$	$
03346.20	Concrete, abutments, rub		sf	$	$	$	$	$	$
03370.20	Concrete, abutments, curing		sf	$	$	$	$	$	$
03370.30	Concrete, deck, curing		sf	$	$	$	$	$	$
			Subtotals	$		$		$	$
	Metals								
05120.00	Steel, structural, place		lb	$	$	$	$	$	$
05520.00	Guardrail		lf	$	$	$	$	$	$
05812.00	Bearing plates		lb	$	$	$	$	$	$
			Subtotals	$		$		$	$

FIGURE 3.8 Project Budget Summary Form.

60

Challenging Construction Project

EMPIRE STATE BUILDING

Although its record has long since been supplanted by other buildings, when the Empire State Building was completed in 1931 it was a marvel of the construction profession. At the time it was the tallest building in the world and the first one to surpass the 1,000 feet tall mark (it was 1,250 feet tall). The construction firm that built the Empire State Building—Starrett Brothers and Eken—was well experienced at undertaking large construction projects. The firm had already constructed office towers, hospitals, banks, and hospitals in the United States and Europe. The Empire State Building was to be the culmination of the careers of the Starrett Brothers, and it turned out to be a fitting monument to their talents.

The Empire State Building is 103 stories tall and—amazingly for the time—was built in just 15 months. Constructing what at the time was the world's tallest building presented numerous challenges including the following: 1) the design and manufacture of specialized equipment made just for that project, 2) supplies delivered to the site in finished or almost finished condition to avoid the need for on-site prep work, 3) hiring of more than 60 different types of trades, 4) suppliers and subcontractors who could adhere to the tightest possible schedule, 5) the first-ever use of construction fast-tracking where in the actual construction is started before the design is completed, 6) simultaneous excavation of the new site and demolition of the building that currently occupied the site, 7) constructing an on-site railway system for moving materials quickly around the site, 8) development of a series of derricks for passing steel girders up to the desired level (derricks at the time could reach no farther than 30 stories, 9) development of a hopper system for delivering bricks to the various stories as the building went up, and 10) establishing a just-in-time system that allowed work to begin precisely when it was needed and in many cases overlap with other work without interfering with it.

The construction firm employed 3,500 people at the site who worked seven million man hours. In addition to all of the steel used in the framework for the building, construction materials included 62,000 cubic yards of concrete, 200,000 cubic feet of limestone and granite, 10,000 square feet of marble, 6,500 windows, The Empire State Building would be a challenge today, even with all of the advances in construction management and technology. In 1931, it was truly a marvel. It remained the tallest building in the world until 1972 when the World Trade Center Twin Towers were constructed.

Source: Based on Bob Moore Construction, "Empire State Building: A Case Study in Successful Commercial Construction Management." www.commercialconstruction.com/historic-construction-projects/empire-state-building

Construction Monitoring Estimates

The construction monitoring estimate is not a new estimate that is prepared after the construction bid estimate. Rather, it is the budget that is used to monitor and control the cost of the project as change orders, cost overruns, and cost savings are factored in. The construction bid estimate is the baseline the construction firm begins with on the project. It becomes the budget. It is updated continually throughout the project as changes of any kind occur.

On most construction projects, but especially on larger projects, there will be change orders. As they occur, whether they increase the cost of the project or decrease it, the budget

---

CONSTRUCTION COST ESTIMATING PROCESS

- Organization of the estimate
- Quantity takeoff
- Unit prices
- Labor costs
- Material costs
- Equipment costs
- Subcontractors costs
- Indirect costs
- Profit

FIGURE 3.9 Specific tasks in the construction cost estimating process.

for the project must be updated and documentation relating to all changes must be filed. In the final analysis, the actual cost of the project is what both the owner and the construction firm need to know. The owner needs this figure because he has to pay the bill. The construction firm needs the figure to add to its historical record for use in doing two things: (1) determining how accurate its estimate was so that weaknesses in its estimating processes can be strengthened and (2) providing accurate data for use in preparing estimates on future projects.

CONSTRUCTION COST ESTIMATING PROCESS

The construction cost estimating process involves several specific tasks (see Figure 3.9). Regardless of whether a cost estimate is prepared manually, with the assistance of estimating software, or using a combination of the two, the tasks in Figure 3.9 must be understood by construction project managers if the estimates are to be comprehensive, accurate, and dependable. These basic estimating tasks are explained in the following sections.

Getting Organized

Only in the smallest of firms will one person prepare a construction cost estimate. Even in cases such as this, the person doing the estimate will receive information from a variety of different sources. With larger firms, the project manager will oversee a team of estimators with each member assigned to complete a different element of the estimate. For example, one member might focus on quantity takeoffs, another on labor, another on material, and so on. With this many people working on the estimate and using information from numerous different sources, getting organized before beginning the estimate is essential. What follows are some guidelines that can be used for organizing the cost estimating process:

- *Make sure that an entire set of the latest editions of all construction documents is on hand (e.g., contract, architectural drawings, engineering drawings, specifications).* All of these documents should be placed in a central location such as a conference room (unless they are available to the team online). The construction drawing for commercial and industrial projects should include at least the following: civil engineering, architectural, structural, mechanical, and electrical drawings. There may also be separate drawings for fire suppression and security systems. Make sure that addenda to construction documents are readily available so that all team members are working with the official, latest, and most accurate information. Make sure that all construction documents are coded in some way and numbered for easy identification and to promote better communication among the members of the estimating team.

- *Establish a file control system so that everyone on the team knows where all construction documents are at all times and can easily access them.* Teams that are using cost estimating software and shareware do not need to worry about this step as the files will be retained and made available electronically.
- *Examine the specifications carefully.* Pay special attention to any unusual or nontypical aspects of the specifications such as special conditions or agreements that will affect the cost of the project. Also determine whether comparable materials can be substituted should the need arise or should value engineering reveal opportunities to save on material costs by substituting materials comparable to those specified.
- *Consider all of the factors that can affect the cost estimate.* These factors include: (1) quality of materials available, (2) productivity of the trades and subcontractors, (3) allowances for overtime, (4) size and scope of the project, (5) physical location of the site, (6) geographic location of the project, (7) season (winter, summer, etc.), (8) weather conditions, (9) building code requirements, (10) safety requirements, (11) union restrictions, (12) availability of materials and labor, (13) availability of water and electricity, and (14) special requests of the owner.

Preparing the Quantity Takeoff

The most fundamental component of the construction cost estimate is the quantity takeoff. The purpose of the quantity takeoff is to determine how much material of all required types will be needed to complete the project. The key to preparing an accurate quantity takeoff is to identify every type of material that will be needed. To assist with this challenge, estimators can use such tools as the CSI 50-division system shown in Figure 3.5. Using this method, the quantity takeoff requires that the estimator identify which of the CSI's elements applies to the project in question and then determine how much will be needed for each element.

For example, assume that the estimator is trying to determine how much concrete will be needed for a retaining wall that is shown on the drawings to be 3'-0" tall, 2'-0" thick, and 150'-0" long. Concrete requirements are expressed in cubic feet. The estimator could manually perform the computation (3'-0" × 2'-0" × 150'-0" = 900 cubic feet). If using estimating software, the estimator simply enters the dimensions and the computer makes the calculation. Either way, the estimator would record 900 cubic feet as being needed for this particular item in the overall estimate. Once all concrete requirements have been calculated, they are totaled and recorded on the quantity takeoff form as shown in Figure 3.10.

One of the minor items of knowledge needed for preparing quantity takeoffs is the measurements used for quantifying different materials. For example, concrete is measured in cubic feet, floor tile is measured in square feet, and liquid coatings such as paint and sealer are measured in square feet per gallon. There are numerous quantity takeoff forms available commercially and online to assist estimators with this task. Of course, teams using estimating software will have the forms built in. Figure 3.10 is an example of a portion of a completed quantity takeoff form showing the concrete requirements for a project.

QUANTITY TAKEOFF: CONCRETE			
Cost Code	**Work Subgroup**	**Unit**	**Quantity**
	Concrete		
03150.10	Footing forms, fabricate	sf	450
03150.20	Abutment forms, prefabricate	sf	1,627
03157.10	Footing forms, place	sf	1,051
03159.10	Footing forms, strip	sf	998
03157.20	Abutment forms, place	sf	4,054
03159.20	Abutment forms, strip	sf	4,054
03157.30	Deck forms, place	sf	1,728
03159.30	Deck forms, strip	sf	1,728
03200.00	Steel, reinforcing, place	lb	85,860
03251.30	Concrete, deck, saw joints	lf	73
03311.10	Concrete, footings, place	cy	127
03311.20	Concrete, abutments, place	cy	287
03311.30	Concrete, deck, place & screed	sy	216
03345.30	Concrete, deck, finish	sf	1,998

FIGURE 3.10 Quantity takeoff for the concrete component of a project.

It is important for estimators to ensure that the dimensions they use in computing quantity takeoffs are accurate and complete. For example, a mistake commonly made by beginning estimators is to just take the stated room dimensions from the architectural plans and use these figures for computing quantities. The problem with this approach is that architects tend to state room sizes such as 12′ × 15′ and 20′ × 30′ on the basis of internal wall-to-wall dimensions. This practice does not take into account the widths of the building's interior walls. This mistake can reduce the square footage of the building substantially and cause errors in the quantity takeoffs for floor materials.

Preparing the Unit Price Summary

Once the material quantities for the project have been determined, each unit in the quantity takeoff must be priced and the cost of that unit determined. The unit price summary is limited to the unit price for each item and its associated cost. Related costs such as labor, equipment, project overhead, and company overhead each have their own unit price and associated cost. For example, go to Figure 3.10 and determine the number of cubic yards of concrete needed for concrete abutments (287 cubic yards). The estimator would multiply this figure times the going rate for concrete. Assume that the rate for concrete at the time is $75 per cubic yard, the computation would be as follows: 287 cubic yards × $75 per cubic yards = $21, 525. The unit price for concrete and the total for that specific item in the quantity takeoff would be recorded on the unit price schedule (see Figure 3.11).

UNIT PRICE SCHEDULE					
Item Number	Item Description	Unit	Quantity	Unit Price	Total Cost
1	Concrete, abutments	cubic yards	287	$75	$21,525

FIGURE 3.11 Preparing the unit price schedule.

Preparing the Labor Cost Estimate

Estimating the cost of labor for a construction project can be one of the most difficult esti-mating tasks because so many factors can affect labor rates. These factors include unions, differences in specific trades, geographic influences, duration of the project, the types of work to be done, National Construction Estimator efficiency rating of the labor, weather conditions, other concurrent projects that compete for the time of the various trades, and the availability of skilled labor in the area where the project will be built. Labor rates—union and nonunion—are not usually difficult to determine. Of course, estimators must ensure that they are working with loaded labor rates (i.e., wage rate plus other factors such as social security, federal unemployment insurance, state unemployment insurance, workers' compensation, and fringe benefits). The number of hours to allow for each unit of work in the estimate can be a challenge.

Determining the labor cost for each unit of work is a matter of multiplying the esti-mated rate times the number of hours the unit is estimated to require to complete. There are numerous tools and aids available for determining how many hours to allot to a given unit of work and the wage rate to apply to a given trade. These tools and aids take much of the guess work out of estimating labor costs on construction projects. However, they tend to provide national averages. This means that estimators must localize labor data. The better labor estimating tools and aids do this for the estimator.

These tools and aids often use the labor per unit of work. For example, a given tool or aid might recommend a labor cost of $25 per cubic yard for a specific type of concrete work. An example of the type of tools and aids available to estimators is the *National Construction Estimator.* This reference provides labor costs, man-hours, and recom-mended crew as well as the labor costs for installations with geographic adjustments for residential, commercial, and industrial construction projects in all of the areas shown in Figure 3.12.

Preparing the Material Cost Estimate

Estimating material costs is typically less of a challenge than estimating man-hours. As with estimating man-hours, there are tools and aids available to assist estimators. However, the most reliable information material unit prices will come directly from the suppliers of those materials. The key to making accurate material cost estimates is ensuring that prices quoted by suppliers are for the actual material specified in the construction documents.

**AREAS OF COVERAGE FOR LABOR COSTS
NATIONAL CONSTRUCTION ESTIMATOR**

Assemblies—Carpentry, Concrete, Electrical, Masonry, Plumbing, Roofing

Cabinets—Corner, Custom, Island, Metal, Utility, Wall, Wood

Carpentry—Assemblies, Framing, Glulam Beams, Molding, Paneling, Piecework Rates, Sheathing

Concrete—Beams, Finishing, Forming, Foundations, Pouring, Reinforcing, Slabs

Cooling—Air Handlers, Chillers, Condensers, Cooling Fans, Cooling Towers, Heat Pumps

Demolition—Asbestos, Building, Curbs, Flooring, Foundations, Framing, Roofing, Walkways

Doors—Closet, Colonial, Entrance, Fire, French, Glazed, Screen, Sliding

Electrical—Breakers, Conduit, Fixtures, Lighting, Panels, Switches, Wiring

Excavation—By Equipment Type and Hand, Backfilling, Clearing, Grading, Trenching

Finishes—Acoustical Ceilings, Lath and Plaster, Paint, Wallboard, Wallpaper

Flooring—Carpet, Composition, Masonry, Tile, Treads and Risers, Vinyl, Wood

Hardware—Door Closers, Framing Connectors, Locksets, Straps, Weather stripping

Heating—Baseboard, Ducting, Fans, Furnaces, Heat Pumps, Hydronic, Radiant, Steam, Wall

Insulation—Fiberglass, Foil, Perlite, Polystyrene, Urethane, Vermiculite

Lumber—Cedar, Fir, Pine, Plywood, Posts, Redwood, Siding, Spruce, Trim

Masonry—Accessories, Assemblies, Block, Brick, Flagstone, Marble, Pavers, Reinforcing

Metals—Decking, Joists, Pre-Engineered Buildings, Structural Steel, Trusses

Pipe—Cast Iron, Concrete, Corrugated Metal, Clay, Ductile Iron, PVC, Transite

Plumbing—Cleanouts, Drains, Fire Protection, Pumps, Rough-in, Sinks, Tubs, Water Closets

Roofing—Assemblies, Built-up, Corrugated, Elastorneric, Flashing, Shingles, Slate, Tile

Site Work—Clearing, Curb and Gutter, Fencing, Irrigation, Landscaping, Paving, Piping

Windows—Awning, Casement, Double Hung, Louvered Glass, Sliding, Skylights

---

FIGURE 3.12 Estimators use tools and aids such as this to determine labor costs.

Unit prices for materials as quoted by suppliers are typically good for only a specified period of time. For example, a concrete supplier might quote $75 per cubic yard of concrete but with the caveat that the price will hold at that level for only 60 days. Time is a critical issue when estimating material costs for construction projects. Another critical issue with materials is what is known as the *supplier's buying cycle*. Material suppliers have peak and down times in their businesses. Quotes provided on materials that must be delivered during peak times will typically be higher than those for materials that can be delivered during down times. Other issues that estimators must consider when preparing material cost estimates include:

- Availability of materials
- Match between project specifications and materials quoted by the supplier

- Quantity (large orders typically come with quantity discounts)
- Delivery price (it should be included in the material price quoted by the supplier)
- Delivery schedule (Can the material be delivered on a just-in-time basis or will it have to be stored?)
- Warranties on the material
- Transporting considerations (Will the roads accommodate delivery vehicles or will special accommodations be required?)
- Site access (Will delivery vehicles be able to get on and off the site without special accommodations?)
- Lead time required on the order of the materials
- Payment terms, discounts, credits
- Exchange and return policies on the materials

These are the types of issues purchasing and procurement personnel deal with all the time. On larger construction projects, purchasing or procurement professionals will handle the determination of unit prices for materials. On small projects, the project manager handles this task.

Preparing the Equipment Estimate

All construction projects require equipment of some type even if it is just small tools. Many projects require cranes, lift trucks, earth movers, and other large equipment (see Figure 3.13). When estimating equipment costs it is necessary to determine four things: (1) what equipment will be needed, (2) how long will each item of equipment be needed, (3) cost of the

FIGURE 3.13 Equipment estimates must cover the cost of the equipment and its operation.

equipment, and (4) cost of operating the equipment. In determining the cost of equipment, estimators must consider the following factors:

- Purchase, lease, or rental costs
- Depreciation costs for owned equipment
- Storage costs
- Insurance costs
- Cost of money (interest on borrowed money to purchase, lease, or rent)
- Taxes
- License fees

The cost of operating the equipment must also be computed. A major operating cost, of course, is labor. However, the labor cost associated with equipment operation is captured in the labor cost estimate. The cost of operating equipment includes the following:

- Fuel
- Maintenance
- Transportation to and from the site
- Safety checks
- Mobilization

Reviewing Subcontractor Quotes

The general contractor for a construction job requests bids from subcontractors for the various work packages that make up a substantial part of the job. The list of subcontractors that might be involved in a major construction project is substantial (see Figure 3.14). Large construction firms will prequalify subcontractors so they know they are working with reputable, dependable partners that can and do their part to complete the project on time, within budget, and according to specifications.

However, it will occasionally be necessary to work with subcontractors that do not meet the prequalification criteria. When this is the case, before accepting a quote, the project manager for the general contractor should ask the following questions:

- How much is known about the subcontractor?
- Can the subcontractor provide appropriate references and proof of the ability to perform?
- Is the subcontractor bonded or bondable?
- Does the subcontractor's quote cover all applicable aspects of the drawings and specifications?
- Did the subcontractor quote the specified materials?
- Did the subcontractor factor all addenda into the quote?
- Can the contractor meet the schedule requirements?
- Did the subcontractor acknowledge the clause covering liquidated damages?
- Does the subcontractor have any inclusions or exclusions?

A subcontractor's quote is a smaller version of the general contractor's overall estimate for the job. It covers material, labor, equipment, overhead (indirect costs), and profit. Bonding costs are also an issue with subcontractors. A comprehensive cost estimate prepared by a general contractor will include estimates for the various subcontractors, not to do their work for them but to have a way to evaluate subcontractor quotes.

**PARTIAL LIST OF
CONSTRUCTION SUBCONTRACTORS**

- Sitework sub (earthwork, utility, paving, curbs, line painting)
- Landscaping and irrigation subs
- Irrigation
- Concrete formwork
- Concrete supplier
- Reinforcing bar
- Masonry
- Structural steel
- Railing and deck
- Framing material and trim
- Framer
- Finish carpenter
- Millwork
- Casework
- Roof
- Siding
- Insulation
- Doors
- Windows
- Glass
- Drywall
- Stucco
- Acoustical ceiling
- Flooring
- Painting
- Fire suppression
- Security

FIGURE 3.14 Project managers in construction work with many different subcontractors.

Computing Indirect or Overhead Costs

Indirect costs, sometimes referred to as *overhead*, consist of the labor, material, equipment, permits, insurance, signs and barricades, photographs, testing fees, office costs, travel, legal fees, administration, and other nonconstruction costs required to support a construction project. Overhead costs represent too substantial a part of the overall estimate for a project to be overlooked or neglected. Figure 3.15 is a checklist of potential overhead costs that might be associated with a construction project.

POTENTIAL PROJECT OVERHEAD COSTS

✓ Organization, management, and administrative support (Project manager, safety director, clerical support, and other support personnel)

✓ Travel

✓ Testing

✓ Job site office/trailer

✓ Temporary utilities

✓ Winter protection (heating stations, etc.)

✓ Summer protection (cooling stations, etc.)

✓ Temporary access (roads and bridges)

✓ Signs and barricades

✓ Photography

✓ Clean up

✓ Temporary sanitation facilities

✓ Permits

✓ Insurance

✓ Taxes

✓ Bonding

✓ Miscellaneous

FIGURE 3.15 Overhead costs constitute an important component in the overall estimate.

Adding in the Profit

Construction firms are in business to make a profit. The bottom line in construction is the same as it is in any business: no profit, no business. Bankruptcy is the eventual result when construction firms consistently fail to make a profit. Of course, the financial stability of different firms will vary, but even the most financially stable construction firms cannot endure unprofitable projects for long. Construction firms have standard profit margins they add to the overall estimate, although with some government projects the allowable profit margin is prescribed. A larger project might have a slightly lower profit margin than a smaller project depending on the philosophy of the construction firm. Higher risk projects almost always have higher profit margins as a way to offset the risks.

Assume that the total estimate for a commercial building is $12,900,578 and the construction firm that prepared the estimate adds a 7 percent profit to its jobs of this size and level of risk. The projected profit for the project would be $903,040 ($12,900,578 × .07). Once this figure is arrived at and added to the overall estimate, a bid can be submitted on the basis of the estimate. Once accepted, the overall figure is considered firm unless approved change orders alter it.

LEED® CERTIFICATION AND CONSTRUCTION COST ESTIMATING

Green building construction is becoming more and more prevalent. As this happens, the various cost factors that go into computing a construction estimate can be affected. Consequently, construction professionals need to be aware of how building in accordance with the *Leadership in Energy and Environmental Design* (LEED) standards can affect their construction estimates. A study is available in which two buildings were modeled to determine what they would cost to build if the construction firm followed the standards for the various levels of LEED certification (certified, silver, gold, and platinum). The study, entitled *GSA LEED® Cost Study* is available from the U.S. General Service Administration (Order Number P-00-02-CY-0065) or online at www/fypower.org/pdf/gsaleed.pdf.

In order to determine the cost implications of building according to LEED standards, the estimator must understand how LEED certification works. Each of the four levels of LEED certification—certified, silver, gold, and platinum—requires design and construction firms to earn a certain number of points. There are seven mandatory points that must be met to achieve any level of certification and 69 elective points. Design and construction firms earn successively higher certifications by earning more of the 69 elective points. For example, the various levels of certification require at least the following number of elective points:

- Certified   26 points minimum
- Silver   33 points minimum
- Gold   39 points minimum
- Platinum   52 points minimum

Generally speaking, there is a cost associated with earning the points required for each successively higher level of certification. In reality, some of the points can sometimes be earned without incurring additional costs. For example, points are earned by proximity to mass transportation access points. A project either is or is not located in such proximity. If it is, the project will receive the points. If not it will not receive them unless it is moved, which is not a likely scenario.

The types of design and construction considerations that earn points toward the various levels of certification—what LEED refers to as *credits*—include energy efficiency, use of underfloor air delivery systems, dedicated ventilation systems, recycled content in construction materials, types of refrigerants used in HVAC and refrigeration (HCFCs are not allowed) and suppression agents used in fire suppression systems (halons are not allowed), and green power (electricity generated by alternative sources such as wind, solar, or biomass). As interest in green construction continues to grow, estimators will be increasingly engaged in factoring such things as LEED certification into their construction cost estimates.

COMPUTERS AND SOFTWARE IN CONSTRUCTION COST ESTIMATING

Computers and computer software play a major role in construction cost estimating. As with anything else, the computer is a tool to make the process more efficient and more accurate. However, computers and software do not decrease the need of construction professionals to understand any of the various aspects of construction cost estimating (anymore than handheld calculators relieve Math or Engineering students of the need to understand how to perform computations and what the computations mean once they are performed).

Building Information Models (BIMs) are emerging as realistic tools for use by architects, engineers, and construction firms. A BIM is a digital representation of a building or other structure showing every aspect of it and the functionality of all aspects. In other words, a BIM is theoretically a digital three-dimensional model of a building to be constructed. Ideally, a BIM will enhance communication and information sharing between and among architects, engineers, construction firms, and owners.

One obvious use of a BIM is in construction cost estimating. If the digital model contains all of the information traditionally available in architectural drawings, engineering drawings, specifications, and the contract as well as cost information from other similar projects in similar geographic locations, the BIM can be a powerful tool for the construction cost estimator. The ultimate goal of BIM advocates is to have the concept become sophisticated enough that once a digital model of a building has been developed, obtaining an accurate cost estimate will amount to little more than pressing a button.

The list of construction estimating software available commercially is long and getting longer. Software, depending on its quality, can be powerfully helpful for construction cost estimators, but the same problem exists with it that exists with all softwares: *garbage in/ garbage out*. The usefulness of construction estimating software is dependent on the talent and experience of the cost estimator. An abbreviated list of construction estimating softwares includes the following:

- *Bid4Build Construction Estimating*
- *CPR GeneralCOST Estimator*
- *Craftsman National Construction Estimator*
- *Estek Fast Estimate Easy*
- *Nomitech CostOS*
- *Priosoft Estimating and Construction Management*
- *Taskmaster Estimating and Scheduling*
- *WinEstimator*
- *Building Systems Design CostLink*
- *Craftsman CD Estimator Heavy*
- *Craftsman Earthwork & Heavy Equipment Estimator*

This is only a partial list. The key for construction project manager to understand is that although software can help organize an estimate and minimize the amount of labor required to prepare it, software should never be viewed as a substitute for knowledge and experience in estimating.

For example, with well-designed estimating software much of the material the estimator has to look up is preloaded into the software. This convenience can save steps for the estimator and, in turn, reduce the time required to prepare a construction cost estimate. Advocates of construction estimating software look forward to the day when architectural design, cost estimating, scheduling, budgeting, tracking, and monitoring software will all be merged into one powerful package that will reduce the time required by many of the more time-consuming project management tasks, such as cost estimating.

Even as this is happening, the quality of the construction cost estimates generated using computer software is in direct proportion to the quality of the software. As with all computer software, some is better than others. Consequently, it is still a good idea for construction students to learn how to manually prepare cost estimates.

CONSTRUCTION PROJECT MANAGEMENT SCENARIO 3.2

I Don't Need to Learn How to Prepare Estimates—I'll Just Buy the Software

Martha Miller is a good student. She has made mostly As and Bs in her college classes. Her major is Construction Engineering Technology, and she is enrolled in the required project management class this semester. Unfortunately, Miller is not applying herself in this class. While her classmates struggle to learn the details for quantity takeoffs and estimating labor costs, Miller is using her time to focus on other classes. When one of her classmates questioned her about ignoring the estimating lessons, Miller responded: "I don't need to learn how to prepare estimates—I'll just buy the software."

Discussion Questions

In this scenario, Martha Miller is convinced that the construction cost estimating software will prepare her estimates for her—that she does not need to understand how to prepare the estimates herself. If Miller expressed this point of view to you, how would you respond? What problems might Miller run into if she does not learn how to prepare estimates manually before attempting to use the cost estimating software?

SUMMARY

Construction cost estimating skills include the ability to build a structure mentally, work systematically using replicable methodologies, document work clearly and logically, think critically and investigate to verify data, computer labor costs, identify cost optimization strategies, use reference materials, and analyze completed estimates. There are six levels of estimates in construction: (1) screening or rough order of magnitude, (2) preliminary or conceptual or schematic, (3) design or detailed or definitive, (4) construction documents, (5) construction bid, and (6) construction monitoring or control. Each level of estimate is progressively more accurate and detailed.

Commonly used formats for preparing construction cost estimates are the unit price model and the systems or assemblies model. The unit price model makes use of the 50–division format developed by the Construction Specifications Institute (CSI). It is a product-based model. The systems or assemblies model makes use of the seven elements of the UNIFORMAT II format.

UNIFORMAT II is a systems or assemblies-based model. In the unit price format, construction materials are grouped together for the entire job on the basis of the CSI's 50 divisions. In the systems or assemblies format, construction materials are spread over the various systems that make up the building exactly as they will be used in the actual construction of it.

The construction estimating process is made up of the following tasks: (1) get organized, (2) prepare the quantity takeoffs, (3) prepare the unit price summary, (4) prepare the labor cost estimate, (5) prepare the material cost estimate, (6) prepare the equipment estimate, (7) review subcontractor quotes, (8) computing indirect costs/overhead, and (9) add in the profit margin.

Green building construction is becoming more prevalent. Consequently, construction project managers need to be aware of the cost implications of green building and how to computer them. *Leadership in Energy and Environmental Design (LEED)* provides standards for green building construction and levels of certification for buildings constructed according to these standards. The types of design and construction considerations that earn points toward LEED certification of a building include: energy efficiency, use of underfloor air delivery systems, dedicated ventilation systems, use of materials with high recycled content, the types of refrigerants used in HVAC and refrigeration systems and the kinds of suppression agents used in fire suppression systems, and green power (electricity generated by alternative sources, e.g., wind, solar, and biomass).

Building Information Models (BIM) are digital representations of a building or other structure showing every aspect of it and the functionality of each aspect. The use of a BIM as aids to design, estimating, and construction planning is increasing. In addition, there are numerous commercially available construction cost estimating software packages. BIM and estimating software can be invaluable tools for construction cost estimators, but they should not be used as crutches. Construction professionals need to learn the basics of preparing estimates before using either BIM or estimating software.

KEY TERMS AND CONCEPTS

Screening or rough order of magnitude estimates
Preliminary, conceptual, or schematic design estimates
Design, detailed, or definitive estimates
Construction documents estimates
Construction bid estimates
Construction monitoring or control estimates
Unit price model
Systems or assemblies model
CSI 50–Division MasterFormat®

UNIFORMAT II System Format
Quantity takeoff
Unit price summary
Labor cost estimate
Material cost estimate
Equipment estimate
Subcontractor quotes
Indirect or overhead costs
LEED® Certification
Building Information Models

REVIEW QUESTIONS

1. List and briefly explain the generic skills needed by construction cost estimators.
2. What are the six levels of construction cost estimates? Provide a brief explanation of each.
3. Compare and contrast a Level 1 and a Level 5 construction cost estimate.
4. Explain the unit price model for preparing construction cost estimates.
5. Explain the systems or assemblies model for preparing construction cost estimates.
6. Compare and contrast the unit price and systems or assemblies models.
7. What are the seven "systems" in the UNIFORMAT II model?
8. List and briefly explain the various tasks in the construction estimating process.
9. What are the different factors that make up the cost of construction equipment?
10. What are the factors that make up the cost of operating construction equipment?
11. List the most common overhead costs in a construction project.
12. Explain the concept of LEED® certification of buildings and how it can affect the cost of the building.
13. List the various factors that give construction firms LEED® points or credits when constructing a building.

14. What is a BIM and what is its relevance to construction cost estimating?

15. Explain what is meant by the following statement: construction cost estimating software should be used as a tool not a crutch.

APPLICATION ACTIVITIES

1. Contact construction firms in your region to determine the square foot costs for residential, commercial, and industrial construction. Use these figures to prepare rough order of magnitude estimates for the following buildings:
- Single-story ranch style home of 2,200 square feet
- Office building of 15,000 square feet
- Manufacturing building of 25,000 square feet

2. Identify a building on your college campus. Examine the building inside and out and check each division in Figure 3.5 that appears to apply. In some cases, you may not be able to tell without seeing the drawings and specifications for the building. However, note all divisions that observably apply. Now, reexamine the building according to the seven systems of the UNIFORMAT II model. Write a description of each of the seven systems of the building.

3. Refer to Figure 3.6 to complete this activity. A commercial building will use precast double "T" beams with no topping for its floor. The span for the beams will be 60 feet, the superimposed load will be 100 per square foot, the beams will be 32 inches deep and 10 feet wide, the concrete type will be regular weight, and the total load will be 164 per square feet. What are the square foot costs of the material and installation? If the dimensions of the floor are to be 60 feet × 60 feet, what is the estimated cost of the floor installed?

4. Refer to Figure 3.7 to complete this activity. An industrial building will use tilt-up concrete panels for the walls. The panels will have a steel trowel finish, be 5 ½" thick, and 3,000 pounds per square inch (psi). What are the square foot costs of the material and installation? If the four walls are 60 feet × 12 feet, what is the estimated cost of the walls installed?

5. For this activity, obtain the most recent edition available in your college's library or through another source of the reference book, *R.S. Means Assemblies Cost Data*. Using the national average estimates computed in Activities 3 and 4, make the conversions of your estimates for the following cities (identify the "city cost indexes" in *RSMeans Assemblies Cost Data*):
- Milwaukee, Wisconsin
- Norfolk, Virginia
- Spokane, Washington
- Helena, Montana
- Santa Barbara, California
- Pensacola, Florida

6. Identify a construction company in your region that will allow you to examine a completed estimate for a construction project for a residential, commercial, or industrial building. What estimating model was used? What was the estimated cost of the building? What were the overhead costs (individual items and overall cost)?

ENDNOTE

1. RSMeans, *Assemblies Cost Data,* 36th ed. (Norwell, Massachusetts: Reed Construction Data, 2011), 7.

Planning and Scheduling Construction Projects

A fact that is reiterated throughout this book is that all construction projects have the following success criteria: finish on time, within budget, and according to specifications (quality). Two additional criteria also apply to all construction projects: do the job safely and in an environmentally friendly manner. Some project managers separate safety and the environment out as separate criteria while others subsume them into the first three criteria—time, cost, and quality—because they affect these three criteria. Regardless of which approach is used, it is important for project managers to understand that none of these criteria can be met on a construction project, large or small, without *thorough planning* and *careful scheduling*.

The two principal outputs of the planning and scheduling process explained in this chapter are the project schedule and the policies and procedures manual. The schedule shows the starting and ending dates for the project as well as for all activities required to complete the project, the duration of each activity, project milestones, and the critical path for the activities. The policies and procedures manual explains how certain issues that are inherent in construction projects will be handled (e.g., duties and responsibilities of the owner, architect/engineer, and construction firm; division of labor; progress payments, meetings, inspections, reporting, safety, and other issues).

BENEFITS OF PLANNING AND SCHEDULING

Even a small construction project is a complex undertaking with a lot of moving parts that have to fit together just right and at exactly the right time. The more complex the project, the more difficult this *orchestration* becomes. The most important tool in making all of the moving parts fit together just right and at exactly the right time is a comprehensive, well-planned schedule. Ensuring that a construction project has a comprehensive, well-planned schedule is the responsibility of the project manager. The quality of the construction project's schedule affects all of the success criteria for the project: time, cost, quality, safety, and environment.

Time and Cost-Related Benefits of Scheduling

The old saying that time is money certainly applies to construction projects. Anything that adversely affects one adversely affects the other. For example, poorly planned projects have an uneven workflow. Construction personnel just stumble along until they finally realize that the deadline is drawing near. Then, all of sudden, work crews have to go into the hurry-up mode. Invariably, when construction firms go into hurry-up mode quality and safety begin to suffer. The schedule suffers because hurry-up mode produces bottlenecks that throw required work off the schedule. The budget suffers because unbudgeted overtime must be approved in order to meet the looming deadline. All of these factors have an adverse effect on time and cost. A well-planned schedule, on the other hand, provides an even workflow that preempts emergencies, bottlenecks, unplanned overtime, quality shortcomings, and safety violations.

Another problem with poor scheduling is that the uneven workflow it creates results in tradespeople working on top of each other and getting in each other's way as they try to do their work at the same time. Worse yet, it can result in one trade standing around doing nothing while waiting for another trade to complete his or her work. Often, the work of one trade is dependent on that of another in construction projects. This disorganized situation undermines productivity, and productivity is a key factor in completing a project on time and within budget. An even workflow has the opposite effect. It gives tradespeople the room they need to ply their trades without undue interference and without standing around doing nothing while waiting for another tradesperson to finish his or her work.

A well-planned schedule has the benefit of providing a yardstick for measuring progress on the construction project. There is a reason that the yard lines are drawn on a football field: They allow all stakeholders to know exactly where things stand. In other words, the yard lines provide a mechanism for measuring progress, and the best way to ensure that progress is being made is to measure it.

Another time and cost-related benefit of developing a schedule is that it encourages critical thinking. In the chapter on construction cost estimating, it was mentioned that construction estimators need to be able to mentally construct the building they are estimating the cost of. The same is true when developing a schedule. The project manager must be able to mentally envision the scope of the project and all activities within it. Developing a schedule encourages those involved to mentally construct the project from the ground up.

A final time and cost-related benefit of a well-planned schedule is that it promotes more effective communication among the many stakeholders on a construction project. Consider the following scenario. Even a small construction project has a long list of stakeholders: owner, architects, engineers, subcontractors, suppliers, inspectors, and tradespeople. Ineffective communication among these stakeholders can create disorganization, disruption, and chaos. On the other hand, effective communication will allow all of the various moving parts and players involved in a construction project to come together in the right way and at the right time. A well-developed schedule is as much a tool for the construction project manager as the baton is for an orchestra conductor.

Quality-Related Benefits of Scheduling

The owner's expectations are translated into practical terms by the project's specifications. Consequently, completing all work in strict accordance with the specifications—doing quality work—is of paramount importance. A well-developed schedule promotes quality work. To do quality work, subcontractors and trades need time, room to do their jobs, and support from the project manager. They also need the work that must be completed before their work can start to be completed on time and done properly. This can happen with a well-developed schedule. However, without such a schedule the work environment becomes disorganized and rushed. People cannot do their best work in an environment of disorganization and unrealistic time frames.

Safety and Environment-Related Benefits of Scheduling

Just as quality goes down when trades and subcontractors have to rush unrealistically, safety and environmental concerns fall by the way side. When workers are rushed unrealistically they respond by taking shortcuts, cutting corners, and neglecting proper work practices. When this happens, safety and environmental concerns are put aside for the sake of making up time and getting the work done by the deadline. For example, when working at heights it can take time to get properly rigged. Workers who are rushed might simply leave off their safety harness and rigging as a way to complete their tasks quicker. Another example is the proper disposal of toxic materials. It can take time to go through all the steps necessary to properly dispose of toxic waste materials at a job site. Workers who are rushed might simply pour the material into the sewer system or throw it in a dumpster to save time. These types of unsafe and environmentally unsound practices can be avoided by having a well-developed schedule that allows trades and subcontractors the time they need to do their work properly.

CONSTRUCTION PROJECT MANAGEMENT SCENARIO 4.1

I Don't Need a Schedule—It's in My Head

Mack Gainer had been building residential and commercial projects for 25 years and he is good at it. Gainer is usually the jobsite superintendent on Ray-Alto Construction Company's most important projects. The firm's latest project is the most important in its history. If Ray-Alto performs well on this project—a chain restaurant—the firm will receive a contract to build 10 additional restaurants for the same chain. Naturally, Mack Gainer will be the jobsite superintendent on the first project. The firm is depending on Gainer to complete the project on time, within budget, and according to specifications, and to make a favorable impression on the owner.

Consequently, the project manager—Gerald Caldwell—was shocked when he asked Gainer to work with him to develop a tight, well-planned schedule for the job. Gainer responded: "I don't need a schedule—it's in my head." Caldwell has been with Ray-Alto only six months, but he was brought in to professionalize the company and its processes. He is concerned that Mack Gainer—in spite of his excellent track record—might not make the desired impression on the owner in this case. He needs to convince Gainer of the value of having a well-planned schedule on paper.

Discussion Question

In this scenario, a talented builder operates by the seat-of-the-pants method. The plans for his project are in his head but not laid out on paper. Gerald Caldwell, the project manager in this case, has been brought in to correct unprofessional practices such as this no matter how talented the people who use them may be. If you were in Caldwell's place, how would you explain the importance of having a well-planned, comprehensive schedule available for all stakeholders to see?

THE PLANNING AND SCHEDULING PROCESS

Like so much of project management, developing a schedule is a process. Processes have three components: (1) inputs, (2) tools/techniques/methods, and (3) outputs. The inputs for the scheduling process are the construction documents and the construction cost estimate. Commonly used tools, techniques, and methods include the Work Breakdown Structure (WBS), bar charts, network logic/critical path diagrams. The output of the planning/scheduling process is a well-developed schedule that encompasses all of the activities that have to be completed to finish the project on time, within budget, and according to specifications and an up-to-date policies and procedures manual.

Project planning and scheduling can be a complex undertaking, but there are numerous scheduling aids available. Scheduling software is readily available and much of it is excellent. However, regardless of whether a construction project is planned and scheduled using appropriate software or by hand, the process has the same basic steps (Figure 4.1): (1) clarify the project's goals, (2) develop the Work Breakdown Structure, (3) put the WBS activities in sequence, (4) compute/estimate and chart the duration of all WBS activities, (5) develop the network diagram and determine the critical path, (6) develop or update the policies and procedures manual, (7) update the schedule as needed throughout the project, and (8) monitor the schedule continually. Once Steps 1 to 6 are completed, milestone events in the schedule should be indicated. Steps 1 to 6 are explained in this chapter. Steps 7 and 8 are covered in Chapter Six.

CLARIFY THE PROJECT'S GOAL

The goal for a project is a brief statement that summarizes why the project is being undertaken in the first place. For example, if an owner initiates a project for the purpose of building a summer cabin on a lake, building the cabin is the goal of the project. If a county or state initiates a project to construct a bridge across a river, constructing the bridge is the goal of the project. Project managers should never assume that stakeholders will instinctively

---

PLANNING/SCHEDULING PROCESS FOR CONSTRUCTION PROJECTS

1. Clarify the project's goal.

2. Develop the Work Breakdown Structure (WBS).

3. Put the WBS activities in sequence.

4. Compute/estimate and chart the duration of all WBS activities.

5. Develop the network diagram and determine the critical path.

6. Develop or update the policy and procedures manual.

7. Update the schedule as needed throughout the project.

8. Monitor the schedule continually.

FIGURE 4.1 Steps in the scheduling process for construction projects—manual planning and computer-assisted.

understand a project's goal. The goal should be stated in simple terms and written down. What follows are sample goals for construction projects:

- Construct a 2,500 square foot ranch style home on time, within budget, according to specifications, and in a safe and environmentally friendly manner.
- Construct a 15-story condominium complex on time, within budget, according to specifications, and in a safe and environmentally friendly manner.
- Construct a 25,000 square foot office complex on time, within budget, according to specifications, and in a safe and environmentally friendly manner.
- Construct 40,000 square foot manufacturing facility on time, within budget, according to specifications, and in a safe and environmentally friendly manner.

The goal of the project should be reviewed by all stakeholders—owner, architect, engineers, trades, and subcontractors—so that everyone has the same understanding before work begins. This seemingly obvious and simple exercise can prevent a number of major problems as construction on the project progresses.

DEVELOP THE WORK BREAKDOWN STRUCTURE

Construction project managers use the concept of deconstruction to break the entire project down into its components parts and then identify all activities that must be performed for each component. Different construction firms approach this task differently. Some use their own internal methodology and break the work down into phases of construction (e.g., site work, foundation, framing). Some create the WBS by following the guidelines used in developing the construction cost estimate. For example, the ASTM UNI-FORMAT II Classification for Building Elements is often used in preparing construction cost estimates. This classification system breaks construction jobs down into elements and levels as follows:[1]

A. Substructure (Level 1: Major Group Element)
    A10 Foundations (Level 2: Group Element)
        A1010 Standard foundations (Level 3: Individual Element)
        A1020 Special foundations (Level 3: Individual Element)
        A1030 Slab on grade (Level 3: Individual Element)
    A20 Basement construction (Level 2: Group Element)
        A2010 Basement excavation (Level 3: Individual Element)
        A2020 Basement walls (Level 3: Individual Element)
B. Shell (Level 1: Major Group Element)
    B10 Superstructure (Level 2: Group Element)
        B1010 Floor construction (Level 3: Individual Element)
        B1020 Roof construction (Level 3: Individual Element)
    B20 Exterior Enclosure (Level 2: Group Element)
        B2010 Exterior walls (Level 3: Individual Element)
        B2020 Exterior windows (Level 3: Individual Element)
        B2030 Exterior doors (Level 3: Individual Element)
    B30 Roofing (Level 2: Group Element)
        B3010 Roof coverings (Level 3: Individual Element)
        B3020 Roof openings (Level 3: Individual Element)

C. Interiors (Level 1: Major Group Element)
    C10 Interior construction (Level 2: Group Element)
        C1010 Partitions (Level 3: Individual Element)
        C1020 Interior Doors (Level 3: Individual Element)
        C1030 Fittings (Level 3: Individual Element)
    C20 Stairs (Level 2: Group Element)
        C2010 Stair construction (Level 3: Individual Element)
        C2020 Stair finishes (Level 3: Individual Element)
    C30 Interior finishes (Level 2: Group Element)
        C3010 Wall finishes (Level 3: Individual Element)
        C3020 Floor finishes (Level 3: Individual Element)
        C3030 Ceiling finishes (Level 3: Individual Element)
D. Services (Level 1: Major Groups Element)
    D10 Conveying (Level 2: Group Element)
        D1010 Elevators and lifts (Level 3: Individual Element)
        D1020 Escalators and moving walks (Level 3: Individual Element)
        D1090 Other conveying systems (Level 3: Individual Element)
    D20 Plumbing (Level 2: Group Element)
        D2010 Plumbing fixtures (Level 3: Individual Element)
        D2020 Domestic water distribution (Level 3: Individual Element)
        D2030 Sanitary waste (Level 3: Individual Element)
        D2040 Rain water drainage (Level 3: Individual Element)
        D2090 Other plumbing systems (Level 3: Individual Element)
    D30 HVAC (Level 2: Group Element)
        D3010 Energy supply (Level 3: Individual Element)
        D3020 Heat generating systems (Level 3: Individual Element)
        D3030 Cooling generating systems (Level 3: Individual Element)
        D3040 Distribution systems (Level 3: Individual Element)
        D3050 Terminal and package units (Level 3: Individual Element)
        D3060 Controls and instrumentation (Level 3: Individual Element)
        D3070 Systems testing and balancing (Level 3: Individual Element)
        D3090 Other HVAC systems and equipment (Level 3: Individual Element)
    D40 Fire protection (Level 2: Group Element)
        D4010 Sprinklers (Level 3: Individual Element)
        D4020 Standpipes (Level 3: Individual Element)
        D4030 Fire protection specialties (Level 3: Individual Element)
        D4090 Other fire protection systems (Level 3: Individual Element)
    D50 Electrical (Level 2: Group Element)
        D5010 Electrical service and distribution (Level 3: Individual Element)
        D5020 Lighting and branch wiring (Level 3: Individual Element)
        D5030 Communication and security (Level 3: Individual Element)
        D5090 Other electrical systems (Level 3: Individual Element)
E. Equipment and Furnishings (Level 1: Major Group Element)
    E10 Equipment (Level 2: Group Element)
        E1010 Commercial equipment (Level 3: Individual Element)
        E1020 Institutional equipment (Level 3: Individual Element)
        E1030 Vehicular equipment (Level 3: Individual Element)
        E1090 Other equipment (Level 3: Individual Element)

    E20 Furnishings (Level 2: Group Element)
        E2010 Fixed furnishings (Level 3: Individual Element)
        E2020 Movable furnishings (Level 3: Individual Element)

F. Special Construction and Demolition (Level 1: Major Group Element)
    F10 Special construction (Level 2: Group Element)
        F1010 Special structures (Level 3: Individual Element)
        F1020 Integrated construction (Level 3: Individual Element)
        F1030 Special construction systems (Level 3: Individual Element)
        F1040 Special facilities (Level 3: Individual Element)
        F1050 Special controls and instrumentation (Level 3: Individual Element)
    F20 Selective building demolition (Level 2: Group Element)
        F2010 Building elements demolitions (Level 3: Individual Element)
        F2020 Hazardous components abatement (Level 3: Individual Element)

The UNIFORMAT II classification system provides excellent guidance for deconstructing architectural and engineering drawings and breaking a construction project down into its building elements. This process, in turn, can be used to develop a WBS. A WBS can be laid out in outline form as shown above or it can be presented in graphic form (see Figures 4.2 and 4.3). Figure 4.2 is a portion of a WBS in graphic form showing the breakdown for basement construction—part of UNIFORMAT II major group element A: substructure. Figure 4.3 is a portion of a WBS in graphic form showing the breakdown for interior construction—part of UNIFORMAT II major group element C: interiors.

    Not every element in the UNIFORMAT II classification system is used in every building, of course. Further, UNIFORMAT II is not designed for use with infrastructure projects, such as roads, bridges, sewers, and water systems. However, it does provide excellent guidance when developing a WBS for residential, commercial, and industrial buildings.

    There is an important point about developing the WBS that project managers should understand. Aids such as the UNIFORMAT II classification system are just that aids. These

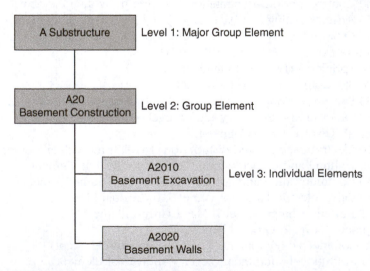

FIGURE 4.2 Portion of a WBS in graphic form.

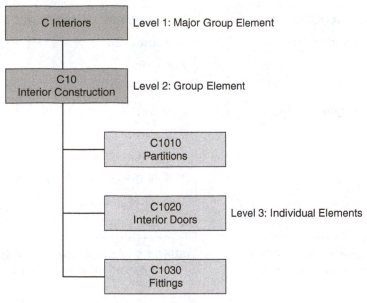

FIGURE 4.3   Portion of a WBS in graphic form.

aids should be used to guide and inform the deconstruction process, not to limit it. Construction companies frequently have their own breakdown guidelines. Some use a standard breakdown methodology and add to or revise it. More recently, construction companies have begun to use the breakdown system that is embedded in the scheduling software they adopt.

PUT THE WBS ACTIVITIES IN SEQUENCE

Once all of the activities in the WBS have been identified, they must be arranged in the order they will be completed. This is called *sequencing*. Many of the activities in a construction project are interdependent, meaning one activity must be completed before another can be started. Others can be undertaken concurrently or in parallel. It is important to determine what activity comes first, second, third, and so on, as well as which activities can be completed concurrently.

For example, assume the project to be built is a ranch style slab-on-grade home. The footings precede the foundation walls. The foundation walls come next and precede the pouring of the floor slab. The walls come next and precede the construction of the roof. While the roof is being constructed, the insulation can be put in the walls as a concurrent activity. The sequencing of all project activities will become important in Step 5 of the planning and scheduling process when the network/critical path diagram is developed. Figure 4.4 is an example of the sequencing for the interior work on a residential construction project. Figure 4.5 is the sequencing for the exterior finish for the same project. There would be a similar sequenced breakdown for the site work, foundation, framing, and finish work activities for this project. All of these subset sequences are arranged into one comprehensive sequence of activities for the entire project.

INTERIOR WORK SEQUENCE
1. Drywall supplies delivered to site
2. Insulation installed
3. Insulation inspected
4. Electrical and HVAC installation prep
5. Drywall installed
6. Drywall inspected
7. Finish package delivered
8. Finish carpentry
9. Paint

FIGURE 4.4 Activities sequence for the interior work on a residential construction project.

EXTERIOR FINISH ACTIVITIES SEQUENCE
1. Housewrap
2. Roofing
3. Masonry
4. Siding
5. Concrete Prep
6. Soffitt and fascia
7. Pour driveway and sidewalk
8. Exterior painting
9. Landscaping

FIGURE 4.5 Activities sequence for the exterior finish of a residential construction project.

COMPUTE/ESTIMATE AND CHART THE DURATION OF ALL WBS ACTIVITIES

How long will each activity in the WBS take to complete? Durations should have been estimated already in the preparation of the construction cost estimate. However, they should be checked for veracity at this point and any adjustments that need to be made should be made at this time. When the activities are put on a schedule, each will be assigned a start date and a completion date. Activity durations may be estimated in several different ways.

Duration Computing/Estimating Methods

When estimating the duration of a project activity, it is important to consider the success criteria that apply to all construction projects: time, cost, quality, safety, and the environment. Each of these factors affect the duration of a given activity. For example, the deadline for project completion can affect the amount of time allotted for individual project activities. An activity that might require three days under normal circumstances might have to be completed in two days in order to complete the project on time.

This type of situation is common in construction projects, but requiring a three-day activity to be completed in two days will affect the next success criteria: cost. To complete an activity in less time than would normally be required, the project manager will have to either authorize the responsible crew or trade to work overtime or add more workers to the crew. In both cases, the cost of the activity increases. A rule of thumb for project managers to keep in mind when estimating the duration of an activity is this: *Shortening the duration typically drives up direct costs whereas lengthening the duration drives up indirect costs.*

Quality, safety, and the environment can suffer also if the planned duration is shorter than the activity would normally take to complete. Pressuring work crews to rush through an activity is a sure way to encourage shortcuts and other expedients that can undermine quality and force workers to ignore safety and environmentally friendly work practices. For

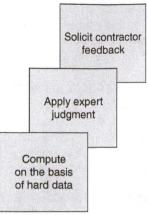

FIGURE 4.6 Commonly used methods for estimating activity durations.

all of these reasons, it is important for project managers to make informed, responsible estimates of activity durations rather than just guessing or engaging in wishful thinking.

The most widely used duration estimating methods are as follows: (1) computing the durations on the basis of hard data, (2) applying expert judgment, and (3) soliciting contractor feedback (see Figure 4.6). Regardless of which of these methods is used, the project manager must remember to consider weather as a factor in making duration estimates. These methods are applied as explained in the following paragraphs.

COMPUTING DURATIONS ON THE BASIS OF HARD DATA. One of the major steps in preparing the construction cost estimate is to complete the quantity takeoff. Retrieving the quantity takeoff for the project is the first step in this approach. The next step is to determine the productivity rate of the crew. This can be done in two ways: (1) using the construction firm's own historical data or (2) using standardized data from a commercial estimating reference such as the *RSMeans Estimating Handbook*. When using standardized data from a commercial source, remember to localize the data before making the actual computations. With this information in hand, the following equation is used to compute activity durations:

Activity duration = Quantity of work/Productivity rate (localized)

Assume that the activity is to lay a concrete sidewalk that will encompass a shopping center. The quantity of work for this activity is 1,600 linear feet of concrete. The data source used shows that a crew of four can prep, form, and lay 300 feet of sidewalk per day. Dividing the quantity of work (1,600 linear feet) by the productivity rate of the crew (300 feet per day) results in an activity duration of 5.33 days that is rounded up to the next whole day or six days. The project manager would schedule six days for this activity.

ESTIMATING DURATIONS ON THE BASIS OF EXPERT JUDGMENT. If the activity in question is of the type that the construction firm has completed many times using the same crew, the project manager may be able to simply apply his experience and determine the activity's duration on the basis of expert judgment. Another approach is to ask for the expert judgment of the crew chief or the construction superintendent. Although expert judgment is a legitimate method for estimating durations, it is important to make sure that the judgment being applied is truly expert judgment. A better use for expert judgment is in verifying and validating duration estimates that were computed on the basis of hard data.

SOLICITING DURATION ESTIMATES FROM SUBCONTRACTORS.   Another way to determine the duration of a given activity is to solicit the information from the subcontractor that will do the work. This can be an especially effective method when the subcontractor is well-known and has an established record with the construction firm from past projects. Even when this is the case, a way to strengthen this approach is to compute the activity duration from hard data and use that information to verify and validate the subcontractor's estimate.

Factoring Weather into Activity Duration Estimates

A factor that almost always has an effect on construction projects is the weather. Weather that is too hot or too cold can slow down work on the job site. Rain, snow, sleet, and ice can slow down work. Consequently, a project manager must factor in the weather when estimating activity durations. There are a number of different ways for factoring weather into a construction schedule. The most common of these are as follows:

- Increase the duration of all activities by the estimating effect of the weather on them.
- Add an estimated number of weather days to the end of the schedule extending the project duration by that many days.
- Put weather contingency days in the schedule as they are activities and assign them an estimated number of days like all other activities.
- Add nonwork days for weather at selected points throughout the project schedule.

Weather days are typically negotiated with the owner, who receives assistance from the architect. The negotiated days become part of the construction contract or an addendum to it. Local data from the weather bureau is a good starting point for estimating weather days. However, local construction experience over a period of years can also be invaluable in making accurate estimates. For example, if a construction firm in the south knows from experience to expect a certain number of rain days during a given period of the year, this is helpful knowledge. Correspondingly, if a construction firm in the north knows from experience to expect a certain number of days with freezing temperatures, snow, or ice, this knowledge can be put to good use in estimating weather days.

Determine the Critical Path

When all activities have been sequenced and assigned durations, they form *paths*. The critical path is the longest path. By totaling the durations of all activities along the critical path, the overall duration of the project can be determined. Shorter paths outside of the critical path have extra time known as *float* or *slack*. Critical path and float/slack are important scheduling concepts that are explained in greater detail later in this chapter.

Charting the Duration of WBS Activities

Bar charts provide a simple and easy-to-understand tool for graphically displaying WBS activities and their respective durations. With the bar chart, all activities can be listed in the proper sequence along with a starting and ending date. The bar between these two extremes represents the duration of the activity in question. In addition, the bar chart can show which activities are interdependent and must, therefore, be performed sequentially as well as which can be performed concurrently. Figure 4.7 is a bar chart showing a portion of the schedule for construction of a residential project. This figure is only a portion of a larger bar chart provided for the sake of illustration. The full chart would contain all of the calendar dates, work days, activities, and activity durations for the entire project.

Notice that the bar chart contains a good deal of information on just one page. For example, the month and calendar dates are shown. The project in question will begin on January 7. Under the calendar dates are the construction days. Down the left-hand side of the bar chart the activities from the WBS are listed in sequential order. In the body of the chart, the estimated duration of each activity is displayed in days in bar form.

Notice that construction day one on the chart is calendar day seven (January 7). On construction day one, two activities will occur: crew prep and kickoff meeting. Notice that

Month	January																					
Calendar Date	7	8	9	10	11	12	13	14	15	16	17	18	19	20	21	22	23	24	25	26	27	28
Construction Day	1	2	3	4	5	6	7	8	9	10	11	12	13	14	15	16	17	18	19	20	21	22
Crew Prep	■																					
Kickoff meeting	■																					
Site Work		■																				
Footings/Fdn.			■	■																		
WP Fdn.					■	■																
Sub Slab							■															
Bsmt. Slab								■														
Back Fill/Rough Grade									■	■												
Windows											■	■	■									
Exterior Concrete														■								
Roofing														■	■							
Masonry														■								
Rough Plumbing																■						
Rough HVAC																	■	■				
Rough Electrical															■							
Wall Insulation																		■				
Exterior Trim																			■			
Drywall																			■	■		
Exterior Paint																				■	■	■

FIGURE 4.7   Portion of a bar chart schedule for a house with a basement.

these two activities are occurring concurrently. A more likely explanation is that each of the activities will require half a day. When this is the case, the activities are displayed as shown in this chart as concurrent activities that are one day in duration. Notice that on January 26, 27, and 28 a crew will be painting the exterior of the house while another crew is hanging drywall inside the house. The footings and foundation for the house will be poured on construction days 3 and 4 (January 9 and 10) and waterproofed on construction days 5 and 6 (January 11 and 12).

Popular scheduling software such as Primavera Project, SureTrak, and Microsoft Project offer an additional feature that can help the project manager and other stakeholders when tracking the progress of a project. In the sequencing of construction activities, the project is broken down into successively smaller activities, and these activities can be grouped into elements or phases. For example, the portion of the construction schedule shown in Figure 4.8 is *Phase Four: Exterior Finish* for a dentist's office. Notice in this figure that Phase Four has a bar on the chart showing the overall duration of the phase. Then, each activity in that phase has its own bar.

When reading a construction schedule, such as the one in Figure 4.8, do not make the mistake of just adding up the duration of each individual activity under a broad phase of construction to determine the duration for that phase. Remember, some of the activities on construction schedules are sequential and some are concurrent. For example, in Figure 4.8 siding, concrete preparation, foundation parging, painting, and landscaping are sequential activities while soffit/fascia and pouring the driveway/walk overlap with other activities. When an activity overlaps with another, the concurrent part of its duration does not add to the overall duration of that phase of construction.

DEVELOP THE NETWORK DIAGRAM AND DETERMINE THE CRITICAL PATH SCHEDULE

Some construction firms go no farther than developing bar charts such as those shown in the previous section when scheduling projects. This approach is acceptable for small, simple projects. However, larger and more complex projects require a scheduling methodology that is equal to their complexity. Network diagram/Critical path method (CPM) scheduling in which planners develop a network diagram and determine the critical path for the project is such a methodology. Now that scheduling software is readily available, the labor intensity of developing a network diagram and identifying the critical path has been simplified somewhat. However, construction project managers still need to understand the *why* and *how* behind the concept.

Network diagram/CPM scheduling was not a popular method in the construction industry prior to the advent of powerful personal computers and scheduling software. Many considered it too labor intensive and difficult to learn. However, with the personal computers and scheduling software now readily available, network diagram/CPM scheduling has become more popular. However, project managers—future and present—should understand that even with personal computers and scheduling software mastering network diagram/CPM scheduling still requires a great deal of work, persistence, and experience. It also requires focused study in the area of construction scheduling. Students should not expect to walk across the stage at graduation expecting to be experts at network diagram/ CPM scheduling. It will take some time on the job before that level of competence will

PROJECT: DENTIST'S OFFICE

Task Name	Dur. Days	Start	Finish	April																									May					
				6	7	8	9	10	11	12	13	14	15	16	17	18	19	20	21	22	23	24	25	26	27	28	29	30	1	2	3	4	5	6
Phase Four Exterior Finish	16	4/6	4/21	■	■	■	■	■	■	■	■	■	■	■	■	■	■	■	■																
Siding	2	4/6	4/7	▨	▨																														
Concrete Prep	1	4/8	4/8			▨																													
Soffitt/Fascia	2	4/8	4/9			▨	▨																												
Pour Driveway & Walk	3	4/9	4/11				▨	▨	▨																										
Fdn Parging	1	4/12	4/12							▨																									
Painting	2	4/13	4/14								▨	▨																							
Landscaping	7	4/15	4/21										▨	▨	▨	▨	▨	▨	▨																

FIGURE 4.8 Partial schedule showing the exterior finish activities, durations, and dates for a dentist's office.

89

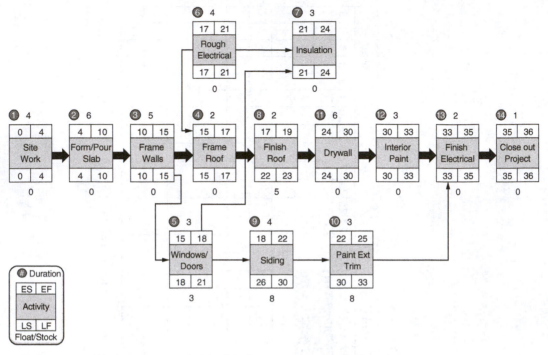

FIGURE 4.9 Network Diagram/CPM Schedule for an accountant's office.

emerge. However, this section will provide sufficient explanation of the concept to give students a good start on their journey. Figure 4.9 is an example of a network diagram/CPM schedule that was developed for the construction of an office for an accountant.

Advantages and Disadvantages of the Network Diagram/CPM Schedule

Network diagram/CPM scheduling offers several benefits to project managers including the following: (1) gives a picture of how the construction of the project fits together, (2) identifies the most critical activities in the project—those around which all other activities revolve, (3) gives the project manager a basis for setting priorities, (4) makes determining the consequences of change orders easier by showing how the change can have a ripple effect throughout the project, and (5) allows the project manager to experiment on paper with different construction sequences to determine the optimum sequence.

The principal disadvantage of the network diagram/CPM schedule is that becoming expert in using this method can take time. One other disadvantage is that this type of schedule can require a lot of space. This final disadvantage can be turned into an advantage. By producing the network diagram/CPM schedule in a large format, it can be taped to the wall of the construction trailer or a bulletin board. In this enlarged format, the project manager and other stakeholders can use it as a manual tracking tool that offers a quick visual of progress. Some project managers actually color in the cells as progress in the project moves across the diagram from left to right.

Developing the Network Diagram/CPM Schedule

Students might find it helpful to make a copy of Figure 4.9 to refer to while reading this section. To understand how to develop such a diagram, it is helpful to begin with a finished product and then break its development into steps. The network diagram/CPM schedule provides a lot of information in a relatively small space. The best place to start with is an understanding of the activity box.

Before attempting to read the actual schedule, examine the key in the lower left-hand corner of Figure 4.9. Figure 4.10 provides a larger version of this key. Figure 4.10 is one of the more commonly used formats for activity boxes in network/CPM schedules. There are other formats, but they all contain the following information or similar information:

- Activity number
- Activity duration in days
- Activity name
- Early start day or date
- Early finish day or date
- Late start day or date
- Late finish day or date
- Float/slack in days
- Name of the person responsible for the activity (optional)

The activity boxes in the center of the diagram with the heavy arrows (boxes 1, 2, 3, 4, 8, 11, 12, 13, and 14) represent the critical path for the project. These boxes represent activities that are sequential in nature and must be completed within the specified duration for each activity. The activity boxes that branch off from the critical path boxes represent activities that can be undertaken concurrently with another activity or activities. Arrows—whether heavy or light—point to a successive activity. For example, box 2 follows box 1 and box 3 follows box 2. These activities must be completed in sequential order and the preceding box must be completed before its successor is started.

Boxes 5, 6, 7, 9, and 10 represent concurrent activities. The work in box 5—windows and doors—can start as soon as the work in box 3 (wall framing) is completed. This means it can be completed concurrently with the roof framing. Also, while the roof is being framed the rough electrical work (box 6) can be done. When the rough electrical work is completed the insulation can be installed (box 7).

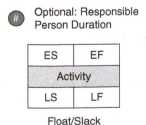

FIGURE 4.10 Commonly used format for the activity box.

Box 1 represents site work. The early starting date is zero because you cannot start before the project begins. The site work is scheduled to take four days. The early start, early finish, late start, and late finish numbers in the activity boxes represent project or construction days. For example, the number 10 in the early start corner of an activity box means day 10 of the project. Notice in activity box 14 that the project is to be completed on project day 36. This day would have been chosen to ensure that the project is completed by the deadline specified in the contract. There is no float (slack) in most of the activities. Only activity boxes 5, 8, 9, and 10 have float. Having examined a completed network diagram/CPM schedule, it will help to now go back to the beginning and develop the schedule one step at a time.

BEGIN BY REVIEWING THE WORK BREAKDOWN STRUCTURE. Before laying out the network diagram/CPM schedule in Figure 4.9, the project manager had already developed a WBS in which all of the activities in the project had been sequenced and for which the durations had been estimated. These steps were explained earlier in this chapter. The project manager may have also developed a simple bar chart to display the WBS and durations graphically—a step that simplifies the development of the network diagram/CPM schedule significantly. However, the WBS in any format—graphic or outline form—that shows the activities sequenced and with durations can serve as the starting point.

IDENTIFY SEQUENTIAL, CRITICAL, AND CONCURRENT ACTIVITIES. By examining the WBS for the project, the project manager is able to determine which activities must be completed sequentially and which can be completed concurrently with other activities as well as which activities must precede which other activities. In the example in Figure 4.9, the sequence of activities turned out to be as follows:

1. Site work (critical)
2. Form and pour the slab (critical)
3. Frame the walls (critical)
4. Frame the roof (critical)
5. Install windows and doors (concurrent)
6. Rough in the electrical work (concurrent)
7. Install the insulation (concurrent)
8. Finish the roof (critical)
9. Install siding on the exterior (concurrent)
10. Paint the exterior trim (concurrent)
11. Hang the drywall (critical)
12. Paint the interior (critical)
13. Finish the electrical work (critical)
14. Close out the project (critical)

LAYOUT, NUMBER, AND LABEL THE ACTIVITY BOXES. In this step, the activity boxes are laid out beginning with the critical path boxes which are placed in order of sequence from left to right as shown in Figure 4.11. As can be seen from the previous step, all activities that will be included in the network diagram/CPM schedule have numbers and the numbers are in sequence. Notice that the critical path activity boxes skip numbers in the sequence as concurrent activities are worked into the sequence at appropriate places.

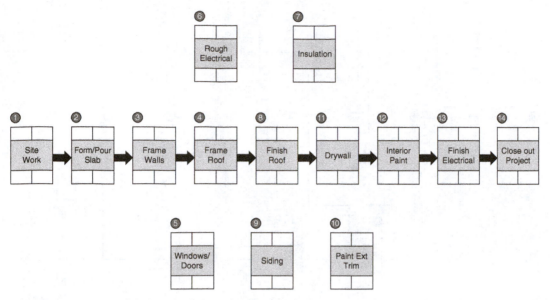

FIGURE 4.11 Layout, label, and number of the activity boxes.

Once the critical activity boxes are laid out, they are numbered and labeled to indicate the activity they represent. Then the concurrent activity boxes are numbered, labeled, and laid out. This step requires some thought. The concurrent activities boxes must be placed on the diagram in locations that will allow the scheduler to indicate that they precede certain activities and follow others. In addition, there must be room to show their relationships—by arrowed lines—not just to the critical path activities but to each other. This step can require some *juggling* of concurrent activity boxes to find the simplest, least cumbersome, and most descriptive arrangement. For example, in Figure 4.11 it was necessary to locate activity boxes 6 and 7 above the critical path boxes and boxes 5, 9, and 10 below the critical path boxes. They could all be placed above or below the critical path provided an appropriate arrangement could be worked out. In this case, placing them as shown in Figure 4.11 provided the most logical arrangement.

ADD RELATIONSHIP LINES/ARROWS. The relational lines/arrows for the activity boxes on the project's critical path were added in the previous step. These lines/arrows are thick and indicate that the critical path for the project consists of activity boxes 1, 2, 3, 4, 8, 11, 12, 13, and 14. The concurrent activity boxes also require lines/arrows to show which activities they precede and follow. These lines/arrows are added in this step. Schedulers use the WBS to determine the sequence for concurrent activities and show the preceding and following activities using lines/arrows (see Figure 4.12).

In Figure 4.12, Activities 1, 2, and 3 are completed sequentially. Then, while the roof is being framed—Activity 4—the windows and doors can be installed (Activity 5) and the electrical work can be roughed in (Activity 6). By dropping a line/arrow from the end of Activity 3—frame walls—the scheduler indicates that as soon as the walls are framed, window and door installation can begin. Installers do not need to wait for the roof to be completed (Activity 4). Also, while the windows and doors are being installed the rough electrical work can

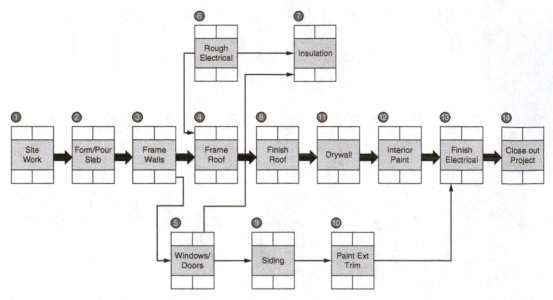

FIGURE 4.12   Add lines/arrows to show sequence in relationships.

begin (Activity 6). As soon as the windows and doors are installed and the rough electrical work is completed, the insulation may be installed in both the walls and the attic (Activity 7).

Once the insulation is installed, the roof can be finished (Activity 8) and the work crews will proceed along the critical path again hanging drywall (Activity 8) and painting the interior (Activity 12). While these activities are going on, the siding can be installed on the exterior of the office (Activity 9) and then the exterior trim of the office can be painted (Activity 10). The last major construction task—finishing the electrical work—occurs as Activity 13. Once the electrical work is completed, the project can be closed out (Activity 14).

All of these preceding, following, and concurrent activities must be indicated by lines/arrows as shown in Figure 4.12. The logic in laying out critical path and concurrent activities is important. On the one hand, any activities that can be completed concurrently should be provided the work can be performed without creating conflict and confusion among subcontractors and trades. Creating a situation in which two or more subcontractors are stepping around and over each other to get their work done. For example, notice that painting the exterior trim of the office (Activity 10) follows installing the siding (Activity 9). This makes good sense because if both subcontractors tried to work at the same time they would be tripping over each other. However, both activities can be completed concurrently with the hanging of drywall (Activity 11) and interior painting (Activity 12) because Activities 9 and 10 occur outside while Activities 11 and 12 occur inside.

ADD START AND FINISH DATES, DURATION, AND FLOAT.   The duration of each activity was estimated in an earlier step in the scheduling process. Hence, the first task in this step is to indicate the duration above each activity box. For example, in Figure 4.13 duration for Activity 1 is four days; the duration for Activity 2 is six days; and so on, through Activity 14, which is one day. Notice that just adding the durations for the critical path will not sum to the total of work days allotted for the project (36 days in Figure 4.13). Rather, the sum is 31.

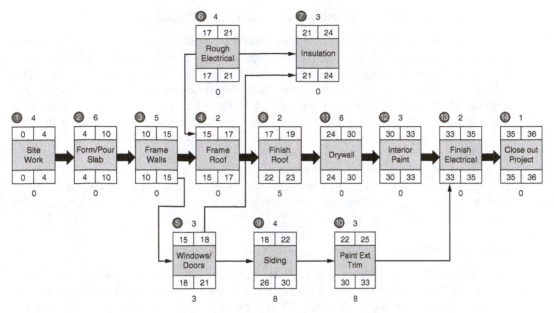

FIGURE 4.13   Start/finish dates, duration, and float.

This is because such an addition process does not account for the five days of float in Activity 8 (finish roof).

The next task in this step of developing the network diagram/CPM schedule is to calculate early start/finish days—also referred to as the *forward pass*—and late start/finish days—also referred to as the *backward pass*. The forward pass is calculated first and the backward pass second.

Calculating the Forward Pass (Early Start/Finish Days)

Beginning with the first activity in the schedule—Activity 1: Site Work in Figure 4.13— place a zero in the early start corner of the box. The zero represents the beginning of the day on the first day of work. Add the duration—four days in Activity 1—to the zero to determine the early finish day for Activity 1. Because Activity 2 can begin as soon as Activity 1 is completed, the early finish day for Activity 1 becomes the early start day for Activity 2.

Adding the duration of Activity 2—six days—to its early start day—four—results in an early finish day of 10 for Activity 2. The process is repeated for Activity 3 resulting in an early start day of 10 and an early finish day of 15. The process is repeated again for Activity 4 resulting in an early start day of 15 and an early finish day of 17. Activity 5 is a concurrent activity. It can begin on the same day as Activity 4. Consequently, its early start day is also 15. However because its duration is three days, its early finish day is 18. This process is repeated right across the network diagram/ CPM schedule. When a single activity has two or more predecessor activities, use the larger of the early finish dates of the predecessor activities as the early start day for the single activity.

Calculating the Backward Pass (Late Start/Late Finish Days)

Calculating the late start/late finish days is called a backward pass because the calculations begin at the end of the project and proceed backwards through the network from right to left (just the opposite of the forward pass). This is the major difference. The calculations are performed in the same way. Beginning with the last activity in the network diagram/CPM schedule—Activity 14 in Figure 4.13—copy the early finish day into the late finish corner of the box. Subtract the duration—one day—from the late finish day to determine the late start. Proceed in this way through the entire schedule using late finish days and the duration to determine late start days. As the scheduler moves backwards through the schedule, the late start day of a successor activity becomes the late finish day of its predecessor. When two or more successor activities back up into a single activity, the earliest late start day becomes the late finish day for the single activity.

Calculating Float

Float is the difference between the early start and the late start days for a given activity in a network diagram/CPM schedule. For example, refer to Activity 8—Finish Roof—in Figure 4.13. The late start day is day 22 and the early start day is day 17. This means that Activity 8 has a total of five days of float ($22 - 17 = 5$). Therefore, when working on Activity 8, the roofing crew has five days of slack to work with.

SCHEDULING SOFTWARE

Computer software has simplified the process of scheduling construction projects markedly over the old days when schedules had to be developed by hand and tacked to the wall of the jobsite office. One of the more advantageous aspects of the better scheduling software is that it will produce schedules in bar chart form that contain more information than the typical bar chart (e.g., project milestones, work package summaries, actual work, remaining work, and critical remaining work). See Figure 4.14 for an example of such schedules. The better software packages can also produce network diagram/CPM schedules and a variety of schedule-oriented reports customized to meet the needs of specific audiences (e.g., owner, project manager, subcontractors). Some of the more widely used software packages for scheduling include the following:

- Plus Series (www.contractorssoftwaregroup.com)
- Maxwell Systems ProContractorMX (www.maxwellsystems.com)
- Procore (www.procore.com)
- Primavera (www.oracle.com/us/primavera/index.html)
- HeadsUp iCPM (www.headsuptech.com)
- PMWeb (www.pmweb.com)
- BuildSoft (www.buildsoft.com)
- BuilderTek Online (www.buildertek.com)
- Microsoft Project (www.microsoft.com)

Like all software, all of these products have their respective strengths and weaknesses. Further, as is often the case with such things as computer software, *beauty is in the eye of the*

Sample High-End Residential Renovation

Activity ID	Activity Name	Original Dur	Start	Finish	Total Float
	Sample High-End Residential Renovation	333	20-Sep-10	17-Jan-12	0
A1020	Notice to Proceed	0	20-Sep-10		0
DESIGN		83	20-Sep-10	18-Jan-11	50
A1000	MEP Design Complete	0		20-Sep-10*	0
A1010	Permit Drawings Complete	0		24-Sep-10	0
A1030	Interior Finish Design Complete	0		29-Oct-10	20
A1040	Furniture, Fixtures, Accessories Design Complete	0		18-Jan-11	50
LONG LEAD PROCUREMENT		251	06-Oct-10	03-Oct-11	50
A1050	Stone Procurement	15	06-Oct-10	26-Oct-10	34
A1090	Kitchen Cabinets Procurement	5	29-Oct-10	04-Nov-10	118
A1080	Art Glass Procurement	5	29-Oct-10	04-Nov-10	176
A1070	Glazing Systems Procurement	15	29-Oct-10	18-Nov-10	20
A1060	Millwork Procurement	15	29-Oct-10	18-Nov-10	67
A2180	Owner FF&E Procurement	180	19-Jan-11	03-Oct-11	50
SUBMITTALS/SHOP DRAWINGS		112	19-Jan-11	17-Mar-11	139
A1980	HVAC Equipment Submittals	5	06-Oct-10	12-Oct-10	114
A1950	Wire Rack Submittals	5	06-Oct-10	12-Oct-10	231
A1920	Elevator Cladding Submittals	10	06-Oct-10	19-Oct-10	241
A2050	Electrical/Lighting Submittals	20	06-Oct-10	02-Nov-10	33
A1100	Stone Submittals	10	27-Oct-10	09-Nov-10	34
A1110	Art Glass Submittals	5	05-Nov-10	11-Nov-10	176
A1120	Kitchen Cabinet Submittals	20	05-Nov-10	03-Dec-10	118
A1140	Glazing Systems Submittals	10	19-Nov-10	03-Dec-10	20
A1130	Millwork Submittals/Shop Dwgs	80	19-Nov-10	17-Mar-11	67
LONG LEAD FABRICATION		168	13-Oct-10	13-Jun-11	110
A1960	Wire Rack Fabrication	10	13-Oct-10	26-Oct-10	231
A1990	HVAC Equipment Lead Time	40	13-Oct-10	08-Dec-10	114
A1930	Elevator Cladding Fabrication	30	20-Oct-10	01-Dec-10	241
A2060	Electrical/Light Fixture Lead Time	40	03-Nov-10	30-Dec-10	33
A1150	Stone Fabrication	80	10-Nov-10	08-Mar-11	34
A1160	Art Glass Fabrication	80	12-Nov-10	10-Mar-11	176
A1180	Glazing System Fabrication	40	06-Dec-10	02-Feb-11	20
A1170	Kitchen Cabinet Fabrication	80	06-Dec-10	31-Mar-11	118
A1190	Millwork Fabrication	100	21-Jan-11	13-Jun-11	46
PERMITS		50	27-Sep-10	06-Dec-10	0
A1200	Permit Approval	50	27-Sep-10	06-Dec-10	0
A1210	Permit Issued	0		06-Dec-10	0

Legend: Actual Work / Remaining Work / Critical Remaining Work / Milestone / Summary

Date Printed: 09-May-11
Data Date: 06-Sep-10
Project Start: 20-Sep-10
Project Finish: 17-Jan-12

Sample High-End Residential Renovation

Project ID No.
Sample Residential
© Primavera Systems, Inc.

Nova Consulting Services, LLC
5680 SE Pot O Gold Place
Stuart, FL 34997
Ph: (772) 781-8672 / Fax: (772) 672-3725

FIGURE 4.14 Computer-generated schedules can present a great deal of information in a relatively small space.
Courtesy of NOVA Consulting Services, LLC.

Sample High-End Residential Renovation

Layout: Classic WBS Layout
Filter: TASK filter: All Activities

Activity ID	Activity Name	Original Dur	Start	Finish	Total Float
CONSTRUCTION		**321**	**06-Oct-10**	**17-Jan-12**	**0**
GENERAL CONSTRUCTION		**294**	**06-Oct-10**	**06-Dec-11**	**0**
A1230	Establish Elevation	2	06-Oct-10	07-Oct-10	0
A1860	Floor Leveling	5	07-Dec-10	13-Dec-10	41
A1220	Layout	5	14-Dec-10	20-Dec-10	0
A2020	X-Ray & Core Drilling	5	17-Dec-10	23-Dec-10	12
A1480	Frame Walls	15	21-Dec-10	12-Jan-11	0
A1490	HVAC Duct Rough	5	07-Jan-11	13-Jan-11	5
A1240	Fire Sprinkler Rough	5	11-Jan-11	18-Jan-11	5
A1800	Field Measure for Millwork Fabrication	5	13-Jan-11	20-Jan-11	46
A1250	Plumbing Rough	10	13-Jan-11	27-Jan-11	25
A1270	Electrical Rough In Walls	15	13-Jan-11	03-Feb-11	25
A1320	HVAC Rough Inspections	3	14-Jan-11	19-Jan-11	6
A1330	Fire Protection Rough Inspections	2	19-Jan-11	20-Jan-11	5
A1870	Demo Existing Balcony Doors	10	27-Jan-11	09-Feb-11	21
A1820	Plumbing Rough Inspection	1	28-Jan-11	28-Jan-11	31
A1260	Frame Ceilings	15	28-Jan-11	17-Feb-11	0
A1880	Install Structural Components @ Balcony Doors	10	31-Jan-11	11-Feb-11	21
A1310	Install Glass Wall Systems	15	03-Feb-11	24-Feb-11	20
A2070	Electrical Wall Rough Inspection	2	04-Feb-11	07-Feb-11	25
A1280	Electrical Rough In Ceilings	15	18-Feb-11	11-Mar-11	0
A2100	Framing @ New Balcony Doors	1	25-Feb-11	25-Feb-11	20
A2120	Framing Inspection @ New Balcony Doors	1	28-Feb-11	28-Feb-11	20
A2110	Drywall @ New Balcony Doors	2	01-Mar-11	02-Mar-11	20
A1290	Fire Alarm Rough	5	07-Mar-11	11-Mar-11	0
A1300	Low Voltage Rough	5	07-Mar-11	11-Mar-11	1
A1500	Low Voltage Rough Inspection	1	14-Mar-11	14-Mar-11	1
A1340	Fire Alarm Rough Inspection	2	14-Mar-11	15-Mar-11	0
A2080	Electrical Ceiling Rough Inspection	2	14-Mar-11	15-Mar-11	0
A1350	Framing Wall Inspection	2	16-Mar-11	17-Mar-11	0
A1970	Spray Insulation @ Wine Room	1	18-Mar-11	18-Mar-11	2
A1360	Hang Drywall One Side	3	18-Mar-11	22-Mar-11	0
A1370	Insulate Walls	1	22-Mar-11	22-Mar-11	0
A1380	Hang Drywall-Walls & Ceilings	7	22-Mar-11	30-Mar-11	0
A1510	Screw Inspection	1	31-Mar-11	31-Mar-11	0
A1520	Finish Drywall	10	01-Apr-11	14-Apr-11	0
A1390	Prime Paint	7	15-Apr-11	25-Apr-11	0

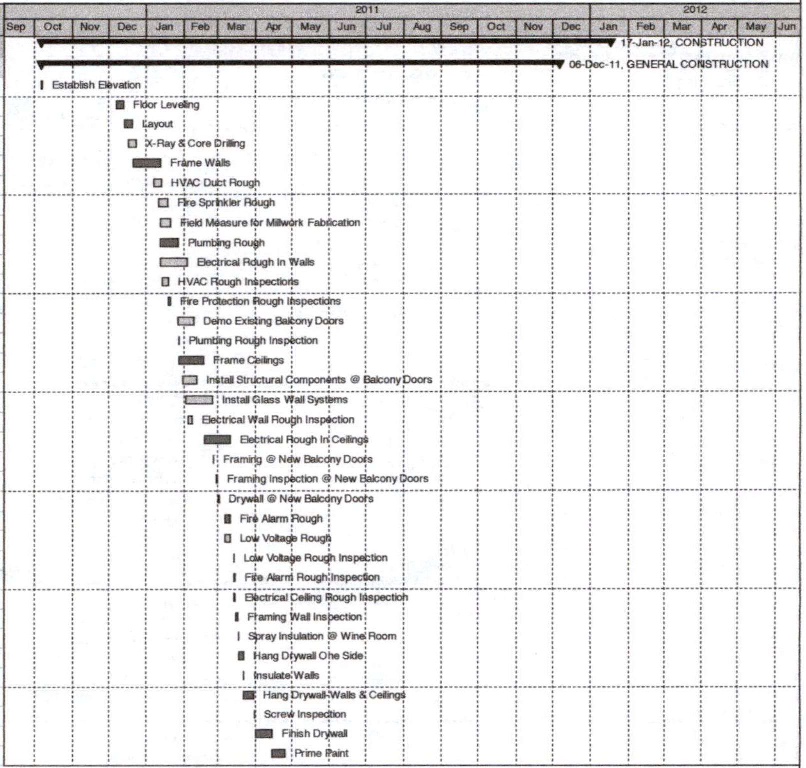

Legend: Actual Work · Remaining Work · Critical Remaining Work · Milestone · Summary

Sample High-End Residential Renovation
Project ID No. — Sample Residential
© Primavera Systems, Inc.

Nova Consulting Services, LLC
5680 SE Pot O Gold Place
Stuart, FL 34997
Ph: (772) 781-8672 / Fax: (772) 672-3725

FIGURE 4.14 (Continued)

Sample High-End Residential Renovation

Layout: Classic WBS Layout
Filter: TASK filter: All Activities

Activity ID	Activity Name	Original Dur	Start	Finish	Total Float
A1910	Install Elevator Door Cladding	3	26-Apr-11	28-Apr-11	142
A2000	Install HVAC Equipment & Final Connections	5	26-Apr-11	02-May-11	20
A2030	Equipment Start-Up & Testing	5	03-May-11	09-May-11	20
A1890	Install Plumbing Fixtures	2	01-Nov-11	02-Nov-11	12
A1810	Measure & Install Shower Enclosures	5	01-Nov-11	07-Nov-11	9
A2160	Final Inspection - Plumbing	1	03-Nov-11	03-Nov-11	12
A1790	Finish MEPF/Trim-Out	2	16-Nov-11	17-Nov-11	0
A1900	Install Kitchen Appliances	1	17-Nov-11	17-Nov-11	2
A2040	Test & Balance	2	18-Nov-11	21-Nov-11	0
A2170	Final Inspection - Fire Alarm	10	18-Nov-11	02-Dec-11	2
A1780	Final Inspection - Fire Protection	10	18-Nov-11	02-Dec-11	2
A2130	Final Inspection - Building	10	21-Nov-11	05-Dec-11	1
A2140	Final Inspection - Mechanical	10	22-Nov-11	06-Dec-11	0
A2150	Final Inspection - Electrical/LV	10	22-Nov-11	06-Dec-11	0
MARBLE/STONE		**111**	**26-Apr-11**	**30-Sep-11**	**34**
A1400	Stone Floors Typical Incl. Foyer	20	26-Apr-11	23-May-11	0
A1410	Master Bath Stone	10	17-May-11	31-May-11	20
A1420	2nd Bath Stone	5	27-May-11	03-Jun-11	28
A1430	3rd Bath Stone	5	09-Jun-11	15-Jun-11	28
A1440	Powder Rooms Stone	5	09-Jun-11	15-Jun-11	28
A1460	Stone Tops	5	29-Jul-11	04-Aug-11	40
A1450	All Other Areas Stone (Wall Inlay)	15	12-Sep-11	30-Sep-11	34
WOOD FLOORS		**30**	**24-May-11**	**06-Jul-11**	**0**
A1470	Wood Floors	30	24-May-11	06-Jul-11	0
MILLWORK/CABINETRY		**113**	**24-May-11**	**02-Nov-11**	**12**
A1530	Kitchen Cabinets	5	24-May-11	31-May-11	81
A1540	Master Bedroom Millwork	5	08-Jun-11	14-Jun-11	0
A1550	Master Hallway Millwork	10	15-Jun-11	28-Jun-11	0
A1560	Master Bath Millwork	5	29-Jun-11	06-Jul-11	0
A1570	Install Closets	10	07-Jul-11	20-Jul-11	0
A1580	2nd Bathroom Millwork	2	21-Jul-11	22-Jul-11	0
A1590	3rd Bathroom Millwork	2	25-Jul-11	26-Jul-11	0
A1600	Powder Room Millwork	2	27-Jul-11	28-Jul-11	0
A1610	Living Room Millwork	10	29-Jul-11	11-Aug-11	0
A1620	Library Millwork	10	12-Aug-11	25-Aug-11	0
A1630	Main Hallways Millwork	15	26-Aug-11	16-Sep-11	0
A1830	Install Wine Rack	2	19-Sep-11	20-Sep-11	6

Legend:
- Actual Work
- Remaining Work
- Critical Remaining Work
- Milestone
- Summary

Date Printed: 09-May-11
Data Date: 06-Sep-10
Project Start: 20-Sep-10
Project Finish: 17-Jan-12

Nova Consulting Services, LLC
5680 SE Pot O Gold Place
Stuart, FL 34997
Ph: (772) 781-8672 / Fax: (772) 672-3725

Sample High-End Residential Renovation
Project ID No.
Sample Residential

© Primavera Systems, Inc.

FIGURE 4.14 (Continued)

Sample High-End Residential Renovation

Layout: Classic WBS Layout
Filter: TASK filter: All Activities

Activity ID	Activity Name	Original Dur	Start	Finish	Total Float
A1840	Other Areas Millwork	10	19-Sep-11	30-Sep-11	0
A1850	Install Glass at Wine Room	2	21-Sep-11	22-Sep-11	6
A2010	Install Art Glass	1	26-Sep-11	26-Sep-11	38
A1640	Install Doors	10	03-Oct-11	17-Oct-11	24
A1940	Install Laminate Counters/Shelving	2	01-Nov-11	02-Nov-11	12
WALL FINISHES		33	03-Oct-11	18-Nov-11	1
A1650	Finish Paint	20	03-Oct-11	31-Oct-11	0
A1670	Wallpaper	20	18-Oct-11	15-Nov-11	0
A1680	Fabric Wall Covering	10	01-Nov-11	15-Nov-11	0
A1660	Specialty Wall Finishes	3	16-Nov-11	18-Nov-11	1
A1690	Install Mirrors				
PUNCHLIST		10	22-Nov-11	06-Dec-11	0
A1700	Punchlist	10	22-Nov-11	06-Dec-11	0
CLOSE-OUT		5	07-Dec-11	13-Dec-11	0
A1750	Final Cleaning	4	07-Dec-11	12-Dec-11	0
A1760	Turnover Apartment	1	13-Dec-11	13-Dec-11	0
A1770	Contract Completion Date = 12/16/11	0		13-Dec-11*	0
FFE		22	14-Dec-11	17-Jan-12	0
A1720	Furniture Delivery	10	14-Dec-11	28-Dec-11	2
A1710	Draperies & Other Fabric	15	14-Dec-11	05-Jan-12	2
A1730	Hang Art	5	29-Dec-11	05-Jan-12	2
A1740	Decoration	5	06-Jan-12	12-Jan-12	2
A2090	Designer FFE Completion = 1/17/12	0	17-Jan-12	17-Jan-12*	0

Legend:
- Actual Work
- Remaining Work
- Critical Remaining Work
- ◆ Milestone
- ▼ Summary

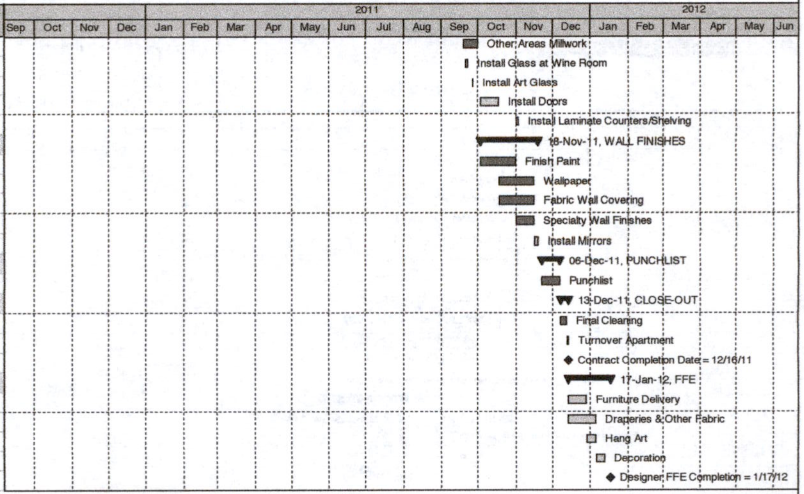

Date Printed: 09-May-11
Data Date: 06-Sep-10
Project Start: 20-Sep-10
Project Finish: 17-Jan-12

Sample High-End Residential Renovation

Project ID No.
Sample Residential

© Primavera Systems, Inc.

Nova Consulting Services, LLC
5680 SE Pot O Gold Place
Stuart, FL 34997
Ph: (772) 781-8672 / Fax: (772) 672-3725

FIGURE 4.14 (Continued)

beholder. Some construction firms like one package while other firms think another package is the best. One firm will recommend a given software package as the best while another will think just the opposite. Regardless, students who become project managers can count on using the computer software to generate, monitor, and update project schedules, even in small construction firms.

FAST TRACK CONSTRUCTION

A concept that is becoming more and more widely used in construction is *fast-track scheduling* or *fast tracking.* Fast tracking involves beginning selected portions of a construction project before the design and/or all of the construction documents have been completed. Fast tracking is not really new. For example, the Empire State Building constructed in the 1930s was fast tracked. What is new about the concept is the frequency of its use. As the design-build and construction management delivery systems have become more common, so has fast tracking because these delivery systems are well-suited to the concept.

The most common approach to fast tracking is to begin excavation and construction of the foundation of a building while the design for the other aspects is still being developed. In addition, advance purchases of critical materials such as the structural steel package for the building are made. While it is necessary for design to stay ahead of construction, it is not necessary for it to be completed before construction can begin. As long as design stays sufficiently ahead of construction, the two can proceed on parallel courses throughout a construction project. This is the basis for fast tracking.

Project managers should be prepared to manage fast tracked projects. Depending on the type of project, as many as 30 to 40 percent of construction projects are now fast tracked. What this means from a scheduling point of view is that the project manager will typically maintain one overall tentative schedule for a project while developing more specific schedules for each successive phase of the construction. This is really just an accelerated version of what project managers do, even when their projects are not fast tracked. Fortunately, the scheduling software that is readily available makes this continual reworking of project schedules a manageable task.

DEVELOP OR UPDATE THE POLICIES AND PROCEDURES MANUAL

Although most of the attention in the planning and scheduling phase of a project is focused on developing the project schedule, there is one additional planning task that gets little or no attention in project management books: *developing or updating the policies and procedures manual.* Most construction firms have a policies and procedures manual that provides guidance for project managers and their teams and promotes better communication among all stakeholders. All firms should have one.

There are a number of topics that will arise with any construction project. These topics grow out of issues that are predictable and recur from project to project. Consequently, construction companies should develop policies and procedures that explain how they handle these topics. Some of the more common of these predictable issues are as follows:

- Duties of the owner
- Duties of the architect/engineers
- Duties of the construction firm

- Coordination
- Correspondence
- Progress meetings
- Progress reports
- Progress payments
- Change order procedures
- Time extensions
- Inspections
- Test and laboratory reports
- Safety
- Acceptance of work
- Project closeout

This list is not all-inclusive. Different construction firms include different topics in their manuals. However, these are the topics that one could expect to find policies and procedures for in most manuals. In fact, these topics should be considered the minimum that should be covered. Having established policies and procedures on these and other topics can prevent issues from arising during the course of the project that become problems. All stakeholders need to understand from the very initiation of a project how the construction firm deals with these topics.

If a construction firm does not have a policies and procedures manual, the project manager should work with higher management to develop one. If there is an existing manual, the project manager should review it to make sure that it is up-to-date. It is not uncommon for language in the project contract to be at odds with one of more of the construction firm's existing policies or procedures. When this is the case, one of the following two things has to happen: (1) the construction firm must negotiate the language out of the contract or (2) the construction firm must update its policies and procedures manual to agree with the contract language—at least for the project in question.

A third option would be for the construction firm to turn down the contract. However, such a drastic step would happen only if the issue in question was of sufficient gravity to warrant it. For example, if it is the construction firm's policy to require the project manager to report on progress once a month and the contract calls for a report every two weeks, the policy would likely be waived for the project in question or even changed. However, if the construction firm's policy is to receive progress payments upon achievement of certain milestones in the schedule and the contract calls for no progress payments the firm would either have to negotiate on this issue or consider turning down the contract.

The actual language in a policies and procedures manual will vary from construction firm to firm, as will the topics covered. However, the language is typically fairly similar as are the topics covered. What follows are examples of the types of language that project managers can expect to find in the policies and procedures manuals of construction firms:

- ***Coordination.*** One of the principal duties of the project manager is to coordinate the work of all stakeholders involved with the project. This includes the owner, architect, engineers, construction firm, subcontractors, and trades. To this end, the project manager will establish practical, dependable communication channels with all stakeholders. Using these channels, the project manager will maintain regular contact with all stakeholders who, in turn, are expected to reciprocate.

- *Correspondence.* Correspondence between any members of the core construction team—owner, architect, engineers, and project manager—will be copied either electronically or in hardcopy form for all other members of the team.
- *Progress meetings.* Progress meetings will be held weekly in the jobsite trailer office and chaired by the project manager. The project manager is responsible for ensuring that accurate minutes are taken. The minutes are to be distributed electronically no later than the end of the day following the meeting. The agenda will normally include the following items but may be revised according to the needs and demands of the project:
  1. Introduction
  2. Review of previous meeting's minutes
  3. Update on action items from the previous meeting
  4. Review of agenda items for the current meeting
  5. Discussion of unanticipated delays and other problems
  6. Discussion of new or anticipated problems
  7. Miscellaneous items
  8. Around the table
- *Safety.* The project manager is responsible for establishing, maintaining, and monitoring a comprehensive construction safety program. The program will be developed in accordance with Title 29 of the Code of Federal Regulations (CFR), Part 1926 of the Occupational Safety and Health Administration of the U.S. Department of Labor.

CONSTRUCTION PROJECT MANAGEMENT SCENARIO 4.2

Let's Just Use the Bar Chart Schedule—CPM is Too Complicated

Danny Forester has just been named project manager for his construction firm's new Value-Tech project. Value-Tech is a technology retailer that specializes in selling used and refurbished technologies, such as personal computers, printers, and other related equipment. The company is thriving and plans to open ten new stores in ten different locations. Danny Forester is the project manager for the first of these ten new stores. Naturally, his company wants to do a good job and make a favorable impression on Value-Tech's management team in hopes of winning the contract to build the other nine stores. Consequently, when Forester's boss—Amanda Parker—asked to see his tentative schedule for the first store, she became concerned when the project manager produced only the bar chart schedule. When Parker asked if he had started laying out the network diagram/CPM schedule, Forester replied: "Let's just use the bar chart schedule—CPM is too complicated."

Discussion Question

In this scenario, Danny Forester does not want to take the time to develop a comprehensive network diagram/CPM schedule. He would rather just use the bar chart schedule. If you were Amanda Parker, would you approve this recommendation? Explain why or why not.

SUMMARY

The success criteria that apply to all construction projects cannot be met without thorough planning and careful scheduling. The two principal outputs of the planning and scheduling process are the project schedule and the policies/procedures manual. Developing a well-planned schedule for a construction project can have time, cost, quality, safety, and environmental benefits.

The planning and scheduling process consists of the following steps: (1) clarify the project's goal, (2) develop the Work Breakdown Structure (WBS), (3) put the WBS activities in sequence, (4) compute and chart the durations of all WBS activities, (5) develop the network diagram and determine the critical path, (6) develop or update the policies and procedures manual, (7) update the schedule as needed throughout the project, and (8) monitor the schedule continually. These last two steps are explained in Chapter Six.

The goal for a project is a brief statement that summarizes why the project is being undertaken. Project manager uses the concept of deconstruction to break a construction project down into its various component parts and then identify all activities that must be performed for each component. The process is used to develop a WBS for the project. Different construction firms do this differently, but all should develop a WBS to use when developing the actual schedule for the project.

Sequencing involves putting the activities identified in the WBS in the order they will be completed on the job site. Once project activities have been identified and sequenced, their durations must be computed or estimated. A rule of thumb to guide project managers when computing/estimating activity durations is this: Shortening the duration of an activity typically drives up direct costs whereas lengthening the duration drives up indirect costs. When using hard data to compute activity durations, the quantity of work is divided by the productivity rate of the crew or individual doing the work. In addition to computing activity durations on the basis of hard data, project managers can estimate them on the basis of expert judgment and input from subcontractors.

Weather is always a factor in construction. Hence, it should be factored in when developing a schedule for a construction project. Weather is compensated for by increasing the duration of activities, adding a certain number of weather days to the end of the project, putting weather contingency days in the schedule, or adding no-work days at selected points throughout the schedule.

The bar chart is a simple and easy-to-understand tool for graphically displaying WBS activities and their respective durations. The bar chart shows all activities listed in sequence with starting and ending dates. A bar chart can also show which activities are interdependent and must be performed sequentially and which can be performed concurrently. The network diagram/CPM schedule is more complex than the bar chart schedule and can require both education and experience to master. The network diagram/CPM schedule offers several benefits: (1) shows how the project fits together, (2) identifies the project's critical activities, (3) gives project managers a basis for setting priorities, (4) makes it easier to see the consequences of change orders, and (5) allows the project manager to experiment with different construction sequences to determine the optimum sequence. The principal disadvantage of the network diagram/CPM schedule is that becoming expert in using this method can take time.

The network diagram/CPM schedule is developed as follows: (1) begin by reviewing the WBS; (2) identify sequential, critical, and concurrent activities; (3) layout, number, and label the activity boxes; (4) add relationship lines/arrows; and (5) add start and finish dates, duration, and float. Float is slack in an activity. Early start/finish days are computed using the forward pass. Late start/finish days are computing using the backward pass. Fast track construction is a concept in which some aspects of a project are undertaken before the design is completed for subsequent aspects of the project.

Although most of the attention in the planning and scheduling phase of a project is focused on developing the project schedule, there is one additional planning task that is important: developing or updating the policies and procedures manual. The manual provides guidance for project managers and their teams and promotes better communication among all stakeholders. The manual explains how certain predicable issues that arise in most construction projects are to be handled. It explains the duties of the owner, architect/engineers, and the construction firm; how coordination, correspondence, progress meetings, progress payments, progress reports, change orders time extensions, and other issues are to be handled.

KEY TERMS AND CONCEPTS

Time and cost-related benefits of scheduling
Quality-related benefits of scheduling
Safety and environmental benefits of scheduling
Work Breakdown Structure (WBS)
Sequencing
Duration
Critical path
Bar chart
Network diagram/CPM schedule

Concurrent activities
Activity box
Relationship lines/arrows
Float
Forward pass
Backward pass
Project milestone
Fast track construction
Policies and procedures manual

REVIEW QUESTIONS

1. What are the time and cost benefits of scheduling?
2. What are the quality-related benefits of scheduling?
3. What are the safety and environmental benefits of scheduling?
4. List and briefly explain the steps in the planning and scheduling process for a construction project.
5. What is a Work Breakdown Structure?
6. Explain the concept of sequencing.
7. Explain how a project manager computes the duration of a project activity.
8. What are other methods that can be used for determining activity durations other than computing them?
9. Explain how project managers can factor weather into their project schedules.

10. Describe the types of information project managers can include on a bar chart schedule.
11. What are the advantages or benefits of the network diagram/CPM schedule?
12. What is the major disadvantage of the network diagram/CPM schedule?
13. What information is typically included on a network diagram/CPM schedule?
14. List the steps in developing a network diagram/CPM schedule.
15. Explain the concept of fast track construction.
16. What is a policies and procedures manual and what purpose does the manual serve?

APPLICATION ACTIVITIES

The following activities may be completed by individual students or by students working in groups:

1. Contact a construction firm in your region that will work with you in completing this activity. Ask to see examples of schedules of projects the firm has completed or that are in progress. Does the firm use bar chart schedules, network diagram/CPM schedules, or scheduling software that can produce both types of schedules?

2. Contact a construction firm in your region that will work with you in completing this activity. Ask if the firm has a policies and procedures manual. If so, review the manual to determine what types of information it covers.

3. Do the research necessary to develop a bar chart schedule and a network diagram/CPM schedule for the following project: Attorney's office that will be 3,000 square feet, wood framed, brick siding, slab-on-grade construction. You may make assumptions and decisions about the project for all information that is required beyond the information that is just provided. You may create the schedules manually or use any scheduling software that is available to you.

4. Select three of the construction scheduling software packages listed in the section "Scheduling Software." Do the research necessary to learn the strengths and weaknesses of your three choices. Then compare and contrast the three software packages you have researched.

ENDNOTE

1. Robert P. Charette and Harold E. Marshall, "UNIFORMAT II Elemental Classification for Building Specifications, Cost Estimating, and Cost Analysis." Retrieved from http://www.fire.nist.gov/bfrlpubs/build99/PDF/b99080.pdf on February 11, 2012.

Procuring for Construction Projects

Because this book is aimed primarily at students who are preparing to work in the construction profession—construction managers, construction project managers, architects, and engineers—this chapter approaches the concept of procurement from the perspective of a construction firm that is serving as the general contractor for a project rather than from the perspective of an owner. What is contained herein is the information that a construction manager, project manager, architect, or engineer needs to know to be a positive participant in or even leader of the procurement process for a construction project. For example, architects use the procurement process to select a general contractor and engineers for the projects they initiate. Construction firms that serve as general contractors and engineering firms use the procurement process to select the subcontractors and materials providers they need for construction projects.

Before construction can begin on a project, the construction firm serving as the general contractor must identify, locate, and obtain the necessary materials and the subcontractors who will do the hands-on work. The process is known as *procurement*. Procurement, like so much of project management, is a process. Hence it has inputs, procedures/methodologies, and outputs. The most common inputs are the construction documents and requests for proposals (RFP) and requests for quotes (RFQ).

The most common procedures/methodologies include: (1) accepting responses submitted on a low-bid, best-value, micro, small, or sole-source basis (depending on instructions from the construction firm serving as the general contractor); (2) evaluating the bids and quotes provided by subcontractors and materials suppliers; (3) determining if a given subcontractor or materials supplier is a responsible provider; and (4) determining if a given subcontractor or materials supplier is a responsive bidder. The most important outputs are the contracts awarded to responsible and responsive subcontractors and materials providers.

This chapter will help construction project managers and other construction-related personnel develop the knowledge and skills they will need to be active and positive participants in the procurement process. As has been explained throughout this book, in smaller firms the project manager might actually have to wear multiple hats—one of which is procurement manager. In larger firms, there will be a procurement professional or even a department of procurement professionals and specialists. Regardless of the project manager's level of involvement in the procurement process, he or she will need to know the basics as presented in this chapter.

PROCUREMENT METHODS

There are a number of different methods used to procure the materials and subcontractors needed in construction projects. The method chosen depends on such factors as the magnitude of the expected price of the materials and/or service, the philosophy of the construction firm procuring the materials and/or service, and the documented record of those who propose to provide the materials and/or service. The most widely used procurement methods in construction are as follows (Figure 5.1):

- Low-bid method
- Best-value method
- Micro-purchase method
- Small-purchase method
- Sole-provider method

Project managers are likely to use all of these at some point in their careers. Consequently, it is important to be familiar with all of these methods.

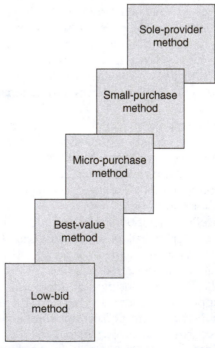

FIGURE 5.1 Widely used procurement methods.

Low-Bid Method

The low-bid method is also called the low-price or best price method. This method is most appropriate for use when the following conditions exist (Figure 5.2):

- A substantial amount of money is involved.
- There are precise specifications for what is needed.
- At least two providers are willing to submit bids.
- Providers have equal capabilities and records of performance.
- Price is the most important consideration (i.e., expectations of other considerations may be lowered to meet the price criterion).

With this method, bidding instructions are prepared and made available to subcontractors or materials suppliers. Bids are often required to be submitted sealed. When sealed bids are called for, representatives of the construction company open them during a bid conference to which all who submitted bids are invited. This approach ensures that all who submitted bids can see that everything about the process is above board and open. Their bids have not been tampered with.

When using the low-bid method, it is important to provide precise instructions to bidders as well as comprehensive and complete specifications. Precisely worded instructions to bidders will help prevent procedural challenges when the bids are opened and the low bidder is named. Comprehensive, complete specifications will ensure more accurate bids. It is also important to provide sufficient time for bidders to develop their proposals. Rushing bidders is likely to result in less accurate bids as key specifications are overlooked, and the price bid is the result of hurried guesswork rather than thoughtful deliberations. Bids should be due no later than a time and date certain that is specified in the instructions to bidders. Bids that come in after the specified deadline should not be returned unopened to the bidder.

With the low-bid method, the bid is awarded to the bidder who meets the following criteria: (1) submits a *responsive* bid (i.e., one that complies with all bidding instructions and meets all specifications, (2) is declared to be a *responsible* bidder (i.e., a bidder that can do the work or provide the materials in strict accordance with the provisions of the contract), (3) submits the lowest price for the products/services needed, and (4) acknowledges that the low-bid price submitted is fixed and final. It is important to understand that all four of criteria apply when choosing the *low-bid* method. If the other three criteria are not applied,

---

Checklist of
CONDITIONS FOR THE LOW-BID METHOD

- ✓ A substantial amount of money is involved
- ✓ There are precise specifications for what is needed
- ✓ At least two providers are willing to submit bids
- ✓ Providers have equal capabilities and records of performance
- ✓ Price is the most important consideration (i.e., expectations of other considerations may be lowered to meet the price criterion)

FIGURE 5.2 The low-bid method is appropriate when these conditions exist.

subcontractors and materials suppliers can simply win the bid by providing purposefully low prices they cannot realistically honor.

Best-Value Method

Best-value procurement is a method that seeks to award contracts to subcontractors and materials suppliers on the basis of their ability to provide the best overall value. Best-value is a combination of best quality, best price, best service, and best record of performance. When all of these factors are considered in combination, the bidder that can provide the best result is considered the best-value provider. It is not uncommon for a best value bid to be higher than its comparable low bid. This is because when considering quality, price, and service, the price might be higher, for example, to achieve the level of quality required. The best-value method is appropriate for use when the following conditions exist (Figure 5.3):

- A substantial amount of money is involved.
- The best construction approach or material for the job is not known.
- More than two providers are willing to submit bids.
- Overall value is more important than the lowest possible price.
- The opportunity to weigh quality, price, service, and past performance according to internally established priorities is important.

With the best-price method, the contract is awarded to the responsive and responsible bidder that proposes to provide the best value for the price. As with the low-bid method, when using the best-price method it is important to provide comprehensive bidding instructions, accurate specifications, and sufficient response time to ensure thorough, thoughtful bids.

Micro-Purchase Method

The micro-purchase method is used when the cost of the material or service is negligible. The construction firm must decide what level of price is negligible and the definition will vary from firm to firm. However, the cutoff point is typically less than $5,000. A micro-purchase is typically a direct purchase without quotes. When using the micro-purchase method, construction firms need to have a fairly accurate idea of what the material or service should cost so there is assurance that the price paid is *fair* and *reasonable*. To know if a

---

Checklist of
CONDITIONS FOR THE BEST-VALUE METHOD

✓ A substantial amount of money is involved

✓ The best construction approach or material for the job is not known

✓ More than two providers are willing to submit bids

✓ Overall value is more important than the lowest possible price

✓ The opportunity to weigh quality, price, service, and past performance according to internal priorities is important

FIGURE 5.3 The best-value method is appropriate when these conditions exist.

price is fair and reasonable, it is necessary to have an accurate price expectation before deciding to use the micro-purchase method.

With this method, there is always a temptation to break up a purchase into component pieces so that the micro-purchase method can be used instead of the other methods. This is rarely justified and should, therefore, be rarely done. Circumventing the appropriate procurement method to compensate for poor planning or other nonemergency factors is unwise. Construction firms that reinforce inefficiency will get more inefficiency.

Small-Purchase Method

The small-purchase method is used when it would cost more to go through a formal bidding process than the service or material needed is likely to cost. Construction firms typically establish procurement levels that allow the formal bidding process to be circumvented for the sake of time and efficiency. The cutoff point will vary from firm to firm depending on size and the philosophy of higher management. However, it is not uncommon for a construction firm to apply the small-purchase method for materials or services that should cost between $75,000 and $150,000.

When using the small-purchase method, the construction firm asks for quotes. The process is informal and may actually be done over the telephone. In fact, it often is when dealing with a known and trusted provider. At least two and occasionally more quotes are solicited to make sure that the laws of competition apply. When using this method, acceptance of the winning quote can be done over the telephone with the contract to follow later.

There is an important point to understand about using the small-purchase method. There will be times when procurement personnel might be tempted to break up a contract into component pieces that do not exceed the established price level for the small-purchase option. This is sometimes done to avoid having to go through the bidding process. When this happens, the culprit is usually time. For example, if due to an emergency there is no time to go through the bidding process. To ensure that the service or material is provided on time, procurement personnel break up the contract into small enough amounts that each piece of the contract can be obtained using the small-purchase method. There may be times—hopefully few—when this approach has to be used. If so, they should be few and far between. Further, this *workaround* should not be used to compensate for poor planning.

Sole-Provider Method

The sole-source method of procurement is noncompetitive in that it seeks to obtain the needed service or material from just one source. This method is used only when the other procurement methods are not appropriate for some legitimate reason. Using the sole-source method to procure construction materials and services is appropriate when one or more of the following conditions exist:

- The needed material or service is available from only one source.
- There is not sufficient time to use a competitive procurement method such as low-bid or best-value.
- There are factors that make the sole-source method legitimate (e.g., the owner specifies that a given provider must be used).
- Only one provider is able to meet the quality criteria in the specifications.
- Special circumstances that could not be anticipated make this method the most appropriate method.

Just as there are conditions that can make the sole-source method a legitimate option, there are also the following conditions that should not be used to justify using this procurement method:

- To compensate for poor planning
- To ensure that the contract in question goes to a favored provider
- To garner favors or even illicit compensation from a given provider
- To gain a competitive advantage based on factors other than performance

CONSTRUCTION PROJECT MANAGEMENT SCENARIO 5.1

I Think We Should Sole Source the Bid and Hire ABC, Inc.

There is disagreement in the procurement team of Construction Management Professionals, Inc. (CMP) concerning what procurement method to use for the new stadium project. CMP is the general contractor for the new football stadium being built by the local university. Most of the structural components for the project will be prestressed concrete. CMP has worked with ABC, Inc., a prestressed concrete firm, on several other projects in which their performance was superb. Consequently, several of the members of the procurement team wants to sole source the prestressed concrete aspects of the project to ABC, Inc. The project manager, Sarah Ferguson, is the leading voice for the sole-source option. She has already stated emphatically: "I think we should sole source the bid and hire ABC, Inc. If we do we can be guaranteed that it will be done right and on time without any problems, excuses, or change orders." CMP's procurement director disagrees. He thinks the bid should be let on a competitive basis using the low-bid method.

Discussion Question

In this scenario, the project manager and the procurement director disagree concerning which procurement method to use in selecting a prestressed concrete subcontractor. Join one side or the other and explain why you think that option is best in this situation.

PREPARING AN RFP OR RFQ PACKAGE

The procurement methods used most frequently by construction firms—low-bid and best-value—require that project managers or project managers working in conjunction with procurement professionals prepare an RFP or an RFQ package for bidders. A comprehensive bid package will contain the following information (Figure 5.4):

- Invitation to bid
- Announcement of the pre-bid conference
- Bidding instructions
- Bidding form

These three components together make up the bid package or RFP/RFQ that is provided to subcontractors and materials suppliers that might wish to submit a bid. The provider selected is then awarded a contract. Contracts are covered later in this chapter.

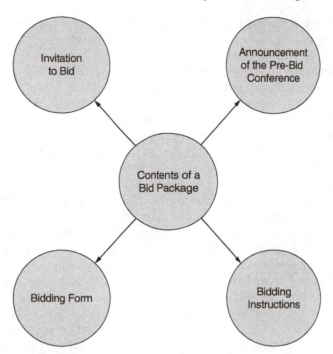

FIGURE 5.4 A comprehensive bid package contains these components.

Invitation to Bid

The invitation to bid alerts subcontractors and materials suppliers that a construction project is being planned that they might want to participate in. The invitation to bid provides potential participants with enough information to know if they would like to submit a bid and if they are capable of providing the service or materials being requested. This latter concern is important because when the bids are evaluated, one of the concerns of the construction company that is hiring subcontractors and suppliers is finding *responsible* providers.

You will recall that a responsible bidder is one that can perform the service or provide the materials in strict accordance with all specifications, requirements, and expectations. To help bidders understand if they can perform as required, the invitation to bid contains the following information:

- ***Project description.*** This is a brief description of the type and size of the project. Is the project residential, commercial, industrial, or infrastructure? Is the project large, mid-sized, or small? This component of the invitation to bid answers these questions. Both questions have a bearing on whether the potential bidder should invest the time and effort to submit a bid. For example, a small contractor that specializes in residential projects may not want to bid on a large industrial project. Correspondingly, a large firm that specializes in infrastructure projects would probably not bid on a residential project. By providing a comprehensive description of the project as the first component of the invitation to bid, construction firms can save themselves time and effort they would otherwise spend evaluating bids submitted by unqualified bidders.

Correspondingly, the project description can save potential bidders the time and effort of preparing a bid on a job they are not able to complete according to specifications, requirements, and expectations.

- *Project location.* This component describes the precise location of the project. This information is important because subcontractors typically have a range of operation within which they work. Some bid only on local jobs, while others are willing to bid on jobs that are regional. Location becomes even more important than usual during difficult economic times. Generally speaking, subcontractors and materials suppliers want to work on projects that are as close to their home offices as possible. However, a rule of thumb in construction is that the worse the economy becomes the farther subcontractors are willing to travel and materials suppliers are willing to ship. Of course, the farther a subcontractor has to travel or the farther a material supplier's goods have to be transported the more their costs will be. These factors will become part of their deliberations as they develop their bids. When economic times are particularly bad, subcontractors and materials suppliers are often willing to cut their profit margins sharply just to have enough work to stay in business.

- *Project start and completion dates.* This component lets bidders know the timeframe within which they will have to provide the service or materials. With materials there will be a more specific date since when materials arrive at the job site is critical. Too early and they have to be stored and secured. Too late and they throw the project off the schedule. For subcontractors, the start and completion dates define how long their personnel will have to be committed to the project in question. This is an important consideration for subcontractors because they are often working on more than one job at a time. If a subcontractor submits a bid containing the proper assurances that its crew will be available for the project, it is making a commitment that will become part of the contract. Violating a contractual obligation can be costly for a subcontractor.

- *Bonds.* This component provides information that automatically eliminates some potential bidders. Consequently, this component is one of the first things subcontractors and materials suppliers will look for in the invitation to bid. The bonding requirement eliminates some bidders at the outset, unless they are willing to undertake the expense of getting bonded. A bond is a financial assurance of performance. Bonds are explained in greater detail later in this chapter.

- *Project documents.* This component explains how the documents that subcontractors need to prepare their bids will be made available. These documents include the specifications, drawings, and any other documents that might apply to a given job. In times past, the documents were made available at specified times on specified days at a central location. Subcontractors would make an appointment to review the documents and come to the central location to do so. In recent times, the ability to provide documents electronically has simplified this aspect of bid preparation.

- *Legal considerations.* This component explains how certain potentially contentious issues will be handled. This component is intended to preclude lawsuits that might arise out of disagreements over the issues covered. For example, this section often contains specific criteria for rejecting a bid or for allowing a subcontractor or materials provider to withdraw a bid with no monetary penalty.

- *Bid deadline.* This component provides a specific date and time by which bids must be received. It will also contain a statement to the effect that bids received after the deadline will not be considered. Often, this will be the first part of the invitation read by potential bidders. Preparing a bid response is a time-consuming process. If the deadline looms too close, some subcontractors and materials suppliers may decide that they simply do not have time to prepare a responsive bid and decline the invitation. Others may decide the work is important enough to work around the clock to respond. This dilemma illustrates why it is so important for construction firms to give subcontractors and materials suppliers sufficient time to prepare responsive bids. Providing less than sufficient time can result in three things happening—all of them bad. First, responsible subcontractors may choose not to submit a bid, thereby reducing the quality of the bidding pool. Second, subcontractors might rush to complete a bid on time and miss important aspects of the specifications or drawings, thereby submitting unresponsive bids. This also reduces the quality of the bid pool. Finally, a subcontractor might suspect that the quick turn around on bids was a ploy by the construction firm to rule out all but a favorite few bidders who were informed of the project prior to the invitation. This situation can lead to challenges to the bidding process and even litigation in cases in which the owner is a government organization.

Announcement of the Pre-Bid Conference

This component provides the date, time, and location of the bid conference. The bid conference is a meeting involving the project manager and potential bidders in which bidders are allowed to ask questions to ensure they understand all aspects of the process, specifications, and instructions. Construction firms do not always hold bid conferences, but they should. Responding to the questions and concerns of bidders at the outset can prevent problems such as unresponsive bids, misunderstandings concerning specifications, and even legal challenges later in the process.

Bidding Instructions

Bidding instructions tell potential bidders specifically what they need to do submit a responsive bid. Some of the material covered in the bidding instructions may also be found in the invitation to bid, a circumstance that is encouraged since redundancy promotes a better understanding on the part of bidders. Information typically contained in the bidding instructions includes the following:

- Date and time bids are due
- Location where bids must be received
- Instructions for completing the bid form
- Unit prices for work/materials to be provided
- How and where to indicate any additional fees that will be charged
- How and when the winning bid will be announced
- How and when the contract will be awarded
- Special instructions (i.e., instruction concerning any aspect of the project that is out of the ordinary)

Bidding Form

To ensure that bids from various contractors can be readily and accurately compared, construction firms typically provide a bid form such as the one shown in Figure 5.5. Providing a standardized bid form helps ensure that all stakeholders are on the same page when it comes to submitting and

BID FORM FOR SUBCONTRACTORS

Job: _____

Date: _____

Time: _____

Location: _____

Bidder: _____

State License #: _____

Address: _____

Telephone: _____ FAX: _____

Email: _____

URL: http:// _____

Addenda:

We acknowledge reviewing all
construction documents _____
                              (Initial)

Bonding:

a. Bonded amount: _____

b. Bonding agent: _____

We propose to complete the following work on this project, as shown on the plans and described in the specifications and addenda as follows:

EXCLUSIONS FROM THIS BID:

PROJECTED SCHEDULE:

BASE BID: $
(including taxes)

Bid price valid until _____
                              (date)

ALTERNATE 1: $

ALTERNATE 3: $

ALTERNATE 2: $

ALTERNATE 4: $

BIDDER'S REMARKS:

Signature: _____

Printed Name: _____

Title: _____

Date: _____

FIGURE 5.5 Standard bid form used for subcontractors.

evaluating bids. Notice on the form in Figure 5.5 that bidders must acknowledge that they have reviewed all addenda to the construction documents, that they are bonded and in what amount and by whom, any part of the required work they will not perform (exclusions), the period of time for which their bids are valid, their base bids plus additional amounts to be added should the construction firm decide to proceed with alternates, and remarks.

Requiring all bidders to submit their final bids on the same form allows the construction firm to make more accurate comparisons when deciding which subcontractor wins the bid and receives the contract. Without this uniformity, bid openings can quickly degenerate into confusions and recriminations. Requiring subcontractors to provide bids on construction alternates is also important. A situation that often occurs is this. The construction firm might want to provide certain additional amenities or a higher quality material or piece of equipment as part of the job. However, its ability to do this will depend on price.

In cases such as the one mentioned above, the base bid represents the minimum the construction firm intends to do. Alternates represent additional work or increases in the quality of the work. For example, masonry subcontractors on a new office building project might be asked to submit a base bid for bricking the exterior walls. In addition, they might be asked to submit an alternate bid to build an in-laid brick sidewalk in front of the building. The construction firm must have the exterior walls bricked. This is the work covered by the base bid. The in-laid brick sidewalk is an amenity the firm would like to add depending on the amount of the base bid and the alternate (the in-laid brick sidewalk).

The construction firm has estimated and budgeted a certain amount for masonry work on the project. If the base bids and alternates are less than that amount or even just slightly more, the firm might decide to award the bid to a masonry firm to do both. However, if the base bids will require all that is budgeted for masonry work on the project, the alternate will likely be discarded. A situation that often arises with alternates is this. A subcontractor's base bid is higher than a competitor, but its base bid plus alternates is lower. In this case, the subcontractor with the lowest total for the base bid and any alternates the construction firm decides to proceed with wins the bid.

Challenging Construction Project

BUILDING THE PANAMA CANAL

Construction of the Panama Canal, completed in 1914, is still considered one of the most challenging construction projects in the history of the world. Attempted at different times by the Scots, Spanish, and French—all of whom failed—the Panama Canal was finally completed by the United States. However, even with American ingenuity and determination, the Panama Canal still required 11 years to complete. In the process, the construction firm in charge had to deal with political intrigue, potential revolutions, tropical diseases, a raging river that continually flooded—carrying away people and equipment, and an environment that rotted clothing, rusted equipment almost overnight, and sapped the strength of workers. And these were just the lesser of the many challenges of building the Panama Canal.

The more difficult challenges—and there were many—all fell into three broad categories: engineering, sanitation, and organization. The engineering problems included: 1) digging through the Continental Divide (envision beginning at the foot of the Rocky Mountains at their highest point in Colorado and digging all the way through to the other side of the state), 2) constructing the largest earthen dam ever built, 3) constructing the largest canal locks ever built, and 4) constructing the largest lock gates ever built. All of this had to be done in the most inhospitable, disease-ridden environment imaginable.

After the first two Chief Engineers—they were the project managers—were unable to contend with the enormity of the project, President Theodore Roosevelt appointed Major George W. Goethals of the U.S. Army Corps of Engineers to take charge of the project. The first project manager—John F. Wallace—could not contend with the mosquito problem and the rampant tropic diseases. He left after just one year on the job. His replacement—an engineer and railroad builder named John Stevens—did make significant contributions to the eventual completion of the project. Stevens established the railroad system that allowed the excavation through the Continental Divide to proceed. At the peak of the excavation, the railroad and steam shovel system established by Stevens was blasting and removing a million cubic yards of rock and dirt every day. He also completed construction of the project's infrastructure including badly needed living quarters for workers and their families. However, when the project moved to the hydraulic phase in which a river had to be dammed and locks built, Stevens was out of his depth. This is when President Roosevelt appointed George Goethals, the project manager who eventually completed the project.

The project that Goethals oversaw involved building three locks on the Pacific Ocean side of the Isthmus of Panama and three on the Atlantic side with a large lake in the middle. It also involved damming the Chagres River. The most successful earlier attempt to build the canal was the French effort to build a sea-level canal without locks. This approach would have required the canal to cross the Chagres River 14 times. In the rainy season, the Chagres flooded and became a raging torrent that would have wreaked havoc on any ships passing through the canal—one of many reasons the French effort failed. Goethals and his team implemented the approach that ultimately was successful: damming the Chagres River and building the six locks that are still used to this day.

Source: Based on "Panama Canal Construction—1903–1914." http://www.globalsecurity.org/military/facility/panama-canal-construction.htm

BONDS, ADDENDA, AND ALTERNATES

In the previous sections, the concepts of bonds, addenda, and alternates were part of the narrative. These concepts are important enough in the procurement phase of project management to warrant their own explanations. The following paragraphs explain the basics of bonds, addenda, and alternates as they relate to the job of the construction project manager:

- ***Bonds for subcontractors.*** Bonding is a way for a construction firm that is serving as the general contractor for a project to manage its risk that a subcontractor might not perform as expected. A bond is a financial guarantee that the subcontractor will perform or forfeit a specified amount of money. In construction there are different types of bonds, although all of them are in essence the same: a financial guarantee of performance. *Bid bonds* guarantee that the subcontractor will follow through and enter into a contract if it is selected during the bidding process. *Performance bonds* guarantee that the subcontractor will do the work contracted for in accordance with all applicable specifications and other requirements and expectations. A *payment bond* guarantees that the subcontractor will follow through and pay all of its bills relating to the project (e.g., materials and labor). This is important because a construction firm serving as the general contractor on a project can be held liable for the unpaid bills of subcontractors in some cases. Consequently, many construction forms will not accept bids from unbounded subcontractors. Subcontractors purchase bonds in the same way that insurance policies are purchased. Then, if they do not perform, the bonding agent must pay the specified amount of money.
- ***Addenda to construction documents.*** Addenda are written changes to the construction documents in response to design changes, to correct errors, or changes of other kinds. Changes to construction projects are sometimes made between when the construction documents are completed and when invitations to bid are sent out. There are

times when changes are made even after invitations to bid have gone out but bids have not yet been received or closed out. In both cases, rather than completely rewrite a certain construction document the construction firm will add an addendum to it. If this happens before the invitations to bid are sent out, the bid form should have a line in it to ensure that subcontractors saw the addendum and factored it into their estimate (see Figure 5.5). If it happens after the invitation to bid has gone out but before the due date for bids, all subcontractors should be notified of the change in writing. Again, bidders are required to indicate on the bid form that they reviewed all applicable addenda and factored them into their bids.

- **Bid alternates.** It is not uncommon for a construction firm that is serving as the general contractor on a project to ask bidders to submit a price for a *base bid* and for *alternates*. The base bid covers the work/materials for the mandatory aspects of the project. The alternates represent discretionary aspects of the project (i.e., things that will be included if the price allows it). Construction alternates typically represent additional construction—another building, wing, story, or major feature. However, they can represent escalating levels of quality. For example, a construction firm might specify a certain approach/materials. Then it might add alternates for upgrading the approach/ materials. Upgrade Level 1 would be alternate one—one level of quality above the base bid. Level 2 would be alternate two—two levels of quality above the base bid. For example, assume that the construction firm is building a coin laundry. A subcontractor to provide and install the washing machines is needed. The base bid represents low-priced washers and driers and standard installation features. The first alternate might be medium-priced machines and additional features. The second alternate might represent high-priced machines and maximum features. The construction firm will select either the base bid or one of the alternates depending on price. Regardless of what the alternates in a construction project actually represent, subcontractors should price them like separate projects and provide a separate price in the bid for each alternate.

CONTRACTS FOR SUBCONTRACTORS AND MATERIALS SUPPLIERS

Once bids have been received and evaluated, the subcontractors and materials suppliers who are selected receive contracts. The contract is an important component in the package of construction documents. There are different forms of contracts, but they all tend to contain the same or similar elements. The elements typically contained in a contract between a construction firm that is serving as the general contractor for a project and its subcontractors are as follows:

- **Participants.** This clause of the contract contains the contract participants—the parties to the contract. The contract is between two parties: the construction firm serving as the general contractor and the subcontractor or material provider. Although the participants may seem obvious, there can be no assumptions when entering into a contract. The full and DBA (doing business as) names of both parties are stated. It is not uncommon for construction firms and subcontractors to have formal corporate names that are different than their DBA names.
- **Description of the work to be provided.** This clause of the contract describes in detail the work the subcontractor will provide or the material the vendor will supply. The more comprehensive and accurate this section is the better. If the description

needs to be too long to fit into the actual contract document, it can be attached to it as an appendix. Ideally, this element will be written in a way that leaves no room for disagreement or creative misinterpretation by either party. If disagreements occur during the course of the project over change orders that are requested, this is the first place the project manager will look to resolve the conflict.

- **Starting date.** This clause of the contract contains the starting date of the subcontractor's portion of the project. Depending on the project schedule, the start date might have some slack in it. For example, it might read as follows: *the subcontractor may begin work on January 5 or as late as—but no later than—January 9.* Starting dates for subcontractors are important for several reasons. First, to complete their work on time subcontractors must start on time. Second, subsequent work on the project might depend on the subcontractor completing its work. Finally, when subcontractors submit a bid, it is valid for only a specified period of time. If the subcontractor's starting date is after this valid period, the bid no longer applies and may have to be renegotiated or even rebid.

- **Completion date.** This clause of the contract contains the completion date—the date by which the subcontractor agrees to have all of its work completed. This date is especially critical since subsequent work in the project may be dependent on the subcontractor completing its work. When one subcontractor falls behind schedule on a project, it can start a chain reaction that throws the rest of the project off the schedule.

- **Contract amount.** This clause of the contract contains the amount the subcontractor submitted to win the bid for the work in question. It is a total of the base bid and any alternates that were accepted during the bid process. This amount will not change unless the subcontractor negotiates a change order with the construction firm that is serving as the general contractor.

- **Progress payments.** This clause of the contract explains the milestones that will trigger progress payments to the subcontractor. Not all subcontractor contracts allow for progress payments. It typically depends on the amount of work to be done and the length of time the work will take to complete. Progress payments provide the subcontractor with cash flow so that it does not have to borrow money to make payroll and pay for materials and equipment it uses in the project. If the subcontractor is borrowing for these purposes, the progress payments allow it to pay the loan off promptly, thereby avoiding interest and other fees. For example, assume that a masonry subcontractor has been awarded a contract to brick all four sides of a commercial building. The subcontractor might receive progress payments at 25, 50, and 75 percent completion. The final payment is made upon 100 percent completion and after passing applicable inspections.

- **Liquidated damages.** This clause of the contract contains an explanation of what failures will trigger liquidated damages and in what amounts. Sometimes referred to as "late fees," liquidated damages are assessed against the subcontractor for any damages suffered by the general contractor as a result of failures on the part of the subcontractor. For example, if the subcontractor does not complete the agreed upon work by the agreed upon date and the project schedule is affected, the general contractor can assess liquidated damages as specified in this element of the contract. For example, assume that the electrical subcontractor on an industrial project does not complete its electrical work on time. This, in turn, means the insulations and drywall subcontractors cannot begin on time. The late completion snowballs through the remainder of the project costing the general contractor not just work and aggravation, but money. In a case such as this, the general contractor would probably elect to invoke the liquidated damages clause of the contract.

- **Retained funds or holdbacks.** This clause in the contract explains how much—usually a percentage of the overall amount—the construction firm serving as the general contractor will hold back from the subcontractor as assurance that the work is completed properly and on time. Subcontractors do not receive their retained funds until the general contractor is satisfied that the work provided is complete and properly done. By retaining funds, general contractors build in a certain amount of protection against a subcontractor completing most but not all of the work, then taking its money, and leaving. The amount of retained funds is typically 10 percent, although this amount can vary depending on the level of confidence the general contractor has in the subcontractor. A subcontractor whose work record is questionable might have to agree to a higher percentage of retained funds than one that has a strong work record and a history of completing its work on time, within budget, and according to specifications.

- **Final payment.** This clause of the contract states the conditions under which the general contractor will release any retained funds to the subcontractor. This is an important clause because by releasing any and all retained funds the general contractor is also releasing the subcontractor from further responsibility on the project, other than any responsibilities contained in the subcontractor's warranty—if applicable—and any that have been put in writing and signed by both parties to the contract.

- **List of construction documents.** This clause of the contract lists all documents—drawings, specifications, addenda, and so on—that are part of the construction documents for the project in question.

- **General conditions.** This clause of the contract contains what is often referred to as *boilerplate* language. Boilerplate language summarizes the standard legal language that should be contained in any contract. The language contained in the general conditions clause of the contract is typically taken from a third-party source that provides wording that has been tested in the courts and over time and proven to be legally valid. The American Institute of Architects (AIA), building industry associations, and professional engineering societies all provide sample language for this clause for their members.

- **Special conditions.** This clause of the contract contains an explanation of any additional conditions of a special nature that apply to the project in question and are not adequately covered by the general conditions clause. Occasionally, there will be special requirements that apply to a project that grow out of the exigencies of that project alone. For example, city regulations might dictate that subcontractors park their vehicles only in certain designated areas. When special conditions apply, they are enumerated and explained in this clause.

- **Bonds.** Bonds were explained earlier in this chapter. This clause in the contract explains what types of bonds are required of the subcontractor and in what amounts.

- **Insurance.** This clause of the contract specifies the types of insurance the subcontractor must carry in order to work on the project in question. For example, general contractors typically require their subcontractors to carry workers' compensation and liability insurance. It is important that this clause and its requirements be included in the contract because the general contractor can be held liable for the accidents, injuries, and damage of a subcontractor that is allowed to work without the appropriate insurance.

EVALUATING BIDDERS AND BIDS

When a construction firm that serves as the general contractor on a project solicits bids from subcontractors and materials suppliers, it needs to know the following two things before accepting a bid:

1. Is the bidder responsible—can it actually perform as specified, required, and expected?
2. Is the bid responsive—does it comply in all aspects with the instructions in the RFP or RFQ?

The answers to these questions are even more important than the price quoted by the bidder. This is because the answers to these questions determine whether or not the price submitted by a bidder is actually valid and whether or not the work proposed can actually be performed. If the bidder submits an unresponsive bid, the price quoted cannot be trusted. If the bidder is not responsible, it may not be able to complete the work as specified. Consequently, construction project managers need to learn how to evaluate bids and bidders.

Evaluating Bidders for Responsibility

A responsible bidder is a subcontractor that can comply with all requirements and expectations in the RFP/RFQ while performing the work in question according to specifications. There are actually two considerations embedded in the responsibility question: (1) *Can* the bidder perform as required? and (2) *Will* the bidder perform as required? Just because a subcontractor can perform as required does not mean it will. Consequently, it is important to answer both aspects of the responsibility question.

The "can" part of the question is answered by evaluating the capabilities of subcontractor from the perspectives of financial resources, organizational infrastructure, skills/experience, and facility/equipment resources. The "will" part of the question is answered by evaluating the subcontractor's record from two perspectives: performance over time and ethics.

To perform as required on the project, a subcontractor must be financially stable. It must have or be able to borrow the money needed to perform on the project. To perform as required, a subcontractor must have the organizational infrastructure required to support its personnel (e.g., operational policies, management/supervisory procedures, an accounting/payroll system). To perform as required a subcontractor must have the personnel available who have the skills and experience to do the required work in the required way. To perform as required a subcontractor must have the equipment called for in the job and appropriate facilities to support its personnel and equipment. Without the appropriate support systems in place, a subcontractor may not be able to perform as required on the project.

The "will" part of the question is answered by examining a subcontractor's performance record and ethics record. Can the subcontractor show that it has performed as expected over a period of time? Can the subcontractor provide the names of other general contractors it has worked with? What do these references say about the subcontractor? Does the subcontractor appear to be an ethical organization? Are there ethical problems in its background? Do references say the subcontractor has integrity or are there questions along these lines? A perfectly capable subcontractor can still fail to perform on a project. Consequently, its records of performance and ethics are important considerations during the evaluation process.

Evaluating Bid for Responsiveness

The bids that should be considered by general contractors are those submitted by subcontractors who appear to be responsible. Once the responsibility question has been answered, the responsiveness of bids can be evaluated. A responsive bid is one that meets the following criteria:

- Received on time
- Complies with all requirements in the RFP/RFQ
- Prices quoted are fair and reasonable

Responsive bids received from responsible subcontractors are the only ones that should be considered for price comparisons. Once the determinations of responsibility and responsiveness have been made, comparing prices is not difficult. Figures 5.6, 5.7, and 5.8 are examples of the types of bid comparison forms used by general contractors for comparing the prices submitted by different kinds of subcontractors. Notice that these forms compare not just the overall price, but the individual components of that price.

Using this approach helps the general contractor make an easy determination of whether subcontractors included all of the work called for in their bids. For example, in

BID COMPARISON FORM				
Framing Subcontractors	1	2	3	4
Base Price				
Framing material				
Trusses				
Lumber/plywood				
Floor framing				
Wall framing				
Roof framing				
Fascia and soffit				
Framing hardware				
Power				
Forklift				
Crane for joist/truss				
Backframing				
Woodstairs				
Other				
Total Price				

FIGURE 5.6 Sample form for comparing the bids of framing subcontractors.

BID COMPARISON FORM				
Masonry Subcontractors	**1**	**2**	**3**	**4**
Base Price				
Brick				
Stone				
Blocks				
Tile				
Rebar				
Rebar install				
Concrete				
Mortar and sand				
Scaffolding				
Forklift				
Precast lintel installed				
Wall cleaning				
Bracing				
Flashing				
Caulking				
Air/vapor barrier				
Insulation				
Fire resistance				
Other				
Total Price				

FIGURE 5.7 Sample form for comparing the bids of masonry subcontractors.

Figure 5.8 assume that the painting subcontractor will be required to apply special wall coatings and that three of the subcontractors—1, 2, and 3—submitted their prices for this part of the job. Subcontractor number 4 did not. This probably means that number 4 overlooked the special wall coatings part of the job and, therefore, has submitted an unresponsive bid.

ETHICS IN PROCUREMENT

No construction-related process has more opportunities for ethical lapses than the procurement process. Unethical players in the process have myriad ways to circumvent the process and make deals that are good for them but bad for the project, the construction firm serving as the general contractor, and the subcontractor/material supplier. Consider the following

BID COMPARISON FORM

Painting Subcontractors	1	2	3	4
Base Price				
Exterior painting				
Interior painting				
Wall coverings				
Columns and beams				
Doors				
Handrails				
Wood trim				
Ceilings				
Stairs				
Special wall coating				
Caulking				
Scaffolding				
Other				
Total Price				

FIGURE 5.8   Sample form for comparing the bids of painting subcontractors.

situations that have occurred numerous times on construction projects. A subcontractor submits an unresponsive bid but is awarded the contract anyway after paying "kickback" to the CEO of the construction firm that solicited the bids. A material provider wins the bid on a large project by knowingly bidding substandard materials that do not meet the specifications while claiming that they do. A subcontractor bribes an inspector to ignore certain deficiencies in its work.

Ethical Procurement in Construction Defined

Ethical procurement in construction is the result of a procurement process that is characterized by integrity, transparency, openness, suitability, and fair competition. When the procurement process operates within an ethical framework, contracts are awarded to subcontractors and materials suppliers on the basis of their ability and commitment to meet all requirements in the RFP/RFQ rather than on such factors, such as favoritism, familiarity, illicit compensation, politics, and intimidation.

Competition and Ethical Procurement

One of the most effective ways for a construction firm to ensure an ethical procurement process is to base the process on fair and open competition. Fair and open competition means that when a construction firm that is serving as the general contractor for a project invites subcontractors to submit bids: (1) the net is cast broadly so as to include as many potential bidders as possible, (2) all subcontractors/materials providers are provided with the same documents on which to base their bids, (3) all bidders and bids are evaluated on the basis of the same criteria, (4) all potential bidders have an equal opportunity to attend bid conferences, (5) all bidders have equal access to construction documents including addenda, (6) all bidders receive the same communications from the general contractor during the bid preparation process, (7) all bidders have an opportunity to be present when bids are opened, and (8) all bids are opened at the same time in the same location.

Other Issues in Ethical Procurement

In addition to having a procurement process that is fair, open, and competitive, construction firms must be cognizant of the ethical behavior of subcontractors and materials suppliers. For example, a construction firm that knowingly awards a contract to a subcontractor who employs illegal aliens, underage workers, drug abusers, or wanted felons is committing an ethical breach by aiding and abetting unethical behavior. A construction firm that knowingly awards a contract to a materials supplier that obtains its materials illegally or abuses its workers is also aiding and abetting unethical behavior. A construction firm that awards a contract to a subcontractor with a record of serious safety violations is aiding and abetting unethical behavior. Consequently, it is important for construction firms that solicit bids to go beyond the walls of their company and consider the ethics of their subcontractors and materials suppliers.

CONSTRUCTION PROJECT MANAGEMENT SCENARIO 5.2

I Don't Think We Should Accept Their Bid

Don Markham is the project manager for the new arena being built by the city. As such, he is one of the members of the procurement team that will evaluate the bids from subcontractors and materials suppliers. Before opening the bids that have been submitted, the procurement team is evaluating the bidders to ensure that they are responsible. A problem has arisen in this regard. The most important subcontractor on this job will be the one that provides and erects the steel. Questions have come up about the ethics of one of the bidders: LMN Steel Contractors. LMN has a long record of completing its projects on time and within budget. However, it also has a reputation for employing illegal aliens and for its shoddy safety practices. Feedback on the company suggests that there are more accidents and injuries than are actually reported. Apparently, when an illegal alien is involved in an accident the company gets away with not reporting it because the illegal worker—for fear of deportation—refuses to take any action.

Some members of the team want to qualify LMN as being a responsible subcontractor on the basis of the firm's record for completing its projects. Some members want to disqualify LMN on the basis of ethical considerations. The most vocal of the latter group is the project manager—Don Markham. Markham began the debate on LMN when he said: "I don't think we should accept their bid." The debate has gone on for more than an hour with nothing resolved.

Discussion Question

In this scenario, the procurement team is evaluating bidders to ensure that they are responsible. Several members think that proven performance is the only criterion they should apply in making the determination. Several other members think that ethics should be an added criterion. Assume that you are a member of this procurement team. Which side of the debate would you take and why?

SUMMARY

Before construction can begin on a project, the construction firms serving as the general contractor must identify, locate, and obtain the necessary materials and the subcontractors who will do the hands-on work. The process is known as procurement. The most commonly used procurement methods are: low-bid, best-value, micro-purchase, small-purchase, and sole-provider. A comprehensive bid package (RFP or RFQ) will contain an invitation to bid, announcement of the pre-bid conference, bidding instructions, and bidding form.

The invitation to bid should contain the following information: project description, project location, start and completion dates, bonds, project documents, legal considerations, and the bid deadline. Bidding instructions should contain the following information: date and time bids are due, location where bids must be received, instructions for completing the bid form, unit prices for work and labor, additional fees, how and when the winning bid will be announced, how and when the contract will be awarded, and special instructions.

A bond is a financial guarantee that a subcontractor will either complete the work as specified or forfeit a specified amount of money. There are three types of bonds widely used in construction: bid, performance, and payment bonds. Addenda are written changes to the construction documents in response to design changes, errors, or changes of other kinds. Alternates are discretionary add-ons to the basic work being bid—extras that will be constructed if funding allows.

A contract for a subcontractor should contain the following information: participants or parties to the contract, a description of the work, starting date, completion date, contract amount, progress payments, liquidated damages, retained funds, final payment, a list of construction documents, general conditions, special conditions, bonds, and insurance.

Bidders should be responsible, meaning they should be able and willing to complete the work in question as specified and in accordance with the construction documents. Bids should be responsive, meaning they should comply with all requirements in the RFP/RFQ, should be submitted on time, and contain prices that are fair and reasonable. Ethical procurement in construction is procurement that is characterized by integrity, transparency, openness, suitability, and fair competition. Fair competition is an effective way to ensure the integrity of the procurement process.

KEY TERMS AND CONCEPTS

Procurement
Low-bid method
Best-value method
Micro-purchase method

Small-purchase method
Sole-provider method
Invitation to bid
Pre-bid conference

Bidding instructions	Contract amount
Bidding form	Progress payments
Bonds	Liquidated damages
Addenda	Retained funds
Alternates	Final payment
Bid bonds	List of construction documents
Performance bonds	General conditions
Payment bonds	Special conditions
Participants	Insurance
Description of the work to be provided	Responsible bidder
Starting date	Responsive bid
Completion date	Ethical procurement

REVIEW QUESTIONS

1. Define the term *procurement* as it is used in construction.
2. List and briefly explain the most commonly used procurement methods.
3. List and briefly explain the contents of a comprehensive bid package.
4. What should be contained in the invitation to bid?
5. What should be contained in the bidding instructions?
6. What is the purpose of a bond in construction?
7. List and briefly explain the types of bonds widely used in construction.
8. What are addenda?
9. What is a bid alternate? Give an example.
10. List the information that should be contained in a contract given to a subcontractor.
11. What is meant by the term *responsible bidder?*
12. How can a project manager determine if a bidder is responsible?
13. What is meant by the term *responsive bid?*
14. How can a project manager determine if a bid is responsive?
15. Define the term *ethical procurement.*

APPLICATION ACTIVITIES

The following activities can be completed by individual students or students working in groups:

1. Identify a construction company in your region that will cooperate with you in completing this project. Ask the procurement director to walk you through how the company solicits bids, evaluates bids, and awards contracts to subcontractors. Ask what the biggest challenges of the procurement process are. Ask to see an invitation to bid, bids that were submitted by subcontractors, and a contract. Determine which procurement method the company uses most often. Report the findings of your research to the class.

2. Do the research necessary to identify at least one case in which there were ethical failures in the procurement process of a construction firm. Write a report on your findings and report the results of your research to the class.

CHAPTER 6

Managing Risk in Construction Projects

Construction projects will occasionally be characterized by what went wrong with them. It is not uncommon for unplanned events and unanticipated problems to be what is remembered most about construction projects. This is why risk management is so important a part of construction project management. There is always risk involved when a firm undertakes a construction project, whether the project is large or small. Consequently, identifying, prioritizing, and minimizing risk is an important responsibility of the construction project manager.

There are three broad success criteria that apply to all construction projects. These criteria relate to completing the project on time, within budget, and according to specifications. Satisfying these three criteria is a goal of the project manager in every construction project. Unfortunately, construction is like everything else in life. Things do not always go according to plan. This fact is the origin of the maxim that claims *anything that can go wrong will go wrong,* sometimes referred to as *Murphy's Law.*

Although the anonymous Mr. Murphy might have overstated his case somewhat, a construction project is a complex undertaking with a lot of moving parts that must come together at the right time and in the right way if the project is to be completed on time, within budget, and according to specifications. Consequently, there is a lot of room for errors and problems in a construction project. Further, there is always the risk that errors that can be made will be made and that problems that can arise will arise. Identifying, assessing, and mitigating risk is part of the project manager's job on construction projects.

RISK DEFINED

Risk is simply the possibility that things will not go as planned and that counterproductive unplanned events will arise during the course of a construction project. In both cases, what is at risk is the completion of the project on time, within budget, and according to

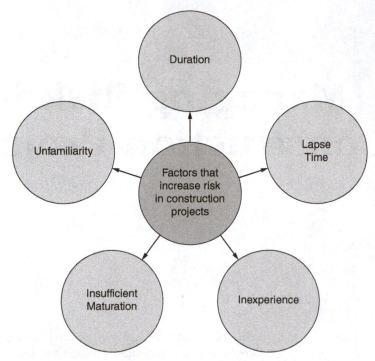

FIGURE 6.1   These factors can multiply the level of risk for a construction project.

specifications. Several factors can increase the level of risk in construction projects (Figure 6.1), which are as follows:

- **_Duration._** The longer it takes to complete a project the more likely it is that something will go wrong. This is like saying the longer you are behind the wheel of a car, the more likely it is that you will have an accident. Increasing the duration of a project correspondingly increases the project's exposure to plans falling apart and unplanned events occurring.
- **_Lapse time._** A lapse between winning a construction contract and beginning construction increases the likelihood of something going wrong. The longer the lapse, the greater the risk. For example, material prices might increase, labor unions might go on strike, material shortages might occur, qualified subcontractors might get tied up with other projects, and other unplanned factors might come into play.
- **_Inexperience._** Generally speaking, the more experience a construction firm has with a given type of project the less risk associated with the project. Inexperience can lead to errors and introduce problems into a project. Of course, there has to be a first time for every type of project, but the risk factor increases as the experience factor decreases.
- **_Insufficient maturation._** As construction processes and technology mature, they become more reliable. On the one hand, construction personnel get better at using them over time. On the other hand, bugs and problems with them get worked out over time. Projects in which new technologies and/or new processes are introduced have a higher level of risk than those that use proven technologies and processes.

- *Unfamiliarity.* Construction projects use a lot of different trades and subcontractors. Larger firms often prequalify subcontractors and trades to ensure they can do the jobs called for in the project. However, it is not uncommon for construction firms to have to work with subcontractors and trades they are unfamiliar with because they have never worked with them before. Unfamiliarity with any members of the construction team—especially the architect/designer, engineers, subcontractors, and trades—increases the level of risk on a construction project.

RISK MANAGEMENT DEFINED

Construction project managers must also be good risk managers. Minimizing and mitigating risk on construction projects is an important responsibility of project managers. From a construction project manager's perspective, risk management is defined as follows:

> Identifying risks that might negatively affect the proper completion of construction projects, assessing their potential impact, developing mitigation plans, and implementing the plans in ways that minimize the risk.

The term "minimize" in this definition is important because that is the best project managers can do. They cannot eliminate risk. There are just too many factors outside of their control that can introduce risk into a construction project. To minimize and manage risk, construction project managers should do the following (Figure 6.2):

- *Be aware that risks exist on all projects.* Project managers should assume that risks are present in any construction project no matter how large or small. Ignoring risk as an issue when planning projects is a mistake that can become costly. The first question to ask in the initiation stage of a project is: What are the risks? The answer to this question is essential to making an informed go/no-go decision concerning whether to pursue the project.
- *Identify project-specific risks.* Although some risk factors apply to many or almost all construction projects, it is important to go beyond the *usual* and look for project-specific risks with every construction project. The same type of project built in a different location or with a different crew can have vastly different levels of risk. Consequently, it is important for construction project managers to identify the risks that apply to every individual project they manage.

RISK MANAGEMENT STEPS

- Be aware that risks exist on all construction projects.
- Identify project-specific risks.
- Assess the potential consequences of the risks.
- Communicate the risks and their potential consequences to stakeholders.
- Develop and implement risk mitigation plans.
- Monitor the effectiveness of risk mitigation strategies.

FIGURE 6.2 Project managers must perform all of the steps in the risk management process.

- ***Assess the potential consequences of the risks.*** One type of risk might be easily dealt with in even its worst-case scenario. However, some risks have the potential to completely undermine the entire project. Consequently, it is important for project managers to carefully assess all risks that apply and determine what might actually happen—from best-case to worst-case scenario—should the risk actually come into play.
- ***Communicate the risks and their potential consequences to stakeholders.*** Once the project manager has identified all applicable risks—at least as best as he or she can—the risks and their potential consequences should be communicated to all project stakeholders (e.g., the construction firm's higher management, architect/designer, engineers, and members of the project team).
- ***Develop and implement risk mitigation plans.*** Since risk cannot be eliminated altogether, the next best course of action is to develop strategies for minimizing it. These mitigation strategies, taken together, become the risk mitigation plan for the project. All construction projects should have a risk mitigation plan.
- ***Monitor the effectiveness of risk mitigation strategies.*** Once the risk mitigation plan is implemented, the project manager should monitor the effectiveness of the mitigation strategies for the duration of the project. In addition to monitoring the risks that were identified earlier in the process, project managers should be vigilant in looking for others that might arise during the course of the project. It is not always possible to identify all project risks on the front end. Consequently, it is not uncommon for other risks to unexpectedly crop up during the course of the project. When this happens, mitigation steps should be taken immediately and the mitigation plan should be updated accordingly.

CLASSIFICATIONS OF RISK FACTORS IN CONSTRUCTION

The risks associated with construction projects may be classified in different ways. One of the more effective ways to classify risk is according to where control lies:[1]

- ***External—Unpredictable.*** These types of risks arise from third parties, acts of God, and other factors over which the project manager has no control. They are completely unpredictable. For example such natural disasters as floods, hurricanes, tornados, and earthquakes can devastate a construction project, but they cannot be predicted.
- ***External—Predictable but Uncertain.*** It can be predicted that these types of risk will occur but not the extent to which they will occur. For example, it can be predicted that the weather will affect progress on construction projects but not the extent of the effect. It can be predicted that the customer might ask for late-in-the-game changes, but not the type or extent of the changes. Market changes and regulatory issues can also be predicted as potential problem causers, but not the extent of the problems they might cause.
- ***Internal—Technical.*** These risks arise primarily from technologies that are used in any phase of a project's life cycle. Risk increases when new and untried technologies are used in construction projects since technologies often have bugs that must be worked out over time. Another factor that increases risk when using new technologies is the ability of people to operate the technologies since they are on the low end of the learning curve at the beginning of the project. Consequently, internal-technical risks are typically tied to such issues as performance, quality, complexity, and specifications.

- *Internal—Nontechnical.* These risk factors grow out of human and organizational issues. For example, an individual who has always performed well on past projects may not perform well on the current project for a variety of reasons. Nontechnical internal risks are typically associated with problems in such areas as teamwork, funding, higher management support, and communication.
- *Legal/Ethical—Civil and Criminal.* These risks grow out of an organization's ethical and legal obligations to project stakeholders. The risk is that the firm, project team, or an individual member of the team might commit some type of ethical, civil, or criminal violation relating to the project and that the violation will lead to mediation, arbitration, or adjudication. For example, a negligent act that is perpetrated by or tacitly approved by the construction firm's on-site superintendent and results in a death on the job could lead to criminal charges. These charges and the tragedy and controversy surrounding them could affect the firm's ability to perform as expected on the job.[2]

Figure 6.3 is a graphic representation of an RBS or Risk Breakdown Structure.[3] Project managers who lead risk management teams should develop an RBS for each project they are involved with and share it with all members of the team. It can be an excellent tool for triggering the thinking of risk team members as they try to identify project risks. On smaller projects in which the project manager undertakes the risk analysis himself or herself, developing an RBS is still an effective first step in the process.

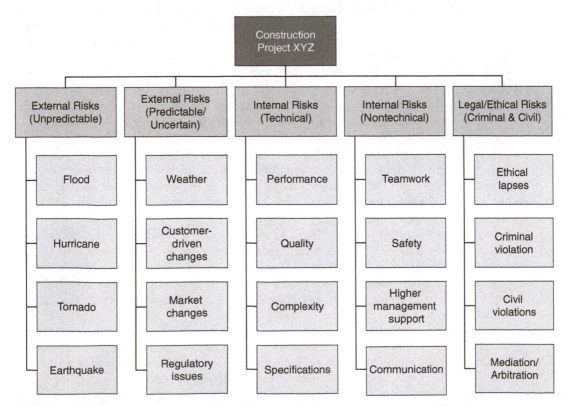

FIGURE 6.3 Template for a Risk Breakdown Structure (RBS) for construction projects.

RISK AS IT RELATES TO PROJECT SUCCESS CRITERIA

Throughout this book, the point is made over and over that there are three basic success criteria that apply to all construction projects. These criteria are to complete the project on time, within budget, and according to specifications. When considering risk, two additional success criteria must be added: (1) complete the project in a manner that is environmentally friendly and (2) complete the project without safety and health violations. One can argue that the latter two criteria are really subsets of the first three. They can be treated in this way—as subsets—or they can be broken out as separate criteria.

The rational for making environmental and safety/health concerns success criteria is simple. A project that is completed on time, within budget, and according to specifications at the expense of the environment or the safety/health of trades, subcontractors, or the general public will be a short-term success but a long-term failure. Fines, negative publicity, and other expenses levied by government regulatory agencies can eventually turn what is a successful project into a failure. In fact, otherwise competent and successful construction companies have been driven into bankruptcy by the litigation and fines associated with regulatory violations.

Hence, it is advisable to look at the risk associated with construction projects as it relates to the following five success criteria: (1) time, (2) budget, (3) specifications (quality), (4) environment, and (5) safety/health. There are specific risk factors associated with each of these five success criteria. These risk factors are as follows:[4]

- ***Time-related risk factors.*** There are a number of factors that can put completing a construction project on time at risk. These factors include: too tight a schedule, poorly planned schedule, design changes, government bureaucracy, material delivery problems, work slowdowns or strikes, weather problems, and natural disasters.
- ***Cost-related factors.*** There are a number of factors that can put completing a construction project within budget at risk. These factors include: too tight a schedule, poorly planned schedule, poor cost estimating, disputes among stakeholders, price inflation of construction materials, government bureaucracy, work slowdowns and strikes, and natural disasters.
- ***Specification/quality-related factors.*** There are a number of factors that can put completing a construction project according to specifications at risk. These factors include: rush induced by too tight a schedule, disorganization caused by poor planning, poor attention to specifications when preparing the cost estimate, incompetence on the part of trades and subcontractors, lack of sufficiently skilled trades, and poor coordination between and among stakeholders.
- ***Environment-related factors.*** There are a number of factors that can put completing a construction project in an environmentally friendly manner at risk. These factors include: questionable practices induced by too tight a schedule, frustration over working with an impenetrable government bureaucracy, insufficient site data from soil tests and surveys, poor management, insufficient attention to noise pollution, and lack of knowledge of environmental regulations.
- ***Safety/health-related factors.*** There are a number of factors that can put completing a construction project without accidents and injuries at risk. These factors include: unsafe practices induced by rushing because of too tight a schedule or a poorly planned schedule, poor supervision, insufficient support of safety/health measures from higher management, inexperienced workers, use of potentially toxic materials, working at heights, worker reluctance to wear appropriate personal protective gear, and lack of knowledge concerning appropriate safe and healthy work practices.

CONSTRUCTION PROJECT MANAGEMENT SCENARIO 6.1

There Are Risks in Every Endeavor

Professor Jones is introducing her construction project management class to the concept of risk. Before getting into risk as it relates specifically to construction projects, she wants to make the point that there are risks in every endeavor. She believes that seeing how a common activity unrelated to construction has risks incumbent in it will help her students better understand risk. Consequently, Professor Jones gives her class the following assignment: Assume that you and several of your friends plan to take a trip to the beach for Spring Break. The nearest beach is 600 miles away and you plan to drive. Your goal is to spend several days relaxing on the beach and having fun. Identify all of the risks inherent in this trip that might keep you from achieving your goal.

Discussion Question

In this scenario, Professor Jones wants her students to understand that there is risk in every endeavor. Assume that you are in this class. Make a list of all risks you can identify that might prevent you from achieving the goal of spending several days relaxing on the beach and having fun.

RISK IDENTIFICATION PROCESS FOR CONSTRUCTION PROJECTS

The first step in identifying the risk associated with a construction project is to get organized. This means the project manager should do the following: (1) form the risk identification team, (2) distribute the Risk Breakdown Structure template, (3) select the risk identification methods to be used, and (4) decide what the output of the risk identification process will be (see Figure 6.4).

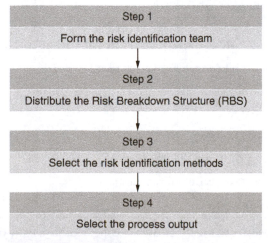

FIGURE 6.4 Risk identification process.

Risk Identification Team

If the construction firm is small and the project is small, the risk management team might consist of only the project manager and one or two other stakeholders. However, in most cases—especially with larger companies and larger projects—the risk management team will consist of the project manager, risk management experts, members of the project team, and other stakeholders who might include the architect, engineers, and owner. Although the risk management team should be kept to a manageable size, anyone from inside or outside the construction firm who can contribute to accurately identifying project risks and planning appropriate responses may be included.

Risk Breakdown Structure Template

Figure 6.3 shown earlier in this chapter is a template for a Risk Breakdown Structure for construction projects. The RBS template is used to trigger the thinking of the risk management team in identifying potential project risks. The team should not be limited by the RBS template. After all, there might be risks that are so specific to the project in question that they do not have a corresponding category on the template. Before beginning discussions and brainstorming activities and even during these activities, members of the risk management team can add categories of risk to the top of the template or specific areas of risk of the existing categories. However, in most cases the team will find that the template covers most of the predictable areas of risk commonly encountered in construction projects.

Risk Identification Methods

There are several different methods that can be used to identify risks for a specific construction project. The most common of these include the following (Figure 6.5):

- Review of the construction documents
- Brainstorming sessions
- SWOT analysis
- Experience review

**CHECKLIST OF
RISK IDENTIFICATION METHODS**

✓ Review of the construction documents

✓ Brainstorming sessions

✓ SWOT analysis

✓ Experience review

✓ Review of professional literature

✓ Survey of experts/Delphi technique

✓ Expert judgment

FIGURE 6.5 Project managers can use these methods to identify risks.

- Review of professional literature
- Survey of experts/Delphi technique
- Expert judgment

REVIEW OF CONSTRUCTION DOCUMENTS. A thorough review of the construction documents by all members of the risk management team is an effective way to identify project-specific risks. Each member of the risk management team has his or her own area of expertise and corresponding perspective. Having people from a variety of backgrounds and perspectives examine the construction documents can reveal risk factors that even the most experienced construction professional might not see.

A review of the construction documents might reveal the following potential problems: (1) the project will take longer to complete than originally thought, (2) there is going to be too much lapse between signing the contract and beginning construction, (3) the construction firm has no experience in building the type of project in question, (4) technologies not previously used will be required to complete the project, (5) it will be necessary to use unfamiliar trades and subcontractors, (6) the specifications are not as well-defined as they should be, (7) the construction cost estimate leaves very little room for error, (8) regulatory issues might be a problem, and (9) the number of rain days planned for the project might not be enough.

BRAINSTORMING SESSIONS. The project manager leading the risk management team through a comprehensive brainstorming session or several sessions is an excellent way to identify project-specific risks. The brainstorming sessions are given structure by the RBS. For large projects, the project manager might conduct one brainstorming session for each of the major risk categories: external—unpredictable, external—predictable but uncertain, internal—technical, internal—nontechnical, and legal/ethical—criminal and civil. For smaller projects, all of the risk categories can be covered in just one brainstorming session. Regardless of the number of sessions held, the goal is to have people of different areas of expertise and different perspectives think creatively about the issue of risk, guided but not limited by the RBS.

SWOT ANALYSIS. A comprehensive SWOT analysis (strengths, weaknesses, opportunities, threats) is an effective way to identify project-specific risks—particularly internal—technical and internal—nontechnical risks. The strengths, weaknesses, opportunities, and threats referred to in this section are those relating to the construction firm's ability to complete the project in question on time, within budget, according to specifications, and in a safe and environmentally friendly manner. The tendency is to think of risk only in terms of the negative (weaknesses and threats), but there is a positive side to risk that should not be overlooked. A construction firm's strengths can sometimes be exploited to help minimize risk as can opportunities.

The project manager conducts the SWOT analysis sessions in a manner similar to the brainstorming sessions. Using the RBS template to provide structure, the project manager poses each SWOT question in order and deals with them one at a time. For example, consider the following questions: (1) What are our strengths relative to the project? (2) What are our weaknesses relative to the project? (3) Are there opportunities that can be exploited with this project? and (4) Are there threats relative to the project?

The construction firm's successful track record in completing similar projects could be a strength. The need to work with unfamiliar trades and subcontractors because the project will be built at an out-of-town location could be a weakness. An opportunity might be the chance to increase the profits by reaching certain milestones ahead of time. A threat might be predictions of an unusually active hurricane season. The advantage of using the SWOT analysis as a risk management method is that it looks at both sides of the risk equation. By so doing it sometimes reveals positives that can be exploited in ways that overcome or, at least, minimize the negatives. For example, a weakness might be inexperience in certain trade areas. An offsetting strength might be a strong construction superintendent who is especially good at monitoring and mentoring inexperienced workers.

EXPERIENCE REVIEW. The experience review is a method available to construction firms that have experience in constructing the type of project in question. Experience is an excellent teacher provided, of course, the student is paying attention. The value of the experience review is increased if it is conducted prior to brainstorming or SWOT analysis sessions. When done in this way, the brainstorming and SWOT analysis sessions are informed by the experience of the risk identification team members.

When conducting an experience review, the project manager will occasionally ask construction professionals and other experienced personnel to *testify* before the risk identification team so that the team gains the benefit of their experience. Another way to conduct the experience review is to assign each member to interview colleagues who have worked on similar projects to determine lessons they have learned and can pass along. One of the key responsibilities of the project identification team is to ensure that the construction firm benefits from its own experience.

REVIEW OF PROFESSIONAL LITERATURE. An effective but often overlooked risk identification method is the review of professional literature. This is another of those methods that should be undertaken before conducting brainstorming or SWOT analysis sessions so that the sessions can be informed by its findings. An effective way to conduct the review of professional literature is to give each member of the risk identification team the assignment of reviewing his or her professional literature to locate articles and other references that explain problems that occurred with other similar projects. These problems, particularly if they surface more than once or even repeatedly, point to potential risks for the current project.

SURVEY OF EXPERTS/DELPHI TECHNIQUE. Surveying experts anonymously for their input can be an effective way to identify project-specific risks. The experts can be internal or they can be professional colleagues or a combination of both. Construction professionals often join professional organizations. Through these associations they meet other construction professionals with expertise in different aspects of construction. These colleagues may be willing to respond to a survey on the basis of problems they have encountered when working on projects similar to the one in question. The survey instrument solicits their input about project risks. This input is summarized, and the summary is circulated among the experts again for additional consideration. The process is repeated until the list of risks has been pared down to those that are most likely to be associated with the project in question. This is another of those risk identification methods that should be undertaken prior to conducting brainstorming or SWOT analysis sessions so that the sessions can be informed by its findings.

EXPERT JUDGMENT. One of the reasons each member of the risk management team was selected to participate in the first place is the expertise he or she brings to the table. Each member of the team has a certain amount of experience, education, and/or on-the-job training relating to the project in question. Consequently, one of the most important assets each team member brings to the table is expert judgment. Expert judgment comes into play in all of the other risk identification methods, but it is especially important when reviewing the construction documents.

To a layperson, the construction documents might appear to be just an attractive set of drawings, an impressive set of specifications, and a potentially lucrative contract. But to a professional with expertise in some aspect of the project in question, the documents can be invaluable sources of information for identifying risk, especially when reviewed within the structure of the RBS. Of course, as is always the case when using the RBS it should be an enabling tool rather than a limited device. Members of the risk management team should begin by thinking inside the box defined by the RBS and end by thinking outside the box.

Process Output

The output of the risk management process is a list of potential risks organized under the broad categories specified in the RBS along with any new categories added to the RBS during the risk identification process. Each entry in the list should be stated in a standard format that will inform the risk response process. A helpful format for risk statements is the cause-and-effect format (e.g., If a certain *cause* occurs, it will have the following *effect*.). Figure 6.6 contains examples of risk statements written in this format and tied to their corresponding risk category from the RBS.

Each of the risk statements shown in Figure 6.6 is written in the cause-and-effect format. Under the broad risk heading *Internal Risks—Unpredictable,* one can see that the risk management team expressed concern about the negative effects an unusually active hurricane season could have on the project. The team also expressed concern about what might happen to work in progress if a hurricane hit during the construction phase of the project. In the first risk statement, the cause is the unusually active hurricane season. The possible effect is schedule delays that might keep the project from being completed on time.

SUMMARY OF RISK STATEMENTS XYZ PROJECT

External Risks—Unpredictable

- If hurricane season is unusually active, the project may not finish on time.
- If hurricane hits during construction, the work in progress might be severely damaged.

Internal Risks—Technical

- If the new material control software does not live up to expectations, the material delivery, unloading and storage process will break down resulting in delays.

Internal Risks—Nontechnical

- Because of the firm's OSHA violations or its last job, if safety procedures are not followed meticulously, heavy fines might be assessed.
- If the new interior wall sound insulation material is not delivered on time, the delay will increase cost and impede the schedule.

FIGURE 6.6 Sample risk statements.

Under the heading *Internal Risks—Technical,* the team expressed concern about a new material control software package that will be used by the construction firm for the first time on the XYZ Project. In this case, the cause is less than optimum performance from the software and the effect is a breakdown in material control and flow. The software might perform at less than optimum levels because the firm's personnel do not yet know how to use it effectively or because there are still bugs to be worked out of it or both. The negative effect if the software does not perform optimally is schedule delays.

Under the heading *Internal Risks—Nontechnical,* the risk identification team was concerned because the construction firm had experienced safety violations on its last job. As a result, the Occupational Safety and Health Administration (OSHA) will be watching the company even more closely than usual on the current project. The cause in this case is noncompliance with safety procedures that results in safety violations. The effect is heavy fines levied by OSHA. OSHA fines can be substantial, especially when the construction firm has been previously cited for safety violations.

Because the risk statements were written in the cause-and-effect format, risk management personnel will find it easier to develop risk response plans to counteract the potential causes. This is important because the goal of the risk management process is to minimize the exposure of stakeholders to risk. Risk cannot be eliminated, but it can be minimized provided the project manager takes the necessary steps.

QUALITATIVE RISK ANALYSIS

Once the risks for a project have been identified, they must then be analyzed. Risk analysis is the process used to assess the qualitative or quantitative value of risk as it relates to specific risk factors. Qualitative risk analysis assesses the *probability* that a given risk factor will have an *impact* on the project and the extent of the impact. Quantitative risk analysis assesses the *probability* that a given loss will occur and the *magnitude* of the loss.

Both types of risk analysis are complex undertakings requiring expertise beyond that typically expected of a construction project manager. Hence, both types of risk analysis require the participation of risk assessment experts as well as construction experts. Of the two—qualitative and quantitative—the more widely used in construction project management is qualitative risk analysis. Qualitative risk analysis is an important step that should be completed prior to developing risk responses because it can inform the risk response process. Risk probability assessment involves estimating the likelihood that a given risk will actually occur. Risk impact assessment involves estimating the potential effect each risk identified might have on the success criteria for the project. Although there may be additional success criteria, there are always the three fundamental criteria of cost (within budget), time (on time), and quality (according to specifications). Safety and environmental risk factors should also be considered, but they can be subsumed into the first three if their principal potential impacts are on cost, time, or quality.

The risk management team, led by the project manager, is responsible for establishing risk probability definitions and corresponding impacts. Probability is a number between 0 and 1. A risk probability of 0.0 means the risk will never happen. A risk probability of 1.0 means it is guaranteed to happen all the time. Impact can also be assigned numerical values between 0 and 1. For example, 0.0 impact means the risk factor will have no effect, but an impact of 1.0 means it will have a devastating effect. By graphing probability on one axis and

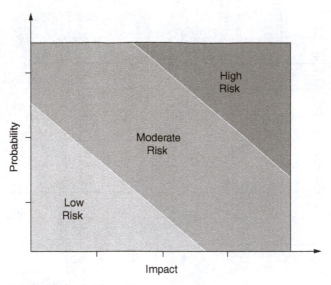

FIGURE 6.7   Plotting risk.

impact on the other, the risk management team can check which of the following levels of risk applies (Figure 6.7):

- Low risk
- Moderate risk
- High risk

Different construction firms will have different levels of risk tolerance. However, generally speaking, risk management teams become concerned when the risk factor falls into the moderate range or higher.

Another approach is to transform the graph in Figure 6.7 into a probability and impact matrix such as the one shown in Figure 6.8. With this approach, the probability and impact scales of the graph are placed along each respective axis to form a grid. The risk value of each intersection on the grid is computed (probability × impact = risk value). The risk management team must decide what it considers the cutoff points for low, moderate, and high risk. In Figure 6.8, the team established the following cutoff points:

- 0 − .19 = Low risk
- .20 − .39 = Moderate risk
- .40 plus = High risk

The probability value chosen by the risk management team is multiplied times the estimated impact value for each risk factor. The resultant risk value determines if the risk of the factor in question is low, moderate, or high. The darker shaded areas in Figure 6.8 represent the high risk zone. The lighter shaded areas represent the moderate risk zone. Any risk factors for the project in question that fall into either of these two zones will receive special attention when the risk management team develops its risk responses.

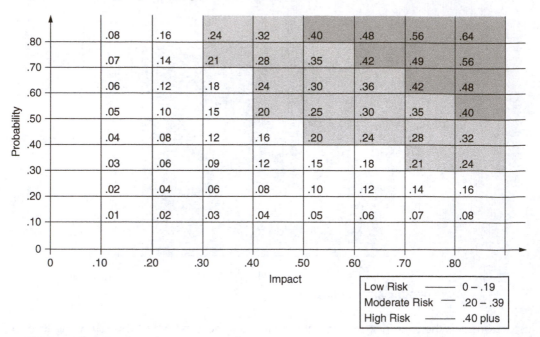

FIGURE 6.8 Computing and plotting risk values.

One of the weaknesses of qualitative risk analysis comes into play at the point where the risk management team plots the estimated probability and impact values on a graph such as the one in Figure 6.7 or 6.8. Expert judgment is required in selecting the probability and impact values. Consequently, the old information technology maxim that says *garbage in/garbage out* applies when performing any form of risk analysis. Hence, qualitative risk analysis is only as good as the judgments made by the members of the risk management team.

Risk Response Strategies

In developing appropriate responses to risk, project management teams have three broad options: (1) elimination, (2) transfer, and (3) minimization.

ELIMINATION. With this option, the risk management team attempts to completely eliminate the risk in question. For example, one of the risks identified in the previous section on qualitative risk analysis was a new insulation material for sound proofing interior walls (see Figure 6.6). The risk was that the material might not be delivered on time. One way to eliminate this risk is to eliminate this new material and substitute another material that can be delivered on time with greater certainty.

TRANSFER. With this option, the risk management team recommends paying a third party to assume some or all of the risk in question. The most commonly used risk transfer method is the purchase of insurance. For example, one of the risks identified in the previous section on qualitative risk analysis was the likelihood of an active hurricane season during the time of the project (see Figure 6.6). A way to transfer the risk associated with hurricanes would be for the construction firm to purchase a special hurricane policy for the project.

MINIMIZATION. With this option, steps are taken to mitigate the impact of the risk in ways that minimize the construction firm's exposure. There are two types of minimization strategies: (1) minimize the chance that the risk will occur and (2) minimize the damage if the risk does occur. For example, one of the risks identified in the previous section on qualitative risk analysis was the likelihood of heavy OSHA fines being assessed if violations occur on the job. The construction firm could minimize the chance that violations will occur by providing mandatory safety training for all personnel who will work on the project. The firm could minimize the amounts of the fines should violations occur by asking OSHA, at the beginning of the project and throughout its duration, to provide assistance in establishing a safe job site. An experienced risk management team would recommend doing both.

Once the risk management team has determined which of the risk statements represent moderate to high risks, the statements are listed in order of priority beginning with the highest risk and risk response strategies—elimination, transfer, or minimization—are planned for each. Figure 6.6 contains five risk statements that were developed earlier in the risk identification phase of the process. Assume that three of these statements had risk values that put them in the moderate to high risk range. The risk statements along with a corresponding risk response strategy are as follows:

- ***Internal Risk—Nontechnical (Risk value of .40).*** Because of the firm's OSHA violations on its last job, if safety procedures are not followed meticulously, heavy fines might be assessed.
  Risk response strategies: (1) Train all on-site supervisors in their responsibilities for ensuring that their personnel work safely, (2) Require all subcontractors and trades to complete a two-day safety seminar before beginning their work, and (3) Request assistance from OSHA in establishing a comprehensive safety program for the project.
- ***Internal Risk—Nontechnical (Risk value of .24).*** If the new interior wall sound insulation material does not arrive on time, the delay will increase cost and impede the schedule.
  Risk response strategies: (1) Work with the supplier to establish an incentive payment for early delivery and a penalty for late delivery, (2) Monitor the delivery schedule daily, and (3) Develop a worst-case scenario plan for working around the material should it not arrive on time.
- ***External Risk—Unpredictable (Risk value of .21).*** If a hurricane hits during construction, the work in progress might be severely damaged.
  Risk response strategies: (1) Work closely with the National Weather Bureau to monitor hurricane activity during the duration of the project, (2) Develop a damage minimization plan that can be implemented on short notice, and (3) Purchase a hurricane damage rider on the insurance policy covering this project.

QUANTITATIVE RISK ANALYSIS

Quantitative risk analysis is another approach used by risk management experts to assess the risks incumbent in a construction project. Quantitative analysis can be a complex undertaking—hence the need for risk analysis experts on the risk management team. Quantitative risk analysis is used to address the following three broad questions:

- How long will the construction project take to complete (Will it go past the deadline)?
- How much will it cost to complete the project on time (Will it go over budget)?
- Will the construction team be able to meet all specifications (Will quality standards be met)?

These questions are just another way to get at the three success criteria that apply to all construction projects (i.e., completing the project on time, within budget, and according to specifications). The short form of these three criteria is time, cost, and quality. Of these three criteria, quantitative analysis is typically applied to determinations of time and cost.

QUANTITATIVE ANALYSIS TOOLS

Quantitative analysis experts use a variety of tools to perform an analysis on a given risk factor as it relates to one of the major success criteria: time, cost, and quality. One of the most commonly used tools is the distribution curve. Figure 6.9 contains examples of three of the most frequently used distribution curves in quantitative analysis: Normal Distribution or "Bell

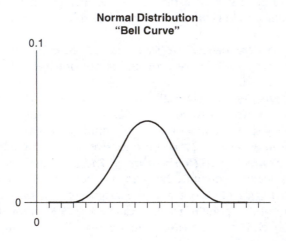

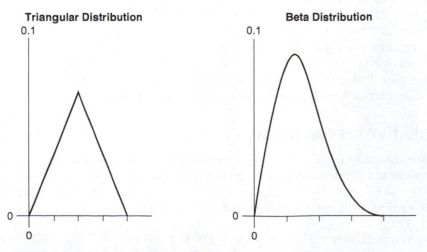

FIGURE 6.9 Widely used distributions in quantitative analysis.

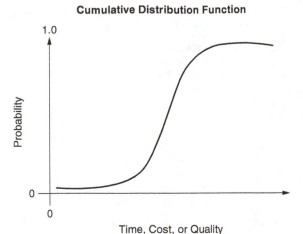

FIGURE 6.10   Shape of the cumulative distribution function.

Curve" (so named because it is shaped like a bell), Triangular Distribution, and Beta Distribution. Another widely used quantitative analysis tool is the Decision Tree.

Distribution Tools in Quantitative Analysis

A normal distribution curve such as the one shown in Figure 6.9 has a probability density function and a cumulative distribution function. The probability density function of the normal curve is shaped like a bell as shown in Figure 6.9. Probability values are plotted on the vertical axis and time, cost, or quality values are plotted on the horizontal. The curve for the cumulative distribution function takes on the shape of an elongated "S" as shown in Figure 6.10. As with the probability density function of the normal curve, the cumulative distribution function has the probability values plotted on the vertical axis and the time, cost, or quality values plotted on the horizontal axis.

The shapes of the curves for the probability density function and the cumulative distribution functions are mathematically prescribed and can be computed by applying standard equations. Changing the variables in the mathematical equations changes their shapes. In the days when these equations had to be computed manually, using distribution curves was a labor-intensive task. However, now that they are embedded in computer software and preloaded in electronic calculators, risk analysis personnel need only supply the variables.

Since this type of assistance is readily available for both types of functions—probability density and cumulative distribution—construction project managers are well-advised to focus on using the distribution curves for analyzing time, cost, and quality risks. For example, Figure 6.11 shows the cumulative distribution function applied to determine the probability that a construction project can be completed in 24 months as specified in the bid package. The plot in Figure 6.11 shows that there is a 90 percent probability that the 24-month goal can be met. Now assume that there is a substantial incentive bonus for completing the

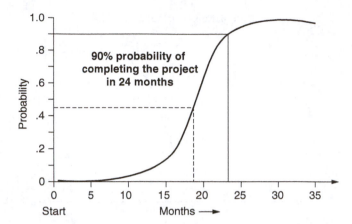

FIGURE 6.11 Using the cumulative distribution function to determine the probability of completing the project on time.

project six months early—in 18 months. The dotted line on the graph shows that the probability of receiving the incentive bonus is approximately 47 percent.

Decision Tree in Quantitative Analysis

The decision tree is a quantitative analysis tool used to analyze cost options in construction projects. For example, assume that a construction firm has received several bids from subcontractors for the electrical work for a large project. Only two of the bids were completed in accordance with instructions. Consequently, the firm must choose between these two bids. The risk management team will use a decision tree to assist in making the decision.

Subcontractors are instructed to provide three prices: a *base price* for completing the electrical work on time, an *incentive* price for completing the project early, and a *penalty* price for completing the project late. Figure 6.12 is a decision tree that was developed by the risk management team. The tree has two main branches—one for each subcontractor—and three branches off of each main branch. The smaller branches represent the on-time, early, and late prices submitted by the subcontractors. The probabilities that were determined for each scenario—on time, early, late—for each subcontractor are shown on the respective branches of the decision tree. For example, the probability that Subcontractor A will complete the project early is .2, on time is .5, and late is .3.

With the decision tree presented as shown in Figure 6.12, the computations for determining the likely price for the project can be completed. In this step, the price on each branch of the tree is multiplied times its probability value as follows:

Subcontractor A:

.2 × $200,000 = $40,000
.5 × $150,000 = $75,000
.3 × $100,000 = $30,000
Total = $145,000

Decision Tree

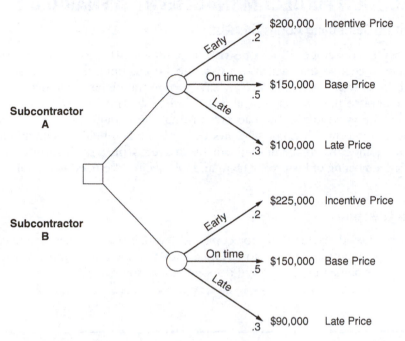

FIGURE 6.12 Decision trees are helpful tools for decision making.

Subcontractor B:

.2 × $90,000 = $18,000

.5 × $150,000 = $75,000

.3 × $225,000 = $67,500

Total = $160,500

These computations show that the construction firm can expect to pay $145,000 with a probability of .7 or a 70 percent chance of having the electrical work completed on time or early if it chooses Subcontractor A. The firm can expect to pay $160,500 with a probability of .8 or an 80 percent change. This clarifies the decision of the risk management team. It can expect to pay $15,500 less with a 70 percent chance of completing the project on time or early or $15,500 more with an 80 percent chance. The determination would be made by deciding which is more important—an 80 percent chance of finishing on time or early or saving $15,500.

There are risks inherent in construction. Those risks relate to completing the job on time, within budget, and according to specifications as well as safely and in an environmentally friendly manner. Consequently, risk must be managed. Risk management is an imperfect process, but the process and methods described in this chapter will allow project managers to manage risks as well as they can be managed on construction projects of any size. The better project managers and others involved in risk management become at this important process the more likely it is that their construction projects will be profitable.

CONSTRUCTION PROJECT MANAGEMENT SCENARIO 6.2
We Need to Do Something About This Risk

David Mackey is concerned about one risk in particular. His construction company has ordered a new machine that will roof trusses to be custom-built on the job site. If the machine works as well as the one Mackey saw demonstrated, his construction team should be able to complete the project well ahead of schedule and earn a substantial incentive bonus. This is the good news. The bad news is that the manufacturer of the truss-making machine has just informed Mackey that production has fallen behind schedule and there is a 60 percent chance that machine will not be delivered on time for the current project. Mackey called a meeting of his project team and told them: "We need to do something about this risk."

Discussion Questions

In this scenario, David Mackey is a project manager with a risk problem. He needs to eliminate, transfer, or minimize the risk. How can he eliminate the risk? Is there a way to transfer the risk? What are some things he could do to minimize the risk?

SUMMARY

Risk is the possibility that things will not go as planned and that counterproductive unplanned events could arise during the course of a construction project. Several factors can increase the level of risk in a project: duration, lapse time, inexperience, insufficient maturation, and unfamiliarity. Risk management is defined as identifying risks that might negatively affect the proper completion of construction projects, assessing their potential impact, developing mitigation plans, and implementing the plans in ways that minimize the risk.

To minimize risks, project managers should: (1) be aware that risks exist on all projects, (2) identify project-specific risks, (3) assess the potential consequences of the risks, (4) communicate the risks and their potential consequences to stakeholders, (5) develop and implement risk mitigation plans, and (6) monitor the effectiveness of risk mitigation strategies. Risk factors can be classified as external—unpredictable, external—predictable, internal—technical, internal—nontechnical, and legal/ethical—civil

and criminal. Risk can be evaluated as it relates to the following success criteria for construction projects: time, cost, quality, safety, and the environment.

The risk identification process includes the following steps: (1) form the risk management team, (2) distribute the Risk Breakdown Structure, (3) select the risk identification methods to be used, and (4) decide what the output of the risk identification process will be. Risk identification methods include the following: review of the construction documents, brainstorming sessions, SWOT analysis, experience review, review of professional literature, survey of experts/Delphi technique, and expert judgment.

Qualitative risk analysis assesses the probability that a given risk factor will have an impact on the project and the extent of the impact. Quantitative risk analysis assesses the probability that a given loss will occur and the magnitude of the loss. Risk response strategies fall into three broad categories: (1) elimination, (2) transfer, and (3) minimization.

KEY TERMS AND CONCEPTS

Murphy's Law
Duration
Lapse time
Inexperience
Insufficient maturation
Unfamiliarity
External—unpredictable
External—predictable
Internal—technical
Internal—nontechnical
Legal/ethical—civil and criminal
Time-related risk factors
Cost-related risk factors
Specification/quality-related risk factors
Environment-related risk factors
Safety/health-related risk factors
Risk management team

Risk Breakdown Structure template
Review of construction documents
Brainstorming sessions
SWOT analysis
Experience review
Review of professional literature
Survey of experts/Delphi technique
Expert judgment
Qualitative risk analysis
Risk response strategies
Elimination
Transfer
Minimization
Quantitative risk analysis
Distribution tools
Decision tree

REVIEW QUESTIONS

1. Explain the concept of Murphy's Law and how it applies to construction projects.
2. Define the term *risk* as it relates to construction projects.
3. How can the following factors affect risk in a construction project: duration, lapse time inexperience, insufficient maturation, and unfamiliarity?
4. Define the concept of risk management.
5. What should project managers do to minimize risk?
6. List and briefly explain four classifications of risk factors in construction.
7. What are three time-related risk factors in construction?
8. What are three cost-related risk factors in construction?
9. What are three quality-related risk factors in construction?
10. What are three environment-related risk factors in construction?
11. What are three safety-related risk factors in construction?
12. List the steps in the risk identification process.
13. List and briefly explain the most commonly used risk identification methods.
14. What is assessed in qualitative risk analysis?
15. What is assessed in quantitative risk analysis?
16. List and explain the three broad options of risk response.

APPLICATION ACTIVITIES

1. Contact a construction firm in your community that will cooperate with you in completing this activity. Ask the firm's representative to explain the most common risk factors the firm has to deal with on its construction projects. Make a list of these risks and discuss them with classmates.

2. Use the Risk Breakdown Structure template in Figure 6.3 to complete this activity. Construction of the Hoover Dam was profiled as this chapter's "Challenging Construction Project." Using the RBS template, identify as many risks associated with that project as possible if it were going to be built today. Do not be limited by the brief

profile presented in this chapter. Do additional research on the project to assist in the completion of this assignment. Hint 1: More than 90 workers died during the construction of the project. What are the risks of OSHA violations, fines, and even criminal prosecutions if this happened on a construction project today? Hint 2: Use the risk identification methods summarized in Figure 6.5. Hint 3: Write your risk statements in the format illustrated in Figure 6.6.

3. Using the probability/impact matrix in Figure 6.8, assign probability and impact values for each of your risk statements and plot them to determine if the risk is low, moderate, or high.

4. Use the cumulative distribution graph in Figure 6.11 to complete this activity. What is the probability that a certain construction project can be completed in 10 months? 20 months? 30 months?

5. Use the decision tree in Figure 6.12 to determine which supplier's bid to accept. Two suppliers have submitted bids to provide all of the steel for your construction project. Each has submitted an on-time, early, and late-bid price. Your risk management team has determined that the probability of on-time, early, and late delivery for each supplier and the corresponding bid amounts are as follows:

Supplier A

On time (.4): $1,525,000

Early (.1): $2,212,000

Late (.3): $998,785

Supplier B

On time (.2): $1,058,000

Early (.1): $1,678,906

Late (.4): $1,000,000

Determine the likely price for each supplier and decide which supplier will receive the contract to supply steel for your project. Explain your decision.

ENDNOTES

1. International Marine Contractors Association, "Identifying and Assessing Risk in Construction Contracts: An IMCA Discussion Document," 5. Retrieved from www.imca-int.com on February 1, 2012.
2. Ibid.
3. Project Management Institute, *A Guide to the Project Management Body of Knowledge*, 4th ed. (Newtown Square, Pennsylvania: Project Management Institute, 2008), 280.
4. Patrick X. W. Zou, Guomin Zhang, and Jia-Yuan Wang, "Identifying Key Risks in Construction Projects: Life Cycle and Stakeholder Perspectives." Retrieved from www.pres.net/Papers/Zou_risks_in_constructionprojects.pdf on January 30, 2012.

CHAPTER 7

Project Construction, Monitoring, and Closeout

Project managers play several different roles during the course of a construction project. At the beginning of the preconstruction phase of the project, they are cost estimators. Once the contract for construction is awarded project managers transition into scheduling mode, then procurement mode, and then risk management mode. These preconstruction processes fall under the broad heading of planning. Once the planning processes are completed or even before in the case of *fast-track* construction, the plans must be implemented. In other words, the project must be constructed. In the construction phase, the project manager's role transitions from planning to monitoring and control.

The construction phase is the most critical phase in the project—the phase where the rubber hits the road so to speak. It is during the construction phase that all of the risk factors associated with the project come into play. It is during the construction phase that Murphy's Law—*anything that can go wrong will go wrong*—comes closest to being true. It is also during the construction phase that the project manager and other stakeholders will find out how effective, accurate, and comprehensive the plans they have developed actually are. Consequently, during the construction phase, all of the critical success factors—time/schedule, cost/budget, specifications/quality, safety, and environment must be carefully monitored and controlled. The principle concerns of the project manager during the construction phase are as follows (Figure 7.1):

- Jobsite layout plan
- Documentation and recordkeeping
- Project meetings
- Conflict resolution
- Labor relations and supervision
- Jobsite safety
- Monitoring of quality
- Schedule control

**CHECKLIST: PROJECT MANAGER'S
CONCERNS CONSTRUCTION PHASE**

✓ Jobsite layout

✓ Documentation and recordkeeping

✓ Project meetings

✓ Conflict resolution

✓ Labor relations and supervision

✓ Safety

✓ Quality

✓ Schedule control

✓ Cost control

✓ Change orders

✓ Project closeout

FIGURE 7.1 Monitoring and control concerns during the construction phase.

- Cost control
- Change orders
- Project closeout

JOBSITE LAYOUT PLAN

Although the construction phase of the project is considered the implementation phase, the project manager does have one additional planning task to perform before hands-on construction can begin: developing the jobsite layout plan. The jobsite plan is developed using the site plan provided by the project's civil engineer. This site plan will show the site with the outline of the building or structure superimposed on it, making it an excellent tool for use when developing the jobsite layout plan. The jobsite layout plan has the following elements (Figure 7.2):

- Material storage/placement areas
- Jobsite office facilities
- Worker facilities
- Tool storage facilities
- Sanitation facilities
- Temporary utilities
- Jobsite security
- Jobsite access
- Jobsite drainage
- Signs and barricades

**CHECKLIST: ELEMENTS OF THE
JOBSITE LAYOUT PLAN**

✓ Material storage/placement areas

✓ Jobsite office facilities

✓ Worker facilities

✓ Tool storage facilities

✓ Sanitation facilities

✓ Temporary utilities

✓ Jobsite security

✓ Jobsite access

✓ Jobsite drainage

✓ Signs and barricades

FIGURE 7.2   Project managers must develop comprehensive jobsite
layout plans.

Material Storage/Placement Areas

Construction projects require a great deal of material, some of which can be placed in designated areas on the jobsite and some of which must be stored away from the elements. Temporary storage facilities must be built or brought in for materials that must be protected from the elements and, perhaps, theft. Placement areas must be designated for materials that can simply be placed on the jobsite without the need for a protective storage facility. When planning for material storage and placement, project managers must ask and answer the following questions:

- *How much storage space will be needed—facilities and placement areas?* There are several tools available to the project manager for answering this question including the project schedule, the bill of materials, and the site plan. The project schedule can be especially helpful because it will show the intervals at which certain materials will be delivered and used. By carefully scheduling the delivery of materials, project managers can limit the amount of space required for storage and material placement. When determining the amount of storage space needed on the jobsite, project managers should factor in the materials needed by subcontractors.
- *Is adequate access available to and from storage facilities/areas?* Material suppliers must be able to access storage facilities and designated placement areas to make their deliveries. Once materials have been stored and placed, the workers and subcontractors who will use them must have safe and convenient access to them. Consequently, careful planning of ingress and egress to storage locations is important. Figure 7.3 shows material that has been temporarily placed in a designated location at the jobsite.
- *Will the storage facilities/areas protect the material from the elements?* Some material can simply be placed on the ground and left. Other material can be placed on the ground and covered with a tarp. For materials placed on the ground it is important

FIGURE 7.3   Material placement area on a jobsite.

to provide adequate drainage to prevent water damage. It is also wise to elevate the material on skids or pallets to allow air to circulate and to prevent moisture build up. Some material must be even more securely protected from the elements. For these materials, facilities are built or brought in (e.g., temporary buildings, trailers). These facilities should be sealed properly against wind, rain, snow, sleet, and sun as appropriate.

Jobsite Office Facilities

The jobsite construction superintendent, engineers, foremen, and support staff will need a facility to serve as their onsite office. This facility should also be able to accommodate meetings with visitors, inspectors, the architect, engineers, representatives of subcontractors, and other personnel affiliated with the project. The jobsite office facility serves as the onsite headquarters throughout the construction phase. It typically contains telephones, a facsimile machine, a computer and printer, a conference room with a large table for spreading out drawings, all of the safety, EEOC, and other compliance posters that must be displayed for personnel, construction records and reports, storage space, a bulletin board or empty wall for displaying the project schedule, an AED, and first aid supplies. The most common solution to the need for a jobsite office is the contraction trailer—a trailer that has been converted to serve this specific purpose. Sometimes construction firms are able to rent existing business space in proximity to the jobsite. On long-term projects, the firm might even build a temporary construction office. Figure 7.4 is an example of a construction office at a jobsite.

Worker Facilities

It is important to provide facilities that give workers a place to take breaks, eat lunch, clean up, and change clothes. The jobsite slang term for worker facilities is *dry shacks*. If the workforce is unionized, providing worker facilities will probably be in the union contract. Like the

FIGURE 7.4   Construction trailer—Jobsite office.

jobsite office, worker facilities can vary in size and amenities. As a general rule, the larger the project and the longer its duration, the more elaborate the worker facility.

Tool Storage Facilities

On every construction site there are specialty tools and equipment that are provided by the construction firm serving as the general contractor including surveying equipment, power tools, saws, ladders, scaffolding, drills, nail guns, pumps, generators, and other items. These tools and equipment must be protected from both the weather and theft. Consequently, most jobsites will require a tool shed that is easily accessible, but also secure. The tool shed must be constructed in such a way that the tools and equipment housed in it can be easily accessed and returned, but also mad secure once inside.

Sanitation Facilities

The construction firm serving as the general contractor must provide toilet facilities, drinking water, and washing water at the jobsite. Toilets should be located so that no worker on the jobsite has to walk more than 200 feet to access one. A good rule of thumb is that one toilet should be provided for every ten workers. On multistory building projects, there should be

FIGURE 7.5   Portable toilets must be provided and placed in convenient locations.

at least one toilet on every third floor. The availability of portable toilets has made satisfying this requirement much easier now than in the past (see Figure 7.5).

Fresh drinking water should be made readily available at the jobsite. This can be done by providing drinking fountains, ice chests full of bottled water, or water on tap. Regardless of how it is provided, contractors should make plenty of fresh drinking water conveniently available in various locations at the jobsite. In warmer environments, water becomes not just a convenience but a life saver. Consequently, providing plenty of fresh drinking water is important. So is providing water for washing. The worker's facility often has sinks with running water and hand or paper towels available. If not, this requirement must be met in some other way.

On unionized jobsites, such things as sanitary facilities, drinking water, water for cleaning and worker facilities will be prescribed in the labor agreement. However, even on nonunionized jobsites, these amenities are important and should be provided. Failing to do so will dampen morale, create resentment, and introduce safety problems. Providing workers with what they need to be safe and productive is the responsibility of the project manager.

CONSTRUCTION PROJECT MANAGEMENT SCENARIO 7.1

I Don't Need a Jobsite Office and They Don't Need a Break Facility

John Clark had been a jobsite superintendent for 10 years before he was promoted to project manager. Clark has a well-deserved reputation for being a tough but fair no-nonsense superintendent. He always runs his jobsites like a Marine Corps drill instructor, but he gets results. His projects are always completed on time, within budget, and according to specifications.

Nobody on the jobsite works longer or harder than John Clark. This is why he was recently promoted to the position of project manager.

Clark has just submitted the jobsite layout plan for his first project: a three-story motel located at a high traffic intersection. Thinking that John has left out two important components of the jobsite layout plan, his supervisor asked Clark, "Where is your jobsite office and what about the break facility for workers?" The supervisor was taken aback by Clark's response. "I don't need a jobsite office and they don't need a break facility." Clark went on to explain that he did not intend to spend any time in an office—he planned to be right in the middle of the activities at the jobsite. Further, he did not think giving workers a place to take breaks, wash up, and change clothes was necessary. "It just spoils them." Clark's response surprised and concerned his supervisor. He began to wonder if he had made a mistake by promoting Clark from jobsite superintendent to project manager.

Discussion Questions

In this scenario, it appears that John Clark might not understand the difference between the responsibilities of the jobsite superintendent and the project manager on a construction project. Do you see any problems with Clark's thinking in this case? If so, what are they? If you were Clark's supervisor how would you explain to him why having a jobsite office and providing a facility for workers are good ideas?

Temporary Utilities

Often electricity and water are not available at a jobsite until the job is close to completion. Because of this it is usually necessary for the contractor to install temporary electricity and water. Temporary electricity is installed by the local electric power company for a specified period of time. When installing temporary electricity, it is important to place the service so as to avoid: (1) the overuse of extension cords, (2) having to move the service before permanent electricity is installed, and (3) locations that might interfere with any aspect of the construction project.

Water is used in so many different construction tasks that it is as important as electricity to the construction project. Like electricity, it will be much later in the project before permanent water lines will be made operable. Temporary water lines should be installed according to the following parameters: (1) as centrally located as possible to accommodate all of the different needs on the site for water, (2) away from ingress and egress routes for trucks and heavy equipment, (3) away from excavation (or at least protected from it), and (4) deeply enough to avoid freezing.

Jobsite Security

Jobsite security is an important consideration for two reasons: (1) the jobsite is susceptible to theft and vandalism and (2) the jobsite is not a safe place for unauthorized visits from members of the general public. The following strategies can be used for securing the jobsite against theft and vandalism and for ensuring public safety (Figure 7.6):

- Installing a perimeter fence with locking gates that encloses the entire jobsite
- Establishing a relationship with local law enforcement officials so that the jobsite is regularly monitored during nonwork hours

---

**CHECKLIST OF
SECURITY MEASURES FOR JOBSITES**

✓ Perimeter fence with locking gates

✓ Monitoring by law enforcement officials during nonwork hours

✓ Jobsite lighting

✓ Locking up equipment, tools, and supplies

✓ Locking up keys to vehicles and equipment

✓ Hiring security guards

✓ Installing motion-activated alarms

✓ Guard dogs

✓ Blocking ingress and egress routes to the jobsite during nonwork hours

---

FIGURE 7.6 Jobsite security is an important planning topic.

- Installing temporary lighting that covers the entire jobsite at night
- Locking equipment, tools, and supplies in a storage facility
- Locking up keys to vehicles and equipment
- Hiring security guards to patrol the jobsite during nonwork hours
- Installing motion-activated alarms
- Employing guard dogs within the perimeter fence during nonwork hours
- Blocking ingress and egress routes to the jobsite during nonwork hours

Jobsite Access

Access to the jobsite for authorized personnel is an important planning consideration as is access within the jobsite. Construction project managers should plan ingress and egress so that authorized personnel have convenient access to parking areas, storage facilities and placement areas, and the worksite itself. The best planned access route is one in which a truck or heavy equipment can enter the jobsite and leave without having to backup or turn around. This is not always possible, but when it is planning for *drive-through* access routes is the recommended approach. Minimizing the instances in which trucks and heavy equipment must backup is essential since this is when they are most accident prone.

It is important to ensure that the surface of access routes can accommodate the types of vehicles that must enter the jobsite. For example, dirt roads may not be able to accommodate heavily loaded concrete trucks or semi-trucks bringing in cranes and excavation equipment. This is especially true in areas that experience a lot of rain. Dirt roads that can usually accommodate trucks and equipment can turn into quagmires when it rains. Consequently, providing for the paving of access routes might need to be part of the planning for jobsite layout.

Access within the site is also an important planning consideration. Tradespeople, subcontractors, inspectors, and other authorized personnel need to be able to move around within the jobsite quickly and safely. For example, the project manager should plan for routes to the tool shed, material storage, portable toilets, the jobsite office, the parking lot, and any other areas in the jobsite that must be accessible to authorized personnel. These routes should be kept open and free of construction debris.

Jobsite Drainage

Jobsite drainage is always an important planning consideration when laying out a jobsite, but it can be a critical consideration in certain regions of the country depending on the water table and the typical amount of rainfall. There are two sides to the issue of jobsite drainage: (1) keeping the jobsite dry so that work can proceed unimpeded and (2) preventing environmental problems from jobsite runoff. On larger jobsites the project manager may be required to submit a storm water pollution prevention plan to the Environmental Protection Agency (EPA). Local and state agencies may also require storm water management plans for the project.

Signs and Barricades

Signs and barriers have two principle uses on jobsites: (1) channeling those who need access to the jobsite onto the proper routes and (2) promoting jobsite safety. The types of signs that are most commonly used at jobsites include: (1) safety signs; (2) a project team sign containing the names of the owner, architect, engineers, and general contractor as well as the number of

---

Challenging Construction Project

BUILDING THE HOOVER DAM

The Hoover Dam is considered one of the seven construction/civil engineering wonders of the modern world. The project took 21 months and 5,000 workers. By comparison, the great pyramid in Egypt—which is smaller in volume than the Hoover Dam—required 100,000 men working 20 years to complete. Even compensating for the advances in construction technologies and practices, constructing the Hoover Dam in 21 months was a monumental accomplishment. Consider the following facts about the construction of the Hoover Dam: 1) the construction site was in the middle of nowhere requiring an entire town to be built to support workers, 2) temperatures could soar to more than 120 degrees Fahrenheit, 3) a human-support infrastructure had to be constructed—including a railroad and highways, and 4) construction-support infrastructure had to be built including a gravel screening plant, concrete-mixing plants, air compressor plants, and a steel plate fabrication plant.

The construction bid for the Hoover Dam was approximately $49 million, an astounding amount for the 1930s. In fact, it was the largest bid let by the federal government up to that time. It was necessary to drill four tunnels to divert the water of the river, build earthen cofferdams above and below the dam site to block water, excavate the site within the cofferdams, and build the dam and the power generating plant. A number of construction advances were developed as part of the Hoover Dam project including the hard hat, the motor driven jumbo drill, and on-site fabrication of steel pipe. The dam itself required 3,250,335 cubic yards of concrete, which is 6.6 million tons. More than 90 workers died on the job during the project.

Source: Based on "Hoover Dam Construction." http://www.onlinenevada.org/hoover_dam_construction

the building permit; (3) financing sign containing the name of the organization providing financing for the project; (4) directional signs; (5) restricted access signs; and (6) delivery signs (directing material suppliers where to make their deliveries). Barricades are used to channel traffic onto the correct routes, prevent unauthorized personnel from entering restricted areas. Consequently, they serve several purposes including traffic control, security, and safety.

DOCUMENTATION AND RECORDKEEPING

A good rule of thumb for construction project manager is this: *document everything and keep accurate records*. It is a sign of the times that documentation and recordkeeping have become so important a responsibility of project managers. But in a litigious society with trial attorneys always waiting to pounce, the construction firm's first line of defense is comprehensive, accurate records. Records kept during the construction phase of the project should meet several criteria as follows:

- *Trustworthy.* It is important that stakeholders be able to trust that what they read in reports and records kept during the construction phase of the project. This means the reports and records must be unbiased, factual, and complete. Bias by omission is just as unethical as bias by commission.
- *Standardized.* There are various kinds of reports prepared and records kept during the construction phase of a project. Each kind of report and record should have a standard format that makes the information contained in it easy to find and understand. Standardizing formats ensures that all stakeholders know what to expect when they receive a report or retrieve a record, what report or record to retrieve when certain information is needed, and where to look in the report/record for the needed information.
- *Timely in preparation and distribution.* Reports and records should keep up with events on the jobsite as they occur. Reports that are overtaken by events before they are read by stakeholders will do little good. In fact, they may even cause problems when decision makers act on the basis of old information. Further, reports should be distributed to any and all stakeholders who need the information contained in them. Consequently, it is important to ensure that all stakeholders—owner, architect, engineers, the contractor's higher management team—have input into who is on the distribution list for all reports.
- *Accessible.* Few things are more disturbing to stakeholders in a construction project than being unable to put their hands on reports and records they need in a timely manner. Reports and records should be easily accessible and readily available when needed.
- *Comprehensive.* When choosing a standardized format for reports and records, it is important to choose formats that demand comprehensiveness. Report and record formats should be designed in ways that require personnel to enter complete, comprehensive information—everything a stakeholder might need to know to thoroughly understand the reported or recorded aspect of the project.

Information to Document and Report

The rule of thumb for construction project managers on documentation and reporting is *document everything and keep accurate records*. This is good advice. But, it is not specific enough

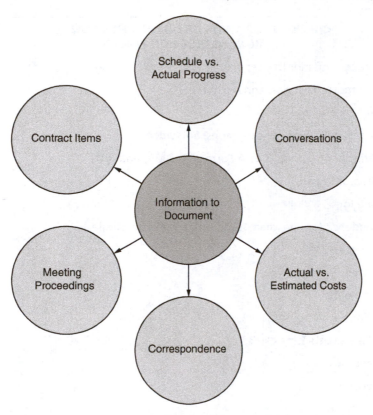

FIGURE 7.7 Accurate comprehensive documentation is a must with construction projects.

to illustrate what actually should be documented over the course of a construction project. The recommended documentation requirements as shown in Figure 7.7 are as follows:

- Conversations between and among stakeholders
- Actual versus estimated costs
- Correspondence
- Meeting proceedings
- Items specified in the contract
- Schedule versus actual progress

By making contemporaneous notes of conversations—whether they take place face to face or by telephone—can prevent *you-said-I-said* disputes or confusion over who said what and when. Monitoring actual costs against estimated costs is a must for budget control. All correspondence of any kind—letters, memorandums, e-mails, facsimiles—should be kept for future reference. The proceedings of meetings, along with the names and titles of all participants, should be kept. Items specified in the contract such as progress payments, change orders, the punch list, and the substantial completion notification should be kept.

**CHECKLIST OF KINDS OF DOCUMENTATION FOR
THE CONSTRUCTION PHASE**

✓ Daily, weekly, and monthly reports

✓ Schedule reports, changes, and updates

✓ Document control logs

✓ Photographs, time-lapse photographs, and video

✓ Accident/incident reports (OSHA Form 300, 300A, and 301)

✓ Subcontractor logs

✓ Telephone logs

✓ Correspondence (letters, memorandums, e-mails, facsimiles)

✓ Purchase orders

✓ Bill of materials

✓ Delivery receipts

✓ Request for payment

✓ Payroll/Time sheets/Time cards

✓ Equipment logs

✓ Rental records

✓ Labor records

✓ Progress payments

✓ Minutes of meetings

✓ Punch list and substantial completion

FIGURE 7.8 Most common kinds of documentation for the construction phase of a project.

Records of actual progress made versus scheduled progress should be made and kept up to date regularly.

Figure 7.8 lists the most frequently used kinds of documentation for recording and reporting on the information listed above. Of course, not all of these kinds of documentation are completed by construction project managers. The project manager typically prepares daily, weekly, and monthly reports (Figure 7.9) as well as schedule updates and reports. Other kinds of documentation are often delegated to specific team members or support staff. For example, the contractor's safety director will complete accident reports and Occupational Safety and Health Administration (OSHA) forms. Various personnel generate and receive correspondence. Photographs, time-lapse photographs, and videos are typically provided by a professional on contract. Telephone logs are kept by all who use telephones. However, the project manager is responsible for ensuring that all necessary documentation is prepared, distributed as appropriate, filed, and made readily accessible.

MONTHLY REPORT: June 1–June 30
PROJECT NUMBER 2012–15: Parking Deck, City of Campbellton
PREPARED BY: Mike Jones–Project Manager

Summary of Progress

During the reporting period, the site work was completed, footings and the foundation were completed, and prestressed concrete columns and beams were delivered to the site. In addition, the crane for erecting columns, beams, and floor members was put in place.

Summary of Problems

The project is two days behind schedule because of the discovery during excavation of a protected species nesting on the site. The appropriate government agencies were contacted, the animal was relocated, and work on the project recommenced. The two days lost will be made up in the month of July with the cooperation and assistance of the prestressed concrete subcontractor with no additional expense.

Cost Update

The discovery of a protected species nesting on the site added $9,000 to expenses. However, this amount can be absorbed in the budget category of *miscellaneous expenses*. Therefore, the project is operating within budget.

Subcontractor Relations

The prestressed concrete subcontractor has been very cooperative and helpful as we have dealt with the protected species issue. There are currently no subcontractor problems to report.

Change Orders

No change order to report in this period.

Date Submitted for Distribution: June 30

FIGURE 7.9 Sample monthly report for a construction project.

PROJECT MEETINGS

Construction project managers are called on to conduct various kinds of meetings over the course of a project. At the outset of the project there will be meetings with team members involved in developing the construction cost estimate, schedule, risk management strategies, and procurement plan. As part of the construction phase of the project, project managers conduct the following types of meetings:

- ***Construction team meeting.*** The construction team consists of the owner, architect, engineers, jobsite superintendent, and project manager. The first meeting of the construction team occurs before construction begins. Its principle purpose is teambuilding (see Chapter Nine). During this meeting, the project manager attempts to gain the commitment of all participants to the team's mission: to complete the project on time, within budget, and according to specifications. Other areas of discussion include: (1) establishing a regular meeting schedule, (2) deciding how problems will be dealt with, (3) determining how communication among team members will be handled, (4) deciding how correspondence will be distributed, and (5) developing a team charter. The team charter contains the team's mission statement and a list of ground rules defining how team members will interact with each other while working together to accomplish the team's mission. Developing the team charter is explained in Chapter Nine.

- ***Organizational/Kickoff meeting.*** The organizational/kickoff meeting occurs once it has been determined who the key players for the general contractor will be for the project in question. These individuals typically include the project manager, jobsite superintendent, safety director, jobsite foremen/supervisors, heavy-equipment operators, key tradespeople, field engineers, and support personnel. They are sometimes referred to as the *project staff team*. During this meeting the project staff team discussed everything necessary to ensure that the project is well-organized, properly staffed, and provided with the necessary resources. The meeting is both informational and planning oriented. Main topics of discussion in this meeting include: (1) overview of the project; (2) explanation of the jobsite layout; (3) responsibilities of each member of the team; (4) documentation, recordkeeping, and reporting; (5) how communication will be handled; (6) key contacts and their contact information (e.g., city and county officials for permits, licenses, local ordinances, and inspections; utility providers including water and electricity); (7) telephone, Internet providers, and mail providers; (8) waste disposal/sanitation providers; (9) equipment rental; (10) schedule; (11) budget; (12) safety and environmental protection including noise control, air quality, and storm water runoff; (13) security; (14) internal processes; (15) site access; (16) facilities; and (17) assignment of office space at the jobsite office.

- ***Subcontractor meetings.*** Once subcontractors begin their work, project managers will need to meet with their representatives periodically—usually weekly. During these meetings subcontractors provide progress reports, explain problems that have arisen, and future problems that might arise. This meeting is also an excellent setting for helping subcontractors to work together to coordinate their schedules and work out any problems between and among themselves with the project manager serving as the facilitator, mediator, and peace maker.

- ***Periodic construction staff meetings.*** Once construction begins, the construction staff—project manager, jobsite superintendent, safety director, jobsite foremen/supervisors, and support staff—meet on a regular basis for updates, progress reports, and to solve problems. These meetings are typically convened on a weekly basis. All participants are typically asked to give a status report on their respective areas of responsibility. Deliveries that have been made and those that are pending, the schedule, the budget, and problems are discussed. These meetings are very action oriented. They focus on status reports and problem solving.

- ***Special meetings.*** Special meetings are topic of problem-driven meetings that are called when needed to deal with specific issues. Consequently, the problem or issue being dealt with determines who attends as well as the timing of the meeting. Often, such meetings are called to resolve a conflict that has arisen, to deal with a risk factor that has come into play and is worse than anticipated, and to solve unanticipated problems.

- ***Project closeout meetings.*** As the project nears completion, stakeholders begin to prepare for the closeout process. This process works as follows: (1) the general contractor, subcontractors, and tradespeople complete their work; (2) the contractor conducts a thorough assessment and develops a *punch list* containing any remaining items

---

PUNCH LIST
PROJECT 2013–15: COLLEGE SPORTS ARENA

General Items

1. Final cleaning of floors and windows
2. Final removal of trash, waste, and construction debris

Interior Items

1. Reseal area marked on the arena floor
2. Repair door to the equipment room (lock sticks)
3. Paint all lines as specified for basketball on the arena floor
4. Install soap dispensers at the sinks in women's restroom number 3

Interior Items

1. Replace all damaged decorative brick on the east elevation
2. Replace the pump in the water feature at the front entrance of the arena
3. Stripe the west parking lot
4. Reclean all bricks on the south elevation
5. Touch paint in the locations marked on all four elevations

---

FIGURE 7.10 Example of a punch list prepared by the architect for the general contractor.

to be completed and issues to be cleared up (see Figure 7.10); (3) the general contractor works with subcontractors and tradespeople to complete all items on the punch list; (4) the general contractor notified the architect that work on the project is completed; (5) the architect conducts a thorough assessment of the project and develops its own punch list; (6) the general contractor, subcontractor, and tradespeople complete the work on the architect's punch list as appropriate; (7) the architect rechecks to ensure that the work on its punch list has been satisfactorily completed; (8) the architect issues a certificate of substantial completion to the general contractor; (9) the general contractor transfers the keys to the project to the owner and completes any remaining contractual obligations; and (10) the owner issues the final payment to the general contractor. As the contractor proceeds through these various steps, closeout meetings are scheduled to facilitate the process. For example, when the general contractor develops its punch list a meeting is scheduled with any subcontractors that have work on the list to ensure that they understand what needs to be done and by what certain date. When the architect develops its punch list, a meeting is scheduled to ensure that the general contractor understands what work remains and the date by which it must be completed. The architect and general contractor must know when the punch list will be completed and when the certificate of substantial completion has been issued so that the owner and contractor can work out the details of transferring keys and other remaining contractual obligations. A number of issues can be covered in this final closeout meeting. For example, this is a good time for the owner to receive the operator's and maintenance manuals for all equipment and technology items in the project. It is also a good time to ensure that all stakeholders understand such issues as final payment, site cleanup, subcontractor payments, warranties, demobilization, interior and exterior cleaning in the case of buildings, and any other outstanding issues.

• ***Project review meeting.*** Once a construction project has been finally completed, it is a good idea for the project manager to convene a meeting of the construction staff to discuss what went right, what went wrong, and lessons learned. This is an important meeting because no construction project is completed without problems. The best project managers are those that learn from mistakes and experience so that they begin the next project better informed and better equipped. One of the most important tasks to perform during the project review meeting is an assessment of the performance of subcontractors, materials providers, and tradespeople. Subcontractors, materials providers, and tradespeople that perform well will receive positive attention the next time they submit a bid to do work for the general contractor. The same is true for materials providers and tradespeople. The obverse is also true. Poor performing subcontractors and others may be marked down as "not responsible" and eliminated from future bids.

CONFLICT RESOLUTION

No matter how tightly a construction contract is written, there will still be room for disagreement among stakeholders. Consequently, it is important that the general contractor, architect, and owner work closely together to head off situations that might generate conflict. It is the project manager's job to provide the leadership to ensure that this happens. This is just one of the reasons that construction project managers must be effective negotiators. They also need to understand the concepts of mediation and arbitration.

Negotiation and Conflict Resolution

Construction project managers are called on to negotiate with other stakeholders in construction projects all the time. It is often necessary to negotiate with the owner and architect over change orders, with materials providers over delivery dates, with subcontractors over quality issues, with union representatives over the labor agreement, and with the union's shop steward over worker performance issues. These are just a few examples of the many differences instances in which negotiations will be necessary over the course of a construction project. Every issue that needs to be negotiated contains the seeds of conflict.

To prevent conflict and resolve it when it cannot be prevented, construction project managers need to be effective negotiators. Negotiating is a skill that can be learned and improved on continually with practice. The first step in becoming an effective negotiator is to learn the best practices of good negotiators and begin developing the ability to apply these practices. Best practices of effective negotiators include the following:

• Ignore the noise, anger, and emotionally charged languages that are often part of discussions when people are in conflict and focus on identifying the cause of the dispute.
• Solve any problems that come up during discussions immediate if possible—do not let problems linger and fester.
• Focus on solutions, not problems.
• Listen more—talk less.
• Keep an open mind and stress that there is usually more than one way to solve a problem. Propose alternative solutions and encourages others to do the same.
• "Listen" visually—watch for nonverbal cues.
• Think critically—recognize assumptions, rationalizations, justifications, and biased information that is presented as facts.

- Give people room to maneuver during negotiations—do not paint people into corners.
- Maintain a sense of humor and a positive attitude during negotiations.
- Consider the issues discussed during negotiations from the other person's point of view.
- Never go into a negotiation unprepared. Prepare, prepare, and prepare.

PREPARING TO NEGOTIATE.   Of the various best practices of effective negotiators, none is more important than thorough preparation. Construction project managers should answer as many of the following questions as possible before beginning negotiations with owners, architects, engineers, subcontractors, materials providers, labor representatives, and other stakeholders:

- What do I want out of this negotiation?
- What does the other party want out of the negotiation?
- What am I willing to give up to get what I want from the negotiation?
- What is the other party willing to give up to get what it wants from the negotiation?
- What is at risk for me if the negotiation fails?
- What is at risk for the other party if the negotiation fails?
- Do I have any "hot-button" issues in the negotiation?
- Does the other party have any "hot-button" issues in the negotiation?
- Are there factors that might affect the outcome of the negotiation over which I have no control?

By using these questions to prepare rather than simply negotiating off the cuff and by applying the best practices presented earlier, construction project managers can prevent and resolve much of the conflict that often arises over the course of construction projects. However, in spite of the best efforts of talented project managers, some conflicts cannot be resolved by negotiating. When this is the case, projects managers have the options of mediation and arbitration.

Mediation and Conflict Resolution

It is not uncommon for general contractors and unions to fall into disputes that, in turn, result in work slowdowns or strikes. It is also not uncommon for the owner in a project to fall into disputes over such things as change orders and other issues concerning the construction. It is always best when the parties in dispute can come to an understanding through negotiations. However, this is not always possible. When negotiations between parties in dispute fail to bring satisfactory results, the next step on the conflict resolution ladder is *mediation*.

Mediation is a legally recognized process for resolving disputes outside of court that involves the parties in dispute placing their cases before an unbiased third party. This third party is an individual who facilitates negotiations between and among the parties in dispute. This facilitator is called a *mediator*. Mediation is like negotiating with a referee in the room— a referee who enforces rules that give structure to the negotiation while protecting the rights of the parties in dispute. The mediator's goal is to keep the negotiations moving in a positive direction, to encourage dialogue, and to guide the parties toward a settlement agreeable to both or all in the case of multiple parties in dispute.

The ultimate result of the process often depends on the knowledge and skills of the mediator. Consequently, mediators are required to complete a specified course of training. It is not the job of the mediator to find a solution or even propose one (although with the permission of all parties in dispute the mediator is allowed to make suggestions). The mediator is a facilitator with no power to decide or enforce decisions. However, if all parties in a dispute come to an agreement during the mediation process, the signed agreement that results is binding.

Before getting involved in mediation, project managers should study the mediation procedures of the American Arbitration Association (AAA). The AAA provides guidelines for initiating mediation, selecting a mediator, explains the duties and responsibilities of the mediator, sets forth the confidentiality and privacy requirements of mediation, and explains the process and procedures. Project managers should be well-versed on all aspects of mediation before proceeding with the process.

Arbitration and Conflict Resolution

When mediation fails to bring a resolution or when the parties in dispute refuse to participate in mediation, arbitration is the next step on the conflict resolution. Arbitration is another process for settling disputes outside of the courts, but unlike mediation arbitration is binding. The arbitrator is a decision maker not a facilitator. Whereas the mediator facilitates in hopes of helping the disputing parties arrive at a resolution, the arbitrator listens to each of the disputing parties and makes a decision that is legally binding.

Before sending a dispute to arbitration, project managers should become knowledgeable of the rules governing the arbitration process. The AAA provides guidelines for arbitration of disputes covering such topics as filing of complaints, appointment of an arbitrator, communication with the arbitrator, attendance at hearings, exchange of information, oaths, recording of proceedings, evidence, conduct of proceedings, and other topics of importance. Project managers should become knowledgeable of the rules and procedures that govern arbitration.

LABOR RELATIONS AND SUPERVISION

Construction project managers must be prepared to lead projects that have union and non-union workforces. Unionized labor is very structured. The distinctions between management and labor are well-defined. For example, the various tradespeople in a unionized workforce are divided into distinct categories: foreman, master craft workers, journeyman craft workers, and apprentice workers. Workers who fall into one of these categories are classified as labor and are paid hourly wages. Foremen have the highest level positions in the labor hierarchy. Helpers have the lowest positions in the hierarchy and are considered unskilled workers. One step above helpers are apprentices in the various trades. Their training consists of both on-the-job practice and formal classroom training of a specified number of hours.

Management personnel in a construction job consist of the project manager, jobsite superintendent, assistant superintendents (depending on the size of the job), and field engineers. Management personnel are typically paid salaries. Subcontractors are just that—companies that are expected to provide their own labor and supervise that labor. There is one caveat here though. In the overall hierarchy on the jobsite, subcontractors fall under the supervision of the jobsite superintendent or one of the assistant superintendents. Figure 7.11 is an organizational chart that shows the typical structure for the personnel who work on a construction job.

Labor Agreements

In some areas of the country, much of the labor needed on construction projects is unionized. The labor unions most commonly associated with construction projects include carpentry,

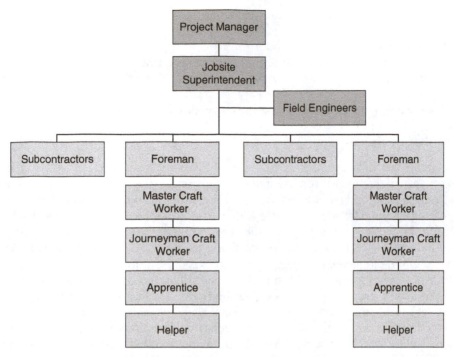

FIGURE 7.11   Labor/management structure of a unionized job.

electricians, masonry, iron and steel, plumbing, and teamsters. When general contractors employ union labor, there must first be a labor agreement—a contract between the applicable labor union and the contractor. The agreement covers three broad areas of concern: wages, benefits, and working conditions. Once the contractor (management) and the union (labor) agree to the various provisions in each of these broad categories, the labor agreement is signed. Once signed both, both management and labor are bound by the agreement. Legally, the labor agreement is considered a contract.

Some general contractors join third-party organization such as the Associated General Contractors of America, Inc., and take advantage of their services in negotiation with unions. A common practice is to ask the third-party organization to conduct the negotiations on behalf of the general contractor. However, some contractors choose to negotiate on their own. These contractors often make use of the guidelines provided by such organizations as the AAA. Regardless of how construction firms choose to approach the development of labor agreements, the type of information contained in these agreements includes the following:

- Purpose of the labor agreement
- Rights and responsibilities of management and labor
- The jurisdictional reach of the labor agreement (scope of its coverage)
- Wage rates for the various levels of workers to be employed on the job
- A clause governing the use of subcontractors (unions will want all subcontractors to sign the labor agreement to ensure that management does not attempt to use subcontractors to preempt the agreement)
- Hiring procedures

- Grievance procedures and dispute resolution
- Substance abuse clause
- Safety and health clause
- Specifications for paydays, holidays, the workweek (days), workday (hours with starting and stopping times), authorization of overtime, shifts (hours with starting and stopping times)
- Rules governing work stoppage (strikes, slowdowns, lockouts)

Jobsite Supervision

The principal supervisor at the jobsite is the jobsite superintendent. The superintendent is the project manager's on-site representative for ensuring that the work of the project gets done on time, within budget, and according to specifications. On an *open-shop* jobsite (non-union), the superintendent works directly with foremen and the representatives of subcontractors to keep the work flowing on schedule, within budget, and according to specifications. On a unionized jobsite, the superintendent works with the union's *shop steward.* The shop steward is a union employee assigned to the job to represent the interests of union workers.

On a unionized jobsite, if the superintendent determines that union workers are not making satisfactory progress, are exceeding the budget, or are not maintaining the quality standards set forth in the specifications, he gets the union's shop involved rather than working directly and independently with the workers or workers in question. Typically the labor agreement will explain how such situations are to be handled. The key is for managers in the construction firm to avoid trying to unilaterally deal with underperforming union workers since to do so might violate the labor agreement.

The jobsite superintendent and shop steward can develop a mutually supportive relationship rather than one that is adversarial and characterized by conflict. Of course, this is not always possible. Personalities, egos, and other elements of human nature often get in the way of mutually supportive relationships between superintendents and union shop stewards. This is one more reason why it is important to ensure that specific procedures for dealing with nonperforming and underperforming union workers are made part of the labor agreement.

BEST PRACTICES OF EFFECTIVE SUPERVISORS. Construction project managers rely on jobsite superintendents and assistant to provide hands-on, day-to-day supervision of the work on construction projects. Since this is the case, it is important for project managers to help ensure that jobsite superintendents and assistant superintendents have strong supervision skills. Project managers can improve the quality of day-to-day supervision on their jobsites by helping superintendents and assistant superintendents learn to apply the following best practices of effective supervisors:

Provide effective leadership. People perform better when they are led than when they are coerced, pushed, and bossed. Inspiring by dint of example will bring better results in the long run than coercing by dint of authority. Effective leaders establish high expectations and then set a consistent example of doing what they expect of others. For example, jobsite supervisors who expect their crews to arrive at work at 7:00 a.m. and to get started immediately should arrive at work at 6:45 a.m. or earlier and be working when the crews arrive. The supervisor's example should extend to ethics. Supervisors who expect work crews to behave in an ethical manner must be seen by the members of those crews setting a consistent example of ethical behavior.

Be a positive change agent. One of the most common causes of conflict on a construction project is change. Change can be traumatic and conflict inducing or it can be positive and result in improvements depending on how it is handled. When change is necessary on a construction job, supervisors can be positive change agents by doing the following: (1) developing a *word picture* that describes the *five Ws and one H*—who, what, when, where, why, and how of the change; (2) communicating the change to the appropriate stakeholders (those who will be affected by it and those who will have to carry it out); (3) giving stakeholders affected by the change an opportunity to ask questions, vent their frustrations, and offer suggestions for implementing the change; (4) make specific assignments and set timetables for implementing the change; and (5) monitor closely and adjust as necessary.

For example, if a change requires taking down a masonry wall that has just been completed because the bricks do not meet the specifications, the supervisor in question should begin by calling a meeting of the masonry crew. Prior to the meeting, the supervisor should develop a word picture that explains the *five Ws and one H* concerning the change. During the meeting the supervisor should: (1) communicate the *five Ws and one H* of the change to the crew; (2) give crew members an opportunity to ask questions, vent their frustrations, and offer suggestions for pulling down the existing wall and replacing it; and (3) make specific assignments and set a timetable for pulling down the existing wall and building the new one. At this point the supervisor begins to monitor the work of making the change and makes adjustments as necessary to keep the work on schedule.

Communicate often and well. Effective supervisors never keep their crews in the dark. The people who do the hands-on work on a construction project need to be fully informed about expectations, progress, problems, schedules, budget, quality, and anything else that will help them contribute to getting the project completed on time, within budget, and according to specifications.

Listen intently to worker complaints and suggestions. No one on the jobsite is closer to the day-to-day work than those who actually do it. It is not uncommon for factors beyond the control of workers to have a negative effect on their ability to do their jobs as expected by the supervisor and higher management. In fact, it is not uncommon for a management policy to unwittingly have unplanned consequences that negatively affect work crews. Consequently, it is important of jobsite supervisors to listen when workers have complaints or suggestions. An effective method is what the author calls the *five-minute rule.* This rule works as follows: workers may have five minutes of the supervisor's time at any time—within reason—to make a complaint or suggestion. However, the complaint or suggestion must be accompanied by a well-thought-out recommended solution. The goal of the five-minute rule is to solve problems while they are still small and to help workers learn how to solve their own problems.

In addition to these supervision strategies, project managers should help their jobsite superintendents and assistant superintendents develop the various people skills explained in Part Two of this book, including leadership, teambuilding, mentoring, and motivation. Supervision takes on the added dimension of the labor agreement on a unionized jobsite. However, even when dealing with a unionized workforce, jobsite supervisors will experience better results if they apply the supervision strategies presented herein and the people skills covered in Part Two of this book.

JOBSITE SAFETY

Construction typically ranks in the top three industries in the United States with the highest annual death rates (the other two are mining and agriculture). Consequently, whether stated openly or tacitly, jobsite safety is one of the major success criteria that apply in any construction project. There are ethical, performance, financial, and regulatory aspects to jobsite safety. From an ethical perspective, providing a safe and healthy jobsite is the right thing to do. From a performance perspective, workers cannot perform at peak levels when they are continually worried about unsafe work conditions.

From a financial perspective, there are several considerations. These considerations are as follows: (1) construction companies can be named in lawsuits filed against subcontractors and other stakeholders when the courts allow the concept of *shared liability* to be applied, (2) construction companies that serve as general contractors are expected to exercise control over all aspects of construction projects and can be held liable for accidents and injuries that occur at the jobsite, and (3) The OSHA tends to hold both general contractors and subcontractors accountable even if they are not directly involved in a regulatory violation. Consequently, jobsite safety must be a high-priority concern of construction project managers.

The regulations governing jobsite safety for construction projects are provided by the Occupational Safety and Health Administration or OSHA, an organization within the United States Department of Labor. The OSHA Standard governing construction safety is Title 29 of the Code of Federal Regulations (CFR), Part 1926. Because of the potential risks associated with jobsite safety and the complexity and ever-changing nature of OSHA regulations, construction project managers are well-advised to employ a safety director to act on their behalf on the jobsite. This individual should be given responsibility for carrying out the general contractor's commitment to jobsite safety by doing the following:

- Ensuring that all job descriptions contain a clause on working safely.
- Ensuring that all contracts with subcontractors and tradespeople contain a clause on working safely.
- Ensuring that the labor agreement for unionized projects contains a clause on working safely.
- Making safe work practices part of all jobsite evaluations and performance appraisals.
- Rewarding safe work practices and ensuring that unsafe practices are not ignored or condoned.
- Developing and implementing work procedures that emphasize safety.
- Ensuring that the general contractor has a comprehensive jobsite safety program that is included in the policies and procedures manual and that all subcontractors agree to abide by the tenets of the program.
- Keeping the general contractor's jobsite safety policies and program up-to-date with the latest versions of OSHA's construction safety standard (29 CFR 1926) and all other applicable federal, state, and local safety regulations.
- Filing the accident, injuries, and illness reports required by OSHA (OSHA Forms 300, 300A, and 301) (see Figures 7.12, 7.13, and 7.14).
- Communicating safety information and expectations to all workers and subcontractors.
- Making jobsite supervisors mutually supportive partners in ensuring jobsite safety (jobsite superintendents and assistant superintendents sometimes double as the general contractor's safety manager).

OSHA's Form 300 (Rev. 01/2004)

Log of Work-Related Injuries and Illnesses

Attention: This form contains information relating to employee health and must be used in a manner that protects the confidentiality of employees to the extent possible while the information is being used for occupational safety and health purposes.

U.S. Department of Labor
Occupational Safety and Health Administration

Year ____

Form approved OMB no. 1218-0176

You must record information about every work-related injury or illness that involves loss of consciousness, restricted work activity or job transfer, days away from work, or medical treatment beyond first aid. You must also record significant work-related injuries and illnesses that are diagnosed by a physician or licensed health care professional. You must also record work-related injuries and illnesses that meet any of the specific recording criteria listed in 29 CFR 1904.8 through 1904.12. Feel free to use two lines for a single case if you need to. You must complete an injury and illness incident report (OSHA Form 301) or equivalent form for each injury or illness recorded on this form. If you're not sure whether a case is recordable, call your local OSHA office for help.

Establishment name ____

City ____   State ____

Identify the person

Describe the case

Classify the case

(A) Case No.	(B) Employee's Name	(C) Job Title (e.g. Welder)	(D) Date of injury or onset of illness (mo./day)	(E) Where the event occurred (e.g. Loading dock north end)	(F) Describe injury or illness, parts of body affected, and object/substance that directly injured or made person ill (e.g. Second degree burns on right forearm from acetylene torch)

CHECK ONLY ONE box for each case based on the most serious outcome for that case:

Death (G)	Days away from work (H)	Remained at work — Job transfer or restriction (J)	Remained at work — Other recordable cases (J)

Enter the number of days the injured or ill worker was:

Away From Work (days) (K)	On job transfer or restriction (days) (L)

Check the "injury" column or choose one type of illness: (M)

(1) Injury	(2) Skin Disorder	(3) Respiratory Condition	(4) Poisoning	(5) Hearing Loss	(6) All other illnesses

Page totals 0 0 0 0 0 0 0 0 0 0 0 0

Be sure to transfer these totals to the Summary page (Form 300A) before you post it.

| | (1) | (2) | (3) | (4) | (5) | (6) |
| Injury | Skin Disorder | Respiratory Condition | Poisoning | Hearing Loss | All other illnesses |

Page ____ of ____   Page 1 of 1

Public reporting burden for this collection of information is estimated to average 14 minutes per response, including time to review the instruction, search and gather the data needed, and complete and review the collection of information. Persons are not required to respond to the collection of information unless it displays a currently valid OMB control number. If you have any comments about these estimates or any aspects of this data collection, contact: US Department of Labor, OSHA Office of Statistics, Room N-3644, 200 Constitution Ave, NW, Washington, DC 20210. Do not send the completed forms to this office.

FIGURE 7.12 Required report from OSHA.

OSHA's Form 301

Injuries and Illnesses Incident Report

U.S. Department of Labor

Occupational Safety and Health Administration

Form approved OMB no. 1218-0176

This *Injury and Illness Incident Report* is one of the first forms you must fill out when a recordable work-related injury or illness has occurred. Together with the *Log of Work-Related Injuries and Illnesses* and the accompanying *Summary*, these forms help the employer and OSHA develop a picture of the extent and severity of work-related incidents.

Within 7 calendar days after you receive information that a recordable work-related injury or illness has occurred, you must fill out this form or an equivalent. Some state workers' compensation, insurance, or other reports may be acceptable substitutes. To be considered an equivalent form, any substitute must contain all the information asked for on this form.

According to Public Law 91-596 and 29 CFR 1904, OSHA's recordkeeping rule, you must keep this form on file for 5 years following the year to which it pertains

If you need additional copies of this form, you may photocopy and use as many as you need.

Attention: This form contains information relating to employee health and must be used in a manner that protects the confidentiality of employees to the extent possible while the information is being used for occupational safety and health purposes.

Information about the employee

1) Full Name _____

2) Street _____

City _____ State ____ Zip ____

3) Date of birth _____

4) Date hired _____

5) ☐ Male
   ☐ Female

Information about the physician or other health care professional

6) Name of physician or other health care professional
_____

7) If treatment was given away from the worksite, where was it given?

Facility _____

Street _____

City _____ State ____ Zip ____

8) Was employee treated in an emergency room?
☐ Yes
☐ No

9) Was employee hospitalized overnight as an in-patient?
☐ Yes
☐ No

Information about the case

10) Case number from the Log _____ *(Transfer the case number from the Log after you record the case.)*

11) Date of injury or illness _____

12) Time employee began work _____ AM/PM

13) Time of event _____ AM/PM ☐ Check if time cannot be determined

14) **What was the employee doing just before the incident occurred?** Describe the activity, as well as the tools, equipment or material the employee was using. Be specific. Examples: "climbing a ladder while carrying roofing materials"; "spraying chlorine from hand sprayer"; "daily computer key-entry."

15) **What happened?** Tell us how the injury occurred. Examples: "When ladder slipped on wet floor, worker fell 20 feet"; "Worker was sprayed with chlorine when gasket broke during replacement"; "Worker developed soreness in wrist over time."

16) **What was the injury or illness?** Tell us the part of the body that was affected and how it was affected; be more specific than "hurt", "pain", or "sore." Examples: "strained back"; "chemical burn, hand"; "carpal tunnel syndrome."

17) **What object or substance directly harmed the employee?** Examples: "concrete floor"; "chlorine"; "radial arm saw." If this question does not apply to the incident, leave it blank.

18) **If the employee died, when did death occur?** Date of death _____

Completed by _____

Title _____ Date _____

Phone _____

Public reporting burden for this collection of information is estimated to average 22 minutes per response, including time for reviewing instructions, searching existing data sources, gathering and maintaining the data needed, and completing and reviewing the collection of information. Persons are not required to respond to the collection of information unless it displays a current valid OMB control number. If you have any comments about this estimate or any other aspects of this data collection, including suggestions for reducing this burden, contact: US Department of Labor, OSHA Office of Statistics, Room N-3644, 200 Constitution Ave. NW, Washington, DC 20210. Do not send the completed forms to this office.

FIGURE 7.13   Required report from OSHA.

OSHA's Form 300A (Rev. 01/2004)
Summary of Work-Related Injuries and Illnesses

Year _____

U.S. Department of Labor
Occupational Safety and Health Administration

Form approved OMB no. 1218-0176

All establishments covered by Part 1904 must complete this Summary page, even if no injuries or illnesses occurred during the year. Remember to review the Log to verify that the entries are complete

Using the Log, count the individual entries you made for each category. Then write the totals below, making sure you've added the entries from every page of the log. If you had no cases write "0."

Employees former employees, and their representatives have the right to review the OSHA Form 300 in its entirety. They also have limited access to the OSHA Form 301 or its equivalent. See 29 CFR 1904.35, in OSHA's Recordkeeping rule, for further details on the access provisions for these forms.

Number of Cases

Total number of deaths	Total number of cases with days away from work	Total number of cases with job transfer or restriction	Total number of other recordable cases
0	0	0	0
(G)	(H)	(I)	(J)

Number of Days

Total number of days away from work	Total number of days of job transfer or restriction
0	0
(K)	(L)

Injury and Illness Types

Total number of...
(M)

(1) Injury	0
(2) Skin Disorder	0
(3) Respiratory Condition	0
(4) Poisoning	0
(5) Hearing Loss	0
(6) All Other Illnesses	0

Establishment information

Your establishment name _____

Street _____

City _____ State _____ Zip _____

Industry description (e.g., Manufacture of motor truck trailers) _____

Standard Industrial Classification (SIC), if known (e.g., SIC 3715) _____

OR North American Industrial Classification (NAICS), if known (e.g., 336212) _____

Employment information

Annual average number of employees _____

Total hours worked by all employees last year _____

Sign here

Knowingly falsifying this document may result in a fine.

I certify that I have examined this document and that to the best of my knowledge the entries are true, accurate, and complete.

_____
Company executive   Title

_____
Phone   Date

Post this Summary page from February 1 to April 30 of the year following the year covered by the form

Public reporting burden for this collection of information is estimated to average 58 minutes per response, including time to review the instruction, search and gather the data needed, and complete and review the collection of information. Persons are not required to respond to the collection of information unless it displays a currently valid OMB control number. If you have any comments about these estimates or any aspects of this data collection, contact: US Department of Labor, OSHA Office of Statistics. Room N-3644, 200 Constitution Ave. NW. Washington. DC 20210. Do not send the completed forms to this office.

FIGURE 7.14 Required report from OSHA.

MONITORING OF QUALITY

With construction projects, quality is defined by the specifications that are part of the construction documents. Performing quality work means meeting or exceeding the expectations prescribed by the specifications. Quality, as in completing the project in accordance with specifications, is one of the basic success criteria that apply to all construction projects. Consequently, quality must be a concern of construction project managers.

Project managers can play a positive role in ensuring quality at the jobsite by doing the following: (1) setting the tone by establishing high expectations with all stakeholders—refusing to accept so-called normal levels of delay and mistakes; (2) making sure that the jobsite superintendent and assistant superintendents are properly focused on quality and the continual improvement of performance; (3) making sure that subcontractors and tradespeople are properly focused on quality; (4) making sure that the labor agreement with a unionized workforce contains a clause about providing quality work and how work that fails to meet specifications will be handled; (5) make sure that all members of the construction project's staff have the training needed to do their part in ensuring quality; (6) making sure that subcontractors are *responsible* before accepting their bids; (7) eliminating barriers between stakeholders—particularly subcontractors tradespeople; (8) getting out of the office regularly and conducting *quality tours* (jobsite observations); (9) taking the necessary action to correct problems and unacceptable work immediately (and ensuring that jobsite supervisors do the same); (10) making sure that all stakeholders understand their respective roles in ensuring quality; and (11) making sure that quality is a standard agenda item for all meetings with the construction team, subcontractors, and the construction project staff.

SCHEDULE CONTROL

The schedule that was developed earlier in the project is a living document that must be monitored, controlled, updated, and reported on as needed throughout the construction project. Monitoring, controlling, updating, and reporting on the schedule is the responsibility of the project manager. The example in Figure 7.15 is a conceptual schedule for a hotel. It is a proposed schedule showing all of the major activities that must be completed in the construction of a hotel. During the scheduling phase of the project, this broad-brush schedule would be used to develop more specific schedules. For example, Activity A1800—General Construction—would have a comprehensive schedule of its own. However, the project manager would continue to use, monitor, and control this broad-brush schedule as a convenient overview of all major project activities.

Schedules like the one in Figure 7.15 provide the information project managers need to develop the reports they will make throughout the project for the owner, architect, engineers, subcontractors, and construction project staff. These reports will contain the following information: (1) completion dates for selected project milestones, (2) current status of the project and all activities, (3) scheduling problems and any revisions made to the schedule and why, (4) delivery dates for materials and equipment, (5) subcontractor starting and completion dates, and (6) other information that might have been requested by stakeholders.

The project manager will also update the schedule as necessary during the course of the project. For example, assume that the bidding period—A1350 in Figure 7.15—had to

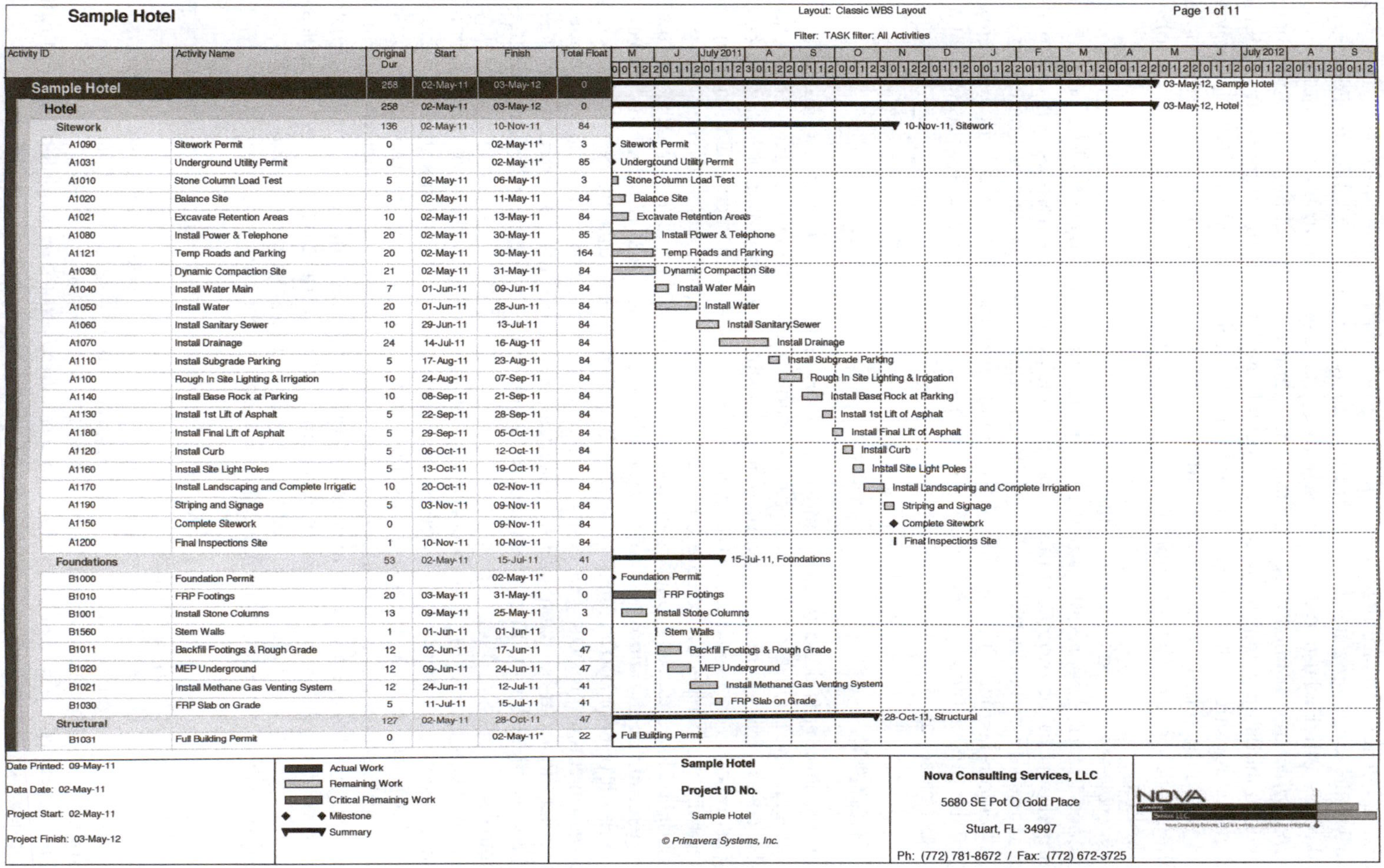

Sample Hotel

Layout: Classic WBS Layout

Filter: TASK filter: All Activities

Page 1 of 11

Activity ID	Activity Name	Original Dur	Start	Finish	Total Float
Sample Hotel		258	02-May-11	03-May-12	0
Hotel		258	02-May-11	03-May-12	0
Sitework		136	02-May-11	10-Nov-11	84
A1090	Sitework Permit	0		02-May-11*	3
A1031	Underground Utility Permit	0		02-May-11*	85
A1010	Stone Column Load Test	5	02-May-11	06-May-11	3
A1020	Balance Site	8	02-May-11	11-May-11	84
A1021	Excavate Retention Areas	10	02-May-11	13-May-11	84
A1080	Install Power & Telephone	20	02-May-11	30-May-11	85
A1121	Temp Roads and Parking	20	02-May-11	30-May-11	164
A1030	Dynamic Compaction Site	21	02-May-11	31-May-11	84
A1040	Install Water Main	7	01-Jun-11	09-Jun-11	84
A1050	Install Water	20	01-Jun-11	28-Jun-11	84
A1060	Install Sanitary Sewer	10	29-Jun-11	13-Jul-11	84
A1070	Install Drainage	24	14-Jul-11	16-Aug-11	84
A1110	Install Subgrade Parking	5	17-Aug-11	23-Aug-11	84
A1100	Rough In Site Lighting & Irrigation	10	24-Aug-11	07-Sep-11	84
A1140	Install Base Rock at Parking	10	08-Sep-11	21-Sep-11	84
A1130	Install 1st Lift of Asphalt	5	22-Sep-11	28-Sep-11	84
A1180	Install Final Lift of Asphalt	5	29-Sep-11	05-Oct-11	84
A1120	Install Curb	5	06-Oct-11	12-Oct-11	84
A1160	Install Site Light Poles	5	13-Oct-11	19-Oct-11	84
A1170	Install Landscaping and Complete Irrigatic	10	20-Oct-11	02-Nov-11	84
A1190	Striping and Signage	5	03-Nov-11	09-Nov-11	84
A1150	Complete Sitework	0		09-Nov-11	84
A1200	Final Inspections Site	1	10-Nov-11	10-Nov-11	84
Foundations		53	02-May-11	15-Jul-11	41
B1000	Foundation Permit	0		02-May-11*	0
B1010	FRP Footings	20	03-May-11	31-May-11	0
B1001	Install Stone Columns	13	09-May-11	25-May-11	3
B1560	Stem Walls	1	01-Jun-11	01-Jun-11	0
B1011	Backfill Footings & Rough Grade	12	02-Jun-11	17-Jun-11	47
B1020	MEP Underground	12	09-Jun-11	24-Jun-11	47
B1021	Install Methane Gas Venting System	12	24-Jun-11	12-Jul-11	41
B1030	FRP Slab on Grade	5	11-Jul-11	15-Jul-11	41
Structural		127	02-May-11	28-Oct-11	47
B1031	Full Building Permit	0		02-May-11*	22

FIGURE 7.15 Project manager must monitor schedules carefully.
Courtesy of NOVA Consulting Services, LLC.

be extended because an insufficient number of responsive bids were received. The project manager would revise the dates shown for this activity (November 17 through February 13) and make any other adjustments to the schedule that this revision might require. This is the *domino effect* in construction scheduling and monitoring. Sometimes schedule changes can be accommodated within the existing schedule and sometimes they create a domino effect that cascades through the rest of the schedule requiring all other dates to be revised. Before revising the schedule, the project manager would work with the other stakeholders to get their concurrence. The schedule would be revised and the new schedule distributed to stakeholders.

This cycle is repeated throughout the course of the contract and right through the project closeout. In fact, the schedule is never really final until the project is closed out and has been turned over to the owner. At this point, the final schedule is used as part of the after-the-fact analysis that should be conducted on all construction projects to determine final costs and lessons learned. Construction firms do not have final cost and schedule determinations until the project is completed or even after that in the case of warranty issues that arise after project closeout.

COST CONTROL

Schedule monitoring is done to keep the project on time. Cost control is done to keep it within budget. The budget for the project was established by the construction cost estimate when the cost of every activity in the project was estimated. Cost control is the process of holding actual costs on the jobsite to the estimated costs or less for each activity. Its principle purpose from the perspective of the construction project manager is to keep the project within budget. However, cost control can also be used for a variety of other purposes including: (1) determining the ultimate profit earned by the project, (2) determining if overruns were the result of poor cost estimating or special circumstances that are not likely to repeat themselves on future projects, (3) identify productivity rates and problems, (4) producing new cost data that can be compared to historical cost data, and (5) for recording lessons learned for use in making future construction cost estimates and establishing more effective cost controls.

Like so many aspects of construction project management, the level of the project manager's involvement in cost control depends on the size of the construction firm and the size and complexity of the project. Larger firms may have a cost control department or, at least, a cost control professional who assists the project manager with this responsibility. Smaller firms may require that the project manager perform this task himself. Also, different construction firms will have different cost control systems and procedures. However, regardless of whether the system is locally developed or purchased from a commercial provider, cost control systems for construction projects should meet the following criteria:

- Simple in format so that it is easy for people who are not finance and accounting professionals to use and understand.
- Provide a specific description of the crew's assignment, size, labor costs per hour, estimated hours, and total estimated cost for each activity.
- Provide both the estimated and actual cost for each activity.
- Have an effective feedback system from work crews so there is immediate notification of problems or unexpected gains that will affect cost in either direction.

COST CONTROL REPORT
PROJECT NUMBER 1555: COLLEGE STUDENT CENTER

Activity	Crew	Labor Cost	Crew Cost/Hr.	Est. Hrs	Act. Hrs	Est. Cost	Act. Cost	Comments
• Excavate footings	2 backhoe operators 6 laborers	$35 $19	$184	40	32	$7,360	$5,888	Under budget $1,472
• Prep and Pour footings	2 carpenters 4 laborers	$31 $19	$138	150	154	$20,700	$21,252	Over budget $522
• Prepare subgrade	4 rebar setters 4 laborers	$28 $19	$188	50	50	$9,400	$9,400	On budget
• Pour Slab	4 concrete operators	$34	$136	60	40	$8,160	$5,440	Under budget $2,720
								Under budget $3,640

FIGURE 7.16 Sample cost control report for constructing footings and slab for a college student center.

Comparisons of estimated versus actual costs should be used by the project manager as the basis for recognizing crews that complete their activities under budget and intercede appropriately when crews go over budget. Figure 7.16 is an example of a cost control report showing comparisons of estimated versus actual costs for the activities that make up the footings and slab work package. This package consists of four activities: excavate footings, prep and pour footings, prepare subgrade, and pour slab.

Each of these activities makes use of a crew. Each type of crew members has an assigned labor rate per hour that is used to determine the cost of the crew per hour. For example, the crew that will excavate the footings consists of two backhoe operators and six labors. The labor rate for the backhoe operators is $35 per hour. The labor rate for the laborers is $19 per hour. Consequently, a crew of tow backhoe drivers and six laborers will cost $184 per hour. The estimated duration of the activity is 40 hours, but it was actually completed in 32 hours. Therefore, the activity was completed $1,472 under budget. The entire work package for pouring the footings and the slab was completed $3,640 under budget.

CHANGE ORDERS

Few things create more consternation in construction projects than change orders. Change orders are required any time a change in the project will result in unplanned work. It is not uncommon for the general contractor to notice a situation in which the architectural drawings are in error, did not properly anticipate a given situation, or leave out pertinent information needed to complete some aspect of the work. When there is a difference between the information contained in the construction documents—most often the drawings or specifications—and reality at the jobsite, the general contractor notifies the architect and asks for guidance.

For example, assume that it becomes obvious to the general contractor that a door will be needed in a certain location, but the architectural drawings show no door. The general contractor cannot just put in the door, no matter how obvious the need for it may be. Instead, the contractor must notify the architect and ask for guidance. The architect must, in turn, let the contractor know what should be done (e.g., ignore the oversight and forget about the door or add the extra door to the work to be completed).

If the architect determines that the door is, in fact, needed a request will be sent to the general contractor to determine what the cost of the extra work will be. If the cost is acceptable to the architect—who will typically confer with the owner—a change order will be issued to the contractor. The contractor is not authorized to do the work or to charge for it without a signed change order from the architect. If the work in question is holding up progress on the project, the architect can issue a *written directive* authorizing the contractor do the work while the official change order works its way through channels. This is often so that more than one change can be collected and included in one large change order to save time and paperwork.

The change order process works as just described when all parties agree what is to be done, how much it will cost, and who will pay. Unfortunately, this is not always the case. A request for a change order is sometimes a trigger for conflict and it is not uncommon for the owner, architect, engineer, and general contractor to dispute what needs to be done and how much it should cost. The best advice for project managers is to use Document A201 provided by the American Institute of Architects (AIA). The provisions set forth in this document can save the owner, architect, engineers, general contractor, and subcontractors time and money when dealing with change orders.

AIA Document A201: General Conditions of the Contract for Construction

AIA Document A201 contains guidelines for all stakeholders in construction projects for writing contracts that minimize the likelihood of disputes among stakeholders. Article 7 of the document provides guidelines for "Changes in the Work." This section also describes the following pertinent concepts as well as how and when they should be used: (1) change order, (2) construction change directive, and (3) order for a minor change in the work. By following the guidelines in AIA Document A201, construction project managers can ensure that changes in the work are handled in a systematic manner that is fair to all parties involved. This, in turn, will reduce the likelihood of disputes that have to be taken to mediation, arbitration, or the courts.

PROJECT CLOSEOUT

Project closeout, like most aspects of project management, is a process. The input into the process consists of a variety of types of documentation attesting to completion of the work. The procedures that make up the project closeout process include: (1) identification of any remaining aspects of the work, any defective equipment and technology items, any defective materials, and any defective workmanship, (2) placement on a punch list—first by the contractor and then by the architect—of all items identified that still require work; (3) start up, testing, and operation of all systems (HVAC, plumbing, electrical, security, etc.) and placement of any problem items on a punch list; (4) completion of all punch list items; (5) issuance of the certificate of substantial completion by the architect; (6) rekeying of all locks for turnover to the owner; (7) review with owner of all operator's manuals and turnover of the manuals; (8) review with owner of warranty materials; (9) remittal of all remaining payments; and (10) release of any retained funds. The output of the closeout process is a completed project turned over to its owner.

The certificate of substantial completion is acknowledgement by the architect that the project has reached a stage in the construction process where it can be occupied by the owner, even though there may still be minor items on the punch list to complete. The certificate of substantial completion will list any remaining punch list items that must be completed. The importance of receiving the certificate of substantial completion to the general contractor is that it established the official date for determining whether or not liquidated damages will be assessed. If the date of issuance of the certificate of substantial completion is on or before the final contracted date for completion of the project, no liquidated damages can be assessed. However, the certificate of substantial completion comes after the contracted completion date, damages can become an issue.

CONSTRUCTION PROJECT MANAGEMENT SCENARIO 7.2

I Think We Need a Safety Director for This Project

Miriam Mantle is concerned that her latest project might be understaffed. The new Technology Park Complex is the largest and most complex project her company has ever undertaken. She won the battle to have not just a jobsite superintendent assigned to the project but also an assistant superintendent, but the construction firm's CEO is balking on assigning a safety director to her. "We don't need a safety director. Let the superintendent or the assistant superintendent take care of safety on this project."

Mantle knows that the jobsite superintendent has always been required to double as the safety director on her firm's construction projects, but none of these projects has been as large or complex as the Technology Park Complex. "I think we need a safety director for this project," said Mantle. "There are just going to be too many hazardous conditions on this project. The superintendent and his assistant are going to have their hands full just keeping the work flowing properly. Besides, neither one of the men in these positions has had any OSHA training."

Discussion Questions

In this scenario, Miriam Mantle is the project manager for the largest construction project her firm has ever undertaken. She is concerned that there are going to be safety issues and wants a full-time safety professional assigned to her staff. The firm's CEO disagrees. Join this debate. Do you agree with Mantle or the CEO? Why? Assume that you have been asked to list the pros and cons of assigning a full-time safety director to the project. Prepare your list.

SUMMARY

Once the planning phase of a construction project is completed, the plans must be implemented. Implementation is the construction phase. The construction phase is the most critical phase in the project. The project manager's principle concerns during the construction phase of a project are: jobsite layout, documentation and recordkeeping, conflict resolution, labor relations and supervision, safety, quality, schedule control, cost control, change orders, and project closeout.

The jobsite layout plan consists of the following elements: materials storage/placement areas, jobsite office facilities, worker facilities, tool storage facilities, sanitation facilities, temporary utilities, jobsite security, jobsite access, jobsite drainage, and signs/barricades. Records kept during the course of a construction project should be trustworthy, standardized, timely in preparation and distribution, accessible, and comprehensive. The types of information to be documented and reported on include conversations among stakeholders, actual versus estimated costs, correspondence, meeting proceedings, items specified in the contract, and scheduled versus actual progress.

There should be regular meetings over the course of a construction project of the construction team and the construction staff. There should also be an organizational kickoff meeting at the beginning of the project and a series of closeout meetings at the end of the project. In between there should be periodic project review meetings, meetings with subcontractors, and special meetings called to deal with problems as they arise.

Project managers should be must negotiators to prevent conflict and to resolve conflict when it cannot be prevented. Preparation is the key to successful negotiations. Mediation and arbitration can also be used to resolve disputes that arise over the course of a construction project. Mediation is a legally recognized process for resolving disputes outside of court that involves the parties in dispute placing their cases before an unbiased third party—the mediator. The mediator serves as a facilitator, not a decision maker. Arbitration is different. The arbitrator is also an unbiased third party, but this individual has the authority to actually decide how the dispute will be resolved. Decisions of the arbitrator are binding on both parties.

On unionized construction projects, the labor agreement prescribes wages, benefits, and working conditions. Unionized workers are represented on the jobsite by a union steward. Regardless of whether the construction project is unionized or open-shop (nonunionized), the jobsite superintendent is responsible for supervising the work and keeping it on time, within budget, and in accord with specifications. The best jobsite supervisors provide effective leadership, are positive change agents, communicate often and well, and listen intently to worker complaints and suggestions.

Jobsite safety is a critical concern on construction projects. From the perspective of performance, workers cannot perform at peak levels when they are continually concerned about unsafe work conditions. From the perspective of finances, construction companies can be named in lawsuits filed against subcontractors and other stakeholders when the courts allow shared liability. General contractors can be held directly liable for accidents and injuries that occur on their jobsites. Finally, OSHA tends to hold both general contractors and subcontractors accountable even if they are not directly involved in a regulatory violation.

Project managers can play a positive role in ensuring quality on their projects by setting high expectations, enlisting the jobsite superintendent on the side of quality, keeping subcontractors and tradespeople focused on quality, putting a quality clause in the labor agreement, providing the proper training for construction project staff, making sure to contract only with responsible subcontractors, eliminating barriers between and among stakeholders, conducting jobsite observation tours, correcting quality problems immediately, helping all stakeholders understand their roles in ensuring quality, and making quality a standard agenda item for meetings of the construction team, subcontractors, and the construction project staff.

The schedule developed during the planning phase of the project is a living document that must be monitored, controlled, updated, and reported on throughout the construction phase of a project. Doing these things is the responsibility of the project manager. Cost control is also the project manager's responsibility. Whether developed internally or purchased commercially, construction firms should have a cost control system that is easy to use and understand. The cost control system is used to determine the ultimate profit of a project, determine

if cost overruns were the result of poor estimating or special circumstances that are not likely to recur, identify productivity rates and problems, produce new cost data that can be compared to historical data, and for recording lessons learned.

Change orders can be problematic if not handled properly. Guidance for properly handling change orders is available to project managers in Article 7 of Document A201 of the American Institute of Architects. This document provides detailed guidance on how to handle change orders, change directives, and minor changes to the work. Project closeout is the process of completing a project and turning it over to the owner. Key concepts in project closeout are the certificate of substantial completion and punch lists.

KEY TERMS AND CONCEPTS

Jobsite layout
Documentation and recordkeeping
Project meetings
Conflict resolution
Labor relations and supervision
Safety
Quality
Schedule control
Cost control
Change orders
Project closeout
Material storage/placement areas
Jobsite office facilities
Worker facilities
Tool storage facilities
Sanitation facilities
Temporary utilities
Jobsite security
Jobsite access

Jobsite drainage
Signs and barriers
Construction team meetings
Organizational/kickoff meeting
Subcontractor meetings
Periodic construction staff meetings
Special meetings
Project closeout meetings
Project review meeting
Negotiation
Mediation
Arbitration
Labor agreements
Jobsite supervision
Shared liability
29 CFR 1926
Domino effect
AIA Document A201
Construction change directive

REVIEW QUESTIONS

1. What are the principle concerns of the project manager during the construction phase of a project?

2. List the elements of a jobsite layout plan.

3. What are the questions a project manager must answer when planning for material and placement at a jobsite?

4. Explain the purposes served by the jobsite office.
5. Explain the types of sanitary facilities that should be available at a construction jobsite.
6. When planning for the installation of temporary electricity at a jobsite, the service should be placed to avoid what three conditions?
7. What are the parameters to remember when planning for the installation of temporary water lines at a jobsite?
8. List five strategies for securing a jobsite against theft, vandalism, and breaches of public safety.
9. Why are *drive-through* access routes to the jobsite the preferred approach?
10. What are the two issues that project managers must deal with when planning for jobsite drainage?
11. What are the two principle uses of signs and barriers on jobsites?
12. What are the types of signs most commonly used at jobsites?
13. Why are documentation and recordkeeping so important for project managers?
14. List the criteria that jobsite records should meet.
15. List the types of things that should be documented by project managers at jobsites.

16. What are the various types of team meetings that project managers are called upon to chair during the construction phase of a project?
17. Explain how a project manager can use negotiation skills to prevent and resolve conflict.
18. Compare and contrast the concepts of mediation and arbitration as they relate to disputes that occur over the course of a construction project.
19. What is a labor agreement, how is it used, and what should it contain?
20. How is supervision different on a unionized jobsite versus an open-shop jobsite?
21. List and explain four best practices of supervisors that can be used by those who supervise construction work.
22. What are the ethical, performance, and financial aspects of jobsite safety?
23. How can project managers play a positive role in ensuring quality at the jobsite?
24. What information should be contained in schedule-update reports prepared by project managers?
25. What are the various uses of a cost control system in construction?
26. Explain the concept of the *change order* in construction.

APPLICATION ACTIVITIES

The following activities may be completed by individual students or by students working in groups:

1. Identify a construction firm in your region that will cooperate in completing this activity. Ask a project manager with the firm to show you how the following issues are handled at the firm's jobsites: jobsite layout, project meetings, labor relations (labor agreement), safety, quality, schedule control, cost control, change orders, and project closeout. Develop a summary of your research and report it to the class.
2. Ask permission to visit a construction jobsite in your region. During your visit, observe how the following aspects of the jobsite layout plan are handled: material storage/placement, jobsite office, worker facilities, sanitary facilities, tool storage, jobsite access/security, and signs/barricades. Develop a summary of your research and report it to the class.
3. Conduct the research to identify a major conflict that has occurred on a construction project. Develop and present a class presentation providing the details of the conflict and how it was resolved.
4. Conduct the research to identify a construction firm that was fined by OSHA for safety violations. Develop and present a class presentation providing the details of the violation, the amount of the fine, and the final resolution of the situation.

CHAPTER 8

Leading Project Teams

Project managers in construction will have opportunities to lead several different types of teams. These teams—depending on the size and type of construction firm—might include the construction cost estimating team, planning and scheduling team, procurement team, risk management team, construction team, and the project staff team. Being able to lead people is important for construction project managers because any time people work in groups to accomplish a specific mission, leadership is required. In fact, the difference between an effective project team and one that is just mediocre is often the quality of the leadership provided by the project manager. As a project manager, it is a good idea to remember that processes can be managed, but people must be led. Therefore, leadership is a critical *people skill* for construction project managers.

Like attitude, leadership is an internal concept that manifests itself in external actions. It is an intangible concept that can produce tangible results. Leadership is both an art and a science. Project teams that are well-led, regardless of their size, are better able to complete their projects on time, within budget, and according to specifications. Consequently, few things will contribute more to a project manager's effectiveness than becoming a good leader.

Even the best estimator, planner, scheduler, and risk manager will be limited as a project manager if he or she lacks good leadership skills because it is the performance of people that will ultimately determine whether a project is completed on time, within budget, according to specifications, safely, and without environmental problems. The best project plan ever developed depends on people to carry it out. The better people are led the better they will carry out the plan.

LEADERSHIP DEFINED

Regardless of the purpose, size, or composition of the project team, leadership is essential to the team's effectiveness. There are a number of different definitions of leadership, primarily because the concept applies to so many different fields of endeavor. The definition presented here applies specifically to leadership for construction project management:

> Leadership is the act of inspiring team members to make a wholehearted commitment to the mission at hand.

The "mission at hand" alluded to in this definition is, of course, to complete the construction project in question on time, within budget, according to specifications, safely, and in an environmentally friendly manner. If this appears to be too simple a definition for such an important concept, look closer. While it is true that some definitions of leadership are longer, do not be misled by the brevity of this one. There is more depth in this brief definition than might be apparent at first. Several aspects of this definition are significant. The first of these is *inspiring team members*.

Other definitions of leadership tend to use the term *motivating* where this one uses the term *inspiring*. While it is certainly true that leaders must be good motivators, in the hierarchy of leadership responsibilities it is a lesser concept than inspiration. Leaders motivate people by providing incentives and rewards that encourage them to perform at their best. This is external motivation. External motivation is an important leadership tool, but it typically provides only temporary results. Consequently, it must be applied regularly. Motivating externally is like filling a car with gasoline. The gasoline will make the car perform, but it burns up quickly requiring frequent refills. Inspiration, on the other hand, can have a long-term or even permanent effect on the performance of people.

Project managers can inspire their team members by being good role models of everything they expect from the team members. Consistency between words and actions is critical to effective leadership. Project managers who fail to live up to the principles they espouse will not be effective leaders. Team members will be inspired by project managers who exemplify traits they themselves would like to have. The most effective way for a project manager to inspire team members is to let them know what is expected of them and to be an exemplary role model of those expectations. Project managers inspire team members by being good at their jobs, exemplifying integrity, being able to make difficult decisions, having the courage to do the right thing in difficult situations, being selfless, helping others, caring about the project as well as the people who do the work to complete it, being fair and consistent, and being a source of calm in the middle of the inevitable storms that arise during the course of a project.

Another important element in the definition of leadership is "wholehearted commitment." There is a quaint saying relating to commitment that project managers should memorize and put to use: *It is easier to ride a horse in the direction it is going.* This saying has direct application for project managers. It is easier to lead team members when they want to follow, when the project goals become their goals, and when it becomes just as important to them as it is to the project manager that the project's purpose be accomplished. This is what is meant by "wholehearted commitment." When team members are as committed to completing the project on time, within budget, and according to specifications as the project manager, everybody wins—the team, the project manager, the construction firm, and the owner.

CONSTRUCTION PROJECT MANAGEMENT SCENARIO 8.1

The ineffective project manager

John's new title is "Project Manager 1." This means he manages residential and light commercial projects for his company, a large construction firm. Level 2 project managers oversee heavy commercial and industrial projects. Level 3 project managers oversee infrastructure projects. When he is given a project, John is responsible for leading a team that consists of the owner, architect, engineers, and jobsite superintendent. He is responsible for securing the commitment of team members; helping develop the construction estimate for the project; developing a schedule for the project; helping minimize all applicable risks associated with the project; helping procure the materials and subcontractors for the project; and monitoring the work at the jobsite to ensure that it is completed on time, within budget, and according to specifications. John is a talented construction professional with excellent credentials, but his former supervisor is concerned that John might not make it as a project manager.

When John was promoted to Project Manager 1, his former supervisor commented to a colleague: "It is as if we have just taken the best hitter on the baseball team and made him the coach. I know John can hit the ball, but I am not sure he can coach the team." This supervisor's concern turned out to be prophetic. As a project manager, John is clearly more comfortable dealing with team members who have the same background as his. When project deadlines create stress, John becomes agitated and panicky. In addition, he struggles with making difficult decisions and often sweeps problems under the carpet rather than dealing with them forthrightly.

Discussion Questions

Assume that you are a friend of John's, one he will listen to. Also assume that you are familiar with John's struggles as a project manager. If he asked you for advice on how to become a more effective project manager, what guidance would you give him?

INFLUENCING TEAM MEMBERS

In positions that have line authority over direct reports, leaders are able to use the authority of their positions to influence those who report to them. They have the authority of their positions to call on when they need people to perform at their best. Project managers, on the other hand, often do not have line authority over members of their project teams. For example, the main team a project manager leads is the construction team consisting of the owner, architect, engineers, and jobsite superintendent. The project manager has no line authority over these individuals. Because this is often the case with the teams they lead, it is important for project managers to understand how to influence team members in ways that contribute to completing the project on time, within budget, and according to specifications. The following strategies will help project managers enhance their ability to influence team members:

- ***Understand the sources of the project manager's ability to influence others.***
  Even in situations where project managers have no line authority, they can still have

influence. The ability of project managers to influence others comes from different sources depending on the type of team in question and its composition. For example, when leading internal teams such as the construction cost estimating team, scheduling team, risk management team, or procurement team, the influence of project managers comes from such factors as their ability to: (1) make recommendations to the supervisors and managers of others concerning salary increases and promotions, (2) make recommendations for assigning individuals to future project teams and other jobs, (3) provide input for the performance appraisals of others, and (4) call on the support of higher management for the project in question. When leading the main construction team—owner, architect, engineers, and jobsite superintendent—the project manager's ability to influence is based on the mutual desire of all members to complete the project on time, within budget, and according to specifications.

• ***Understand the power of the desire of all stakeholders to complete the project on time, within budget, and according to specifications.*** One thing all stakeholders in a construction project—owner, architect, engineers, and construction firm—typically understand and can agree on is the importance of completing a project on time, within budget, and according to specifications. It gives project managers influence when they can tie their recommendations to this overall goal. This means that construction project managers must learn to stay focused on this mutual goal of the project's most important stakeholders and to keep all team members focused on it.

• ***Understand what drives individual team members.*** To influence individual team members, it is necessary for project managers to understand what drives them. Owners in a construction project want their project completed according to all of their expectations. The architectural firm may want a model project to use for marketing purposes—a monument to its talents. The engineers may want to add to their reputation for performing well. And, of course, money is an almost universal motivator applying to all stakeholders. All members of the project team have their individual needs that drive them. Determining what these are and tying recommendations to them will help project managers have influence.

• ***Earn respect, loyalty, and credibility from team members.*** Project managers who have the most influence with their team members—regardless of whether or not they have line authority—are those who have earned respect, loyalty, and credibility from team members. The next section on the Eight Cs of Leadership explains how project managers can earn respect, loyalty, and credibility from team members, regardless of the type of team.

EIGHT Cs OF LEADERSHIP FOR CONSTRUCTION PROJECT MANAGERS

Many people think that leadership is primarily about image, dressing for success, and charisma. This impression is reinforced by the entertainment industry in movies and television. Even news media outlets promote the image-equals-substance theory. For example, watch network news programs during presidential primaries. Even these supposedly serious programs get caught up in discussions about the looks, image, and gravitas of political candidates.

Because of this emphasis on image, many people who want to be leaders contract with image consultants to advise them on such things as dressing for success, how to look taller, compensating for baldness, developing charisma, how to project confidence, and various other image enhancement strategies. One could easily deduce from this fixation on image that the cover of the book is more important than the contents for people who want to be leaders. This is hardly the case for leaders in general and is most certainly not the case for construction project managers.

This is not to say that image does not matter. It does. Image is important because it often takes an attractive cover to convince people to open a book or, in other words, to give leaders a chance to prove themselves. The point to understand here is that the amount of attention devoted to image can be of proportion to its importance. After all, once a book has been pulled off the shelf, the reader's attention quickly moves from the cover to the contents. No matter how attractive the cover, people will not read a book that lacks substance or is poorly written. This same principle applies to leadership in a project management setting. No matter how attractive construction project managers might appear on the outside, their inner substance will soon reveal itself—for better or worse. An empty box, no matter how nicely decorated, is still empty.

Rather than putting too much of their effort into image enhancement, project managers who want to be good leaders should focus on developing the basic characteristics that are essential to effective leadership. The author calls these essential characteristics the *Eight Cs of Leadership*. The Eight Cs of Leadership are leadership characteristics that make people want to follow an individual. These characteristics (see Figure 8.1) are as follows:

- Caring
- Competence

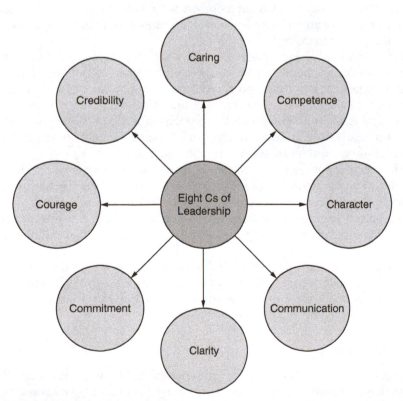

FIGURE 8.1 These characteristics will ensure that a leader has followers.

- Character
- Communication
- Clarity
- Commitment
- Courage
- Credibility

Each of these characteristics is important to project managers who want to be effective leaders.

CARING AND LEADERSHIP

Project managers who are good leaders inspire team members to make a wholehearted commitment to completing the project on time, within budget, and according to specifications. This is important because team members are people and people have agendas, egos, and personality quirks. The only way to keep agendas, egos, and personalities from getting in the way of progress is to gain a wholehearted commitment to the project from all stakeholders.

Few things will inspire team members to follow the lead of project managers more than knowing that they care. Caring is an important ingredient in the formula for inspiring members of project teams to commit fully to the project. When team members know that the project manager cares about both the project and them, they will be more open to following his or her lead. On the other hand, team members who think the project manager cares only about the project and not about them or, worse yet, cares about neither will not wholeheartedly support the project manager.

Of course, in cases where project managers are in positions of authority over team members, they can coerce them into carrying out their demands, but coercion is not leadership. In fact, it is just the opposite and typically results in what the author calls *reluctant compliance*. Reluctant compliance means begrudgingly going along to get along—doing only the minimum required to comply. It is the polar opposite of the wholehearted commitment project managers need from team members. Project managers who get only reluctant compliance from their team members will not be effective.

Wholehearted commitment is important because team members who willingly put their hearts and minds into accomplishing project goals will produce better results than those who just reluctantly go along to avoid trouble. This is why it is so important for team members to know that the project manager cares about them and the project. If a project manager does not care about team members or the project, why should they care about him or her or the project?

Caring leaders consistently display several important traits that make others willing to follow them. These traits are as follows: (1) honesty, (2) empathy, (3) sincere interest, (4) patience, (5) commitment to participatory decision making, (6) a servant's heart, and (7) good stewardship (see Figure 8.2).

Honesty and Caring

Project managers who care about their team members are honest with them. Whether the message they have to convey is good news or bad, effective project managers show they care by telling the truth. Honesty is one of the most fundamental leadership traits. When

---

**Checklist of
CARING TRAITS FOR PROJECT MANAGERS**

✓ Honesty

✓ Empathy

✓ Sincere interest

✓ Patience

✓ Commitment to participatory decision making

✓ A servant's heart

✓ Good stewardship

FIGURE 8.2 Caring leaders engender loyal followers.

communicating with an honest project manager, team members can take comfort in knowing that the message, whether good news or bad, is the truth. What makes this fact so important is that members of project teams will not wholeheartedly support a project manager they do not trust, and nothing dampens trust faster than lies, prevarication, obfuscation, deceit, or failure to follow through on promises.

Team members who think they are being lied to or are having information withheld from them will not be inspired to make a wholehearted commitment to completing the project on time, within budget, and according to specifications. Team members want to be informed, and they want to know that they can trust the message. Whether the news is good or bad, members of project teams want and need the truth. They want to be able to trust that what they are told is reliable, complete, accurate, and up-to-date.

Consequently, effective project managers are honest with all stakeholders: the owner, architect, engineers, subcontractors, material providers, colleagues and higher management in the construction firm, inspectors, public officials, and the public in general. A caveat is in order here. Being honest, especially when the news is bad, does not mean being tactless. Project managers who care about the stakeholders and the project make a point of being tactful when conveying unwelcome news. Using tact can be viewed as driving in the nail without breaking the board. It should not be construed to mean withholding information that might be unwelcome or hurtful. Effective project managers are empathetic and tactful, but they fully and honestly convey all information, including unwelcome messages.

This ability to tactfully deliver unwelcome news is important when trying to lead certain stakeholders to perform at peak levels. For example, project managers often find it necessary to give constructive criticism and corrective feedback to the people who do the hands-on work on a construction project. These are situations where driving in the nail without breaking the board is crucial. Assume that it is necessary to give corrective feedback to a subcontractor that has fallen behind schedule. One way to deliver the message would be to say: "The schedule we established is just that—a schedule. It's not a suggestion. Your contract doesn't say stay on schedule if you feel like." Another way to deliver the message would be to say: "You did outstanding work on our last project. Let's talk about how you can do an equally good job of staying on schedule with this project."

Both versions of the message make the point that falling behind is unacceptable, but the first does so in a tactless manner, a manner likely to break the board. The second is better. It is firm but tactful. Rather than use hurtful sarcasm, the second approach acknowledged that the subcontractor had done well in the past and then offered the constructive criticism about staying on schedule. As a result, the second approach is more likely to be received in a positive manner that will lead to improved performance.

Empathy and Caring

Empathy means identifying with and understanding another person's needs, concerns, fears, and circumstances. Caring project managers are empathetic. They try to put themselves in the shoes of stakeholders when making suggestions, recommendations, and decisions that will affect them. Empathy is about putting oneself in the shoes of the other person and trying to see things from his or her point of view. It should not be confused with sympathy. They are not the same thing. Sympathy is about sharing the sorrow of another person. Empathy is about trying to see things from others' point of view.

Returning to the examples of giving constructive criticism to the poorly performing subcontractor from the previous section, which approach was the more empathetic of the two? The second approach—the tactful approach—was more empathetic because it took into account how the constructive criticism would affect the subcontractor. Being tactful is by definition an empathetic act. Project managers can ensure that they are empathetic by asking themselves the following question before making suggestions, recommendations, and decisions or making statements that affect others: *How would I want this to be said if I were in the other person's place?*

Sincere Interest and Caring

Because they care about stakeholders, the most effective project managers take a sincere interest in them. This means they take the time to find out about their needs concerning the project. In short, they get to know stakeholders. Taking an interest in stakeholders is critical for project managers because it helps them match the goals of stakeholders with those of the project. Further, by getting to know stakeholders project managers can improve their ability to assist them in ways that will enhance their commitment to the project.

For example, if the project manager knows that the owner wants to build a building that will be the most unique structure in the community, he or she can use this information to influence the owner to make the right decisions when difficult choices present themselves. If the project manager knows that the architect wants to build a building that will win state or national design honors, he or she can use this information to help other stakeholders understand when the architect seems to be too demanding. If the project manager knows that the jobsite superintendent wants to make a good impression on the construction firm's higher management team in hopes of earning a promotion, he or she can use this information to motivate the superintendent when the inevitable problems of construction arise.

Patience and Caring

The most effective project managers are patient with stakeholders. This does not mean that they accept inappropriate behavior or condone a lack of commitment. Never forget that the *caring* aspect of leaders means caring about the project as well as the stakeholders. The patience called for in project management means forbearance in dealing with changes and other inconveniences without developing a negative attitude. This kind of patience can manifest itself in numerous ways including staying positive when changes are made during the course of the project or being willing to listen when a stakeholder makes an unwelcome recommendation. It can also mean being willing to mentor a member of the project staff team who, in spite of trying, is falling short of expectations rather than just firing him. Patience requires self-discipline, empathy, and a willingness to maintain a positive attitude when things are not going exactly as planned with the project. Patience does not mean tolerating poor performance or disruptive behavior. Rather, it means staying positive and focusing on solutions when problems arise rather than losing control and becoming angry and uncooperative. For many people, patience comes hard, if at all. But it is a leadership skill that can be learned by project managers and should be learned by them.

Stakeholders who are treated with patience are more likely to maintain their wholehearted commitment to the project. Patience is especially crucial when providing constructive criticism to stakeholders. Project managers who lose their patience with stakeholders on a regular basis will not win the support they need to be effective leaders of project teams. Impatience on the part of project managers often results in reluctant compliance or even purposefully counterproductive behavior on the part of stakeholders.

Commitment to Participatory Decision Making and Caring

Project managers must make many decisions before construction begins and many more during the course of construction. Engaging stakeholders when making decisions is called *participatory decision making*. It does not mean that project managers allow others to make the decisions they should make and it does not mean that they take a vote. Rather, it means that before making a decision project managers solicit input from the stakeholders who will be affected by their decisions. Participatory decision making encourages *ownership* on the part of stakeholders while at the same time leading to better decisions. Different stakeholders will have different perspectives about problems that require decisions. By asking for their input, project managers gain the benefit of their perspectives. Asking stakeholders for their opinions concerning decisions is an excellent way to show that they are part of a team. This simple act shows stakeholders that their opinions matter and, by extension, that they matter. Stakeholders who know they matter are more likely to maintain their commitment to the project.

Participatory decision making is an approach in which the stakeholders who are closest to the problem in question and who will have to implement the decision are included in the decision-making process. For example, when a problem arises that could affect the work of a subcontractor, that subcontractor would be included in the decision-making process. This does not mean that the project manager lets the subcontractor make the decision—far from it. Rather, it means involving the subcontractor in the decision-making process for the purpose of making a better decision and for securing his or her support.

Stakeholders who are given a voice in the decision-making process, even if the eventual decision is not the one they recommended, are more likely to commit to its successful implementation than those who are left out. This phenomenon is known as *buy-in*. Buy-in is one of the benefits of participatory decision making. Another benefit is that it gets the stakeholders who are closer to the problem than the project manager focused on finding the best solution. Project managers are often one or two levels removed from the hands-on aspects of the problems they have to deal with. Consequently, involving stakeholders who are closer to the problem in determining how to solve it can lead to a better solution.

Servant Leadership, Stewardship, and Caring

Effective project managers are servant leaders and good stewards. They know that stakeholders are more likely to commit to the project if the project manager is a good steward and a servant leader. Project managers who are servant leaders put stakeholders and the project itself ahead of their personal agendas. When stakeholders see the project manager working late to solve a problem that has popped up at the last minute or sacrificing his personal time on a weekend to keep the project on schedule, they will know that he is a servant leader who cares about the project. When stakeholders hear that the project manager bought supper for the project staff when its members had to work late, they will know that he is a servant leader who cares about his people.

Project managers who are good stewards take care of the resources—human, physical, and financial—entrusted to them. When stakeholders see the project manager setting an example of fiscal conservatism and wise use of project resources, they will know that he is a good steward who cares about the project. When stakeholders see the project manager making sure that workers subcontractors have the resources they need to do their jobs, they will know that he is a good steward. Nothing shows stakeholders that a project manager cares about them and the project more than servant leadership and good stewardship.

COMPETENCE AND LEADERSHIP

Construction project managers must have three distinct sets of skills in order to be competent. The first skill set consists of construction-specific technical skills. People who want to be construction project managers should be well-versed in the field of construction through formal education, work experience, or both. A good rule of thumb for prospective construction project managers to remember is this: *Before attempting to be a project manager, learn as much as you can about your field.*

The second skill set needed by construction project managers consists of the process skills of project management. These include cost estimating, planning, scheduling, risk management, procurement, jobsite layout, work monitoring/control, and closeout skills. The third skill set consists of the people skills of project management. These skills include leadership, teambuilding, motivating, time management, change management, diversity management, conflict management, and perseverance.

This book is designed to help prospective project managers in the field of construction develop the second and third skill sets—those that relate directly to project management. However, it is important to note that construction knowledge and experience are not just important—they are vital. A construction project manager who does not appear to be well-versed in construction will not inspire confidence in stakeholders.

CHARACTER AND LEADERSHIP

One of the absolute prerequisites of effective leadership is trust. Owners, architects, engineers, jobsite superintendents, the construction firm's higher management, members of the project staff, materials providers, subcontractors, union officials, inspectors, Occupational Safety and Health Administration (OSHA) representatives, government officials, and all other stakeholders in a construction project must know they can trust the project managers they work with. Construction project managers must be trustworthy. Stakeholders will neither follow nor be influenced by project managers they do not trust.

Stakeholders are more likely to trust project managers who consistently exemplify strength of character. Character is what allows project managers to recognize the right course of action in any situation and the courage to do the right thing in spite of pressure and temptations to the contrary. Project managers who consistently tell the truth, follow through on promises, treat stakeholders with dignity and respect, are fair and equitable in human relations, and do the right thing even when it does not serve their personal needs are more likely to be able to influence stakeholders in a positive way. On the other hand, when stakeholders observe or even just sense a lack of character in a project manager, they will not make a wholehearted commitment to the project. Nor will they be influenced by the project manager. Unimpeachable character is the foundation of a construction project manager's credibility with stakeholders.

COMMUNICATION AND LEADERSHIP

Project managers who are effective leaders are good communicators. They have to be. Communication is the oil that lubricates the gears of human interaction. At any given time in a construction project, there are a lot of things happening and a lot of people involved. Effective communication is necessary to keep all of the wheels turning in the right direction and everyone involved fully informed so they can do their parts to keep the wheels properly moving. Effective communication is essential for construction project managers.

The most important communication skill for a project manager is listening. Listening to people empowers them which, in turn, makes them more willing to do their part to complete construction projects on time, within budget, and according to specifications. Listening, verbal communication, and nonverbal communication are such important skills for construction project managers that a chapter is devoted to these concepts later in this book.

CLARITY AND LEADERSHIP

Stakeholders in a project team are just like people on a long trip. They want to know where they are going—where the project manager proposes to take them. When stakeholders are unclear about their direction they feel as if they are peddling a stationary bicycle rather than making progress toward a definite destination. This is why project managers must be clear in establishing the purpose of the project and making sure that stakeholders understand it. Project managers may certainly add to their basic purpose for any construction project—getting it completed on time, within budget, and according to specifications—but they must, at the very least, let all stakeholder know from the outset that this is their basic purpose.

In addition, the hands-on workers in a construction project need to feel that there is meaning in their work. They need to know that the project they have been assigned to is important. If the people who do the hands-on work in a construction project think their work does not matter they might begin to wonder if they matter. Consequently, project managers should develop a clear purpose statement for the project and convey that purpose to all stakeholders including the hands-on workers. The following purpose statement is for a basketball arena that will double as a hurricane shelter when needed.

> The ABC Arena will be the home of the Raiders Basketball team and host to a variety of sporting events during normal times. However, when hurricanes threaten it will serve as an emergency shelter for the community. Our overall purpose in this project is to complete it on time, within budget, and according to specifications.

As this example of a purpose statement for a construction project shows, there are two aspects to the purpose: (1) the purpose of the building being constructed and (2) the purpose of the project team that will build it. Both of these purposes are important and should be stated for clarity for all stakeholders.

COMMITMENT AND LEADERSHIP

To be effective leaders, project managers must be committed to the project and to helping stakeholders do their part to complete the project on time, within budget, and according to specifications. Commitment means more than just trying hard. It means the following: (1) doing everything within the bounds of ethics and the law to get the job done on time, within budget, and according to specifications, (2) striving to maximize the performance of all processes, (3) helping those who do the hands-on work perform at their best, and (4) going the extra mile to ensure that the project is a success.

There is an amusing illustration of the concept of commitment that has circulated over the years in leadership circles. This illustration demonstrates the difference between being just involved and being committed: *With a breakfast of bacon and eggs, the chicken is involved, but the pig is committed.* This humorous illustration makes the point that commitment means more than just trying hard. It means being willing to sacrifice to make the project a success. This is an important distinction for project managers because mediocrity in construction results more often from a lack of commitment than a lack of talent.

When one commits to the success of a project, there will be sacrifices. It might be necessary to sacrifice time that could have been spent doing something else, resources that could have been used for another project, or the comfort of not having to confront factors that undermine the progress of a project. When stakeholders see that the project manager is committed, they are more likely to make a commitment themselves. Of course, the obverse is also true.

It is not uncommon to have people involved in a project who are afraid to make a commitment because they think doing so will require too much of them. Such people tend to sit back and let others do the hard work involved in completing projects. Personnel assigned to a project who are lukewarm in their commitment must be either motivated or removed from the team. Motivating stakeholders is the subject of a chapter later in this book.

COURAGE AND LEADERSHIP

Courage is essential to project managers who want to be effective leaders. In the context of project management, the type of courage required is not physical courage. Rather, it is moral courage—the courage to do the right thing even when it is not convenient or even when it hurts. This is because project managers are often required to deal with difficult situations involving conflict, high expectations, limited resources, pressing deadlines, demanding owners, recalcitrant subcontractors, and delinquent materials suppliers. Further, project managers sometimes have to tell powerful people things they do not want to hear and influence people to do things they do not want to do.

Being a project manager can be a frightening prospect when one considers the level of responsibility and the consequences of failure. It takes courage to face the sometimes daunting responsibilities of the construction project manager. Consequently, project managers need to understand that courage is not a lack of fear. Rather, it is a willingness to do the right thing in spite of fear. It is a willingness to go ahead and do something that is difficult because it is the right thing to do.

For example, a project manager might understandably fear the consequences of failing to complete a project on time, within budget, and according to specifications. He might understandably fear the prospect of telling a demanding owner that what the owner wants cannot be done without delaying completion of the project. Project managers must have the courage to overcome the fear of confrontations and to sacrifice their sense of security to do what is best for the project, even when there is pressure to do otherwise. The most effective project managers are those who learn to put their fears aside and focus on getting the job done rather than being paralyzed by contemplating the consequences of failure.

CREDIBILITY AND LEADERSHIP

Credibility is what project managers have when stakeholders believe in them and see them as leaders who are worthy of confidence. Credibility is earned by exemplifying the first seven of the Eight Cs of Leadership. In addition to exemplifying these characteristics, another important strategy for earning credibility is to continually improve in all aspects of project management—the process skills and the people skills. Project management is like any other endeavor in that those who become good at it will earn credibility from the stakeholders they hope to lead. Credibility, in turn, will make it easier to lead and influence stakeholders.

CONSTRUCTION PROJECT MANAGEMENT SCENARIO 8.2

This team needs some leadership

Jane Evans faces the biggest challenge of her career. The CEO of her company just informed Evans that the project manager for the company's most important project just resigned. She has been selected to replace him. That's the good news. The bad news is that the project is behind schedule and overbudget, the morale of stakeholders is at rock bottom, and the owner is beginning to ask some hard questions. The CEO was frank in telling Evans that ". . . this team needs some leadership."

Discussion Question

In this scenario, Jane Evans has inherited either a mess or a great opportunity. She wants to turn it into an opportunity. How can Jane Evans use the Eight Cs of Leadership to turn the project around and get it back on track? What should she do first? What should she do after that?

SUMMARY

Leadership in construction is the act of inspiring stakeholders to make a wholehearted commitment to the success of a project. The Eight Cs of Leadership are caring, competence, character, communication, clarity, commitment, courage, and credibility. Caring leaders consistently display several important traits that make others willing to follow them. These traits are as follows: honesty, empathy, sincere interest, patience, commitment to participatory decision making, servant's heart, and good stewardship.

Competence is important. Consequently, before becoming project managers, construction professionals need to develop construction-related knowledge or gain construction experience or both. Commitment means more than just trying hard. It means being willing to do everything within the bounds of ethics and the law to complete construction projects on time, within budget, and according to specifications. It also means striving to improve performance all the time and going the extra mile to ensure that the project is a success. Credibility is what project managers have when stakeholders believe in them and when they are seen as being worthy of confidence. Credibility is earned by exemplifying the other seven Cs of the Eight Cs of Leadership.

KEY TERMS AND CONCEPTS

Leadership
Caring
Competence
Character
Communication
Clarity
Commitment
Courage

Credibility
Honesty and caring
Empathy and caring
Sincere interest and caring
Patience and caring
Participatory decision making and caring
Good stewardship and caring
Buy-in

REVIEW QUESTIONS

1. Define the term "leadership."
2. What is the significance of the term "inspiring" in the definition of leadership?
3. How does the adage "It is easier to ride a horse in the direction it is going" apply to leading project teams?
4. How does the concept of caring affect a project manager's ability to lead?
5. List and briefly explain the traits displayed by caring leaders.
6. Why is it important that construction project managers be competent in the field of construction?
7. Why is character so important to those who hope to lead others?
8. Describe the importance of communication skills to project managers.
9. What is meant by the term "clarity" as it relates to leading others?
10. What is meant by the term "commitment"?
11. How does courage apply to construction project managers?
12. Explain how a project manager can earn credibility with stakeholders.

APPLICATION ACTIVITY

This activity may be completed by individual students or by students working in groups. Think of a person you are familiar with who is or was in a leadership position (executive, manager, supervisor, coach, military leader, elected official, historical figure, etc.). This person can be someone you know or someone you have read about. Analyze this person on the basis of the definition of leadership and the Eight Cs of Leadership. Is or was this individual an effective leader? Why or why not? How does this person exemplify or fail to exemplify the Eight Cs of Leadership? Would this person make a good project manager? Why or why not?

Building Project Teams and Managing Conflict

Construction project managers—depending on the size of the construction firm—might be called on to be a member of and lead several different kinds of teams between the initiation and closeout of a project. Prior to construction, these teams include the cost estimating, planning and scheduling, procurement, and risk management teams. During construction, the project manager leads the construction team and the project staff team. Teambuilding and conflict management are important people skills for project managers regardless of the types of team in question, but they are especially important when leading the construction team and the project staff team. The construction team for a given project includes the owner, architect, engineers, jobsite superintendent, assistant superintendents (if there are any), and the project manager. In addition, at different times during the building phase of the project, the team will also include representatives of various subcontractors.

A team is a group of people working together to accomplish a common purpose. In the context of construction project management, a team's common purpose is the successful completion of the project in question. In other words, the common purpose of construction team members is to complete the project on time, within budget, and according to specifications. Construction teams are formed to accomplish this specific purpose. The better the members of the construction team work together and cooperate in achieving the project's purpose, the more effective the project team will be. Molding the owner, architect, engineers, on-site superintendent, and various subcontractors into a mutually supportive team of cooperative members is the responsibility of the project manager. The concept is known as *teambuilding*.

COMMON PURPOSE: THE BASIS OF EFFECTIVE TEAMWORK

Having a common purpose is essential to effective teamwork because common purpose is the foundation of teamwork—it is what brings construction team members together and gives them a reason to be mutually supportive and cooperative. However, being that people are human, individual members of the construction team will have their own agendas, egos, personalities, and perspectives. The project manager's challenge is to take a disparate group of individuals with their own agendas, egos, personalities, and perspectives and help them embrace a common purpose: the successful completion of the project.

Of course, there is more to effective teamwork on a construction project than just committing to a common purpose, but without a common purpose the project is like a house without a foundation. However, once all members of a team have committed to the common purpose of completing the project on time, within budget, and according to specifications, the other ingredients for success will be easier to accomplish.

People, as a rule, are individualistic, and this is especially true of Americans. It is the rights of the individual that are spelled out in the U.S. Constitution and, more specifically, the Bill of Rights—not the rights of the group or team. Respect for the individual is deeply ingrained in the American psyche. America's heroes tend to be rugged individualists who started with nothing and succeeded by triumphing over great odds. Consequently, convincing members of construction teams to put their individualistic tendencies aside for the good of the team can be a challenge. This is one of the reasons that teambuilding never really ends or, at least, not until the project in question is completed. Throughout the duration of a construction project, the project manager must constantly work on teambuilding.

CONSTRUCTION PROJECT MANAGEMENT SCENARIO 9.1

Mack is not a team player

Why don't you want Mack on your construction team? He is the best jobsite superintendent in this company. Nobody manages a jobsite better than Mack. "I know all about Mack's skills. In fact, normally Mack would be my go-to guy. But this project is going to require especially good team players, and Mack is not a team player. He is a my-way-or-the-highway kind of guy. Consider what I am facing as the project manager on this project. The owner is demanding and stubborn. The architect is brilliant but impractical, disorganized, and prone to getting his feelings hurt. Add Mack's sometimes coarse personality into the mix and I will have a potentially volatile situation on my hands. I will have my hands full just trying to keep the owner and architect from drawing their swords. I don't need Mack adding fuel to the fire. Mack would do a good job at the jobsite, but he would clash constantly with this particular owner and architect. The jobsite superintendent for this project is going to have to be a diplomat not a tyrant."

Discussion Questions

In this scenario, the project manager does not want Mack, a talented construction superintendent, on the project team because he is not a good team player. Is this a legitimate concern? Have you ever worked, either in school or on a job, with someone who did a good job individually but did not work well as part of a team? If so, describe the situation. As a project manager, would you ever pass on a potential team member because she is not a good team player?

BUILDING THE PROJECT TEAM

Part of building an effective team is choosing team members wisely. Unfortunately, project managers in construction do not always choose their team members—certainly not all of them. For example, if the project team in question is a construction team, the project manager does not choose either the architect or engineers. The project manager may participate in selecting subcontractors as part of the procurement process, but they are chosen for being responsible and responsive—not for reasons of fit. Project managers may choose or at least have a voice in choosing those members of the project staff team—provided they are not already committed to another project—but for the most part construction project managers do not choose their team members. Consequently, teambuilding in construction can be a challenging undertaking.

Regardless of how project managers get their team members for construction projects, the concept of *fit* should be considered whenever possible. For example, in *Construction Project Management Scenario 9.1* the project manager did not want Mack on his construction team in spite of Mack's talent as a jobsite superintendent. The problem was fit. Because of his personality and approach to the job, Mack would not have been a good fit for the project in question. The project manager thought Mack would clash with the owner and architect. On another project, Mack might have been the best person for the job but on the project in question he wasn't.

The message in this scenario is that, to the extent possible, it is important to consider fit when assigning individuals to teams. Even the most talented individual can undermine the work of the team if she does not fit in as a good team player. During the selection process it is important to consider how well prospective team members will fit in with the rest of the team and if they will be good team players.

Developing a Mission Statement for the Team

Because the team's purpose is obvious to them, some project managers make the mistake of thinking it will be obvious to team members. This is a bad assumption. The purpose of every project team should be defined in writing for team members—even those who have served with the project manager on numerous teams. The team's purpose, when put in writing, becomes its mission statement.

The purpose of any kind of team in construction is to complete the project in question on time, within budget, and according to specifications or to do its part to make sure this happens. If the team is the cost estimating team, its purpose is to contribute to a successful project by producing the most accurate estimate possible for the project. If it is a risk management team, its purpose is to contribute to a successful project by eliminating risk when possible and minimizing it when it cannot be eliminated. For the purpose of illustration, assume in this section that the team in question is a construction team consisting of the owner, architect, engineers, jobsite superintendent, and project manager.

The purpose of the team—the reason its members were brought together in the first place—constitutes the core of the team's mission. This mission should be made clear from the outset and reiterated frequently throughout the duration of the project. This is a critical step in building the project team. When a construction team first comes together, the owner, architect, engineer, jobsite superintendent, and project manager might be complete strangers. The mission statement can be an invaluable tool for helping project managers mold this group of disparate individuals into a team and for keeping them focused on why the team exists in the first place.

A construction team's mission statement should state the name of the project along with a brief description followed by this statement: *The mission of this team is to complete the project on time, within budget, and according to specifications.* Some construction firms prefer to add safety and the environment to the mission statement while others believe that these two factors are covered by the time and budget criteria (i.e., failing to provide a safe and environmentally friendly jobsite can result in fines, medical costs, legal expenses, and schedule delays). What follows are examples of mission statements for construction teams:

- Project XXX is a new governor's mansion for the state of Florida that is to reflect both the importance of the governor's office and the frugality of the state's government. The mission of the construction team is to complete the project on time, within budget, and according to specifications.
- Project YYY consists of 100 miles of four-lane highway and two bridges in Dover County that will connect the two fastest growing cities in the state. The mission of the construction team is to complete the project on time, within budget, and according to specifications.
- Project ZZZ is a 60,000 square foot combination warehouse and manufacturing facility that, when completed, will create 150 new jobs. The mission of the construction team is to complete the project on time, within budget, and according to specifications.

Each of these sample mission statements contains three distinct elements: (1) the name of the project, (2) a brief description of the project, and (3) purpose of the construction team that will build the project. Together, these three elements make up the mission statement for the team that will complete the project.

These examples are intentionally simple, easy to read, and easy to understand. This is how the mission statements for construction teams should be written. Project managers should keep simplicity and understandability in mind when developing their teams' mission statements. A mission statement is a tool for communicating the team's purpose with team members and other stakeholders, some of whom may be laypeople. Consequently, a good rule of thumb is to make the team's mission statement understandable to people who might know little about construction.

Challenging Construction Project
THE BROOKLYN BRIDGE

The Brooklyn Bridge, an infrastructure project connecting Manhattan and Brooklyn in New York City, is one of the most recognized bridges in the world and one of New York City's most famous landmarks. Built between 1869 and 1883, the Brooklyn Bridge took 14 years to build. In the process, some difficult but invaluable lessons were learned about building bridges. Some of the lessons learned were paid for in human lives. In fact, of the 600 people who worked on the bridge 20 died during its construction including the project manager—John Roebling—a Prussian road and bridge builder.

The most challenging aspects of building the Brooklyn Bridge were what the workers called the "caisson disease." This pressure-related malady—much like the bends occasionally suffered by scuba divers who rise too quickly to the surface—afflicted workers who went down in the deep caissons to construct the towers that are the most recognizable feature of the bridge. The other challenge was constructing a mesh of cables that are anchored in the ground and span the length of the bridge. The largest four cables are 11 inches in diameter. These cables support the bridge spans that run from tower to tower.

The Brooklyn Bridge is considered one of the greatest engineering and construction feats of the 19th century. It is 5,989 feet long and was the largest suspension bridge in the world at the time. The towers to which the suspension cables are attached are 276 feet tall. An elevated footpath runs the length of the bridge allowing pedestrians to walk across without having to interact with traffic. Because the footpath provides a breath taking view of New York City, it attracts millions of visitors every year.

Source: Based on A View on Cities, "Brooklyn Bridge." http://www.aviewoncities.com/nyc/brooklynbridge. htm

Understanding the Leadership Responsibilities of Project Managers

Becoming a project manager can be an important step up the career ladder to higher-level leadership positions in a construction firm. In fact, leading project teams is excellent training for higher levels of leadership. Consequently, when an opportunity arises to lead a project team, it is important to get it right. What follows are typical responsibilities of project managers who lead teams:

- Serve as the team's connection with higher management in the construction firm.
- Serve as the leader of the construction team for the project, a team that consists of the architect, engineers, jobsite superintendent, and project manager.
- Serve as the official record keeper for the team. Records include minutes of meetings, correspondence, agendas, and a variety of different kinds of reports. Typically, the project manager will appoint a recorder to take minutes during meetings. However, even if recordkeeping is delegated it is still the responsibility of the project manager. This means the project manager must ensure that minutes are completed in a prompt manner, accurate, distributed to all who should receive them (preferably within 48 hours of the meeting), and kept on file for future reference.
- Participate in team discussions and debates, but take care to avoid dominating.
- Implement team recommendations that fall within the realm of authority—provided they are worthy recommendations—and work with higher management to implement those that fall outside of it.
- Prevent conflict in the team that might undermine its mission and resolve conflict that cannot be prevented.

Developing Positive Working Relationships in Teams

Project teams work most effectively when individual team members form positive, mutually supportive relationships. Positive working relationships among team members can be the difference between having a high-performing team and having one that is just mediocre. Learning how to help team members establish, develop, and nurture positive working relationships is important for project managers. To build positive working relationships in their teams, project managers should do the following:

- Help team members understand the importance of being honest and reliable. Effective teamwork is not possible in an environment of mistrust.
- Help team members develop an attitude of mutual support. Stress that team members are supposed to help each other as they work together to accomplish the team's mission.
- Help team members develop a supportive attitude toward each other as they struggle with the challenges of completing the project.
- Help team members understand that they need to help each other deal with the pressure and corresponding stress involved in getting the project completed on time, within budget, and according to specifications.

These are the basics. Competence, trust, communication, reliability, and mutual support are the foundation on which effective teamwork is built. Any amount of time devoted to improving these factors is a good investment for project managers.

EXPLAINING THE ROLES OF TEAM MEMBERS

It is important that individual team members understand their respective roles in the team. Consequently, project managers must be prepared to explain the roles team members will play in carrying out the work required to complete the project. Specifically, team members need to know: (1) their primary role in the team, (2) their supporting role(s) in the team, and (3) their range of authority (i.e., what they are able to do unilaterally and what requires permission).

Knowing these things will equip team members to do their primary job in the team while also mutually supporting other team members. Knowing these things will also eliminate the risk of conflict that can arise when one team member thinks another is encroaching on his or her territory (primary area of responsibility). Finally, knowing these things will ensure that team members have the latitude to make decisions about their work within clearly defined boundaries without running the risk of overstepping their authority.

EXPLAIN HOW THE TEAM IS SUPPOSED TO OPERATE

Team members need to know how decisions are made and what role they play in the process, how conflict is resolved, how members are to communicate with each other and the project manager, how problems are to be solved, and how the daily work of the team is supposed to be done. Knowing the parameters and how to operate within them will enable team members to give their best to accomplishing the mission of the team without the distraction of wondering if they are overstepping their authority.

FOUR-STEP MODEL FOR BUILDING EFFECTIVE TEAMS

Good teams do not just happen—they must be built. Building effective project teams is best accomplished using a systematic four-step model:

1. Assess
2. Plan

**Four Steps to
EFFECTIVE TEAMBUILDING**

- **ASSESS** the team to identify weaknesses that must be improved and strengths that can be exploited.

- **PLAN** teambuilding activities based on the results of the assessment.

- **EXECUTE** the planned teambuilding activities.

- **EVALUATE** to determine if the teambuilding activities were effective and repeat the cycle continually throughout the project.

FIGURE 9.1   Building effective project teams is a systematic process.

3. Execute
4. Evaluate

This four-step model is applied as follows: (1) assess the team to identify weaknesses that must be improved and strengths that can be exploited, (2) plan teambuilding activities based on the results of the assessment, (3) execute the planned teambuilding activities, and (4) evaluate results (see Figure 9.1). The four-step model for teambuilding presented herein works best for construction teams—owner, architect, jobsite superintendent, and project manager—and the construction project staff team—project manager, administrative support staff, jobsite superintendent and assistant superintendents, safety director, and field engineers.

The work of these teams lasts for the duration of the project. Hence, there is time to actually apply the model and build the team. The work of teams such as the cost estimating, planning and scheduling, risk management, and procurement teams is short-term in nature. Consequently, there may not be time to apply the four-step model to building these teams, unless, of course, the project manager works with them frequently enough on successive projects.

Assessing a Project Team's Strengths and Weaknesses

John is the new coach of a basketball team about which he knows very little. Consequently, the first thing he wants to do is conduct a comprehensive assessment of the team's abilities. He wants to identify specific strengths and weaknesses. With an assessment completed, John will have a better idea of what he needs to do to turn the team into a winner or to improve on an existing winning record. This same approach can be used by construction project managers each time they are to lead a new team.

A mistake commonly made by project managers is trying to begin building their teams before determining where things stand with the team. This mistake can lead to wasted time and resources. Project managers who begin teambuilding activities without first assessing strengths and weaknesses are operating in the dark. Accurately assessing the strengths and weaknesses of teams is an important and necessary first step in teambuilding.

For project teams to be effective and productive, several factors must be present. At a minimum these factors include the following:

- Clear direction that is understood by all team members (e.g., mission, goals, ground rules).

- Team players on the team (e.g., team first—me second).
- Fully understood and accepted accountability measures (e.g., evaluation of performance).

   What follows are specific criteria in each of these three broad areas that can be used for conducting an assessment of a project team's strengths and weaknesses.

DIRECTION AND UNDERSTANDING.   Members of teams need to be clear concerning what the team is supposed to accomplish. They need to have clear direction that is fully understood. The following criteria can be used to determine if a project team has direction and if it understands that direction:

- Does the team have a clearly stated mission?
- Do all team members understand the mission?
- Does the team have a set of goals that translate its mission into more specific terms?
- Do all team members understand the goals?
- Does the team have a schedule for completing the project?
- Do all members of the team know the schedule and intermediate deadlines within it?

CHARACTERISTICS OF TEAM MEMBERS.   Effective teamwork requires that members of teams be good team players. While it is necessary for members of project teams to think independently and critically, once decisions are made they must come together as a team to implement them. The following criteria can be used to determine if a project team's members are good team players:

- Are all team members open and honest with each other all the time?
- Do all team members trust each other?
- Do all team members put the team's mission and goals ahead of their personal agendas all the time?
- Do all team members know they can depend on each other?
- Are all team members committed to accomplishing the team's goals?
- Are all team members willing to cooperate with each other to accomplish the team's mission?
- Do all team members take the initiative to ensure that the project is completed according to specifications, on time, and within budget?
- Are all team members patient with each other when working on solutions to problems?
- Are all team members resourceful in finding ways to get the job done in spite of obstacles?
- Are all team members punctual for work, meetings, assignments, and in meeting deadlines?
- Are team members tolerant of individual differences among members of the team (i.e., intellectual, racial, cultural, gender, and political differences)?
- Are all team members willing to persevere when the work becomes difficult?
- Are all team members mutually supportive of each other?
- Are all team members comfortable stating their opinions, pointing out problems, and offering constructive criticism in team meetings?
- Do all team members support team decisions once they are made?

ACCOUNTABILITY. People in teams need to know how the team's performance will be evaluated and what accountability measures will be used. The following criteria can be used to determine if team members understand the accountability aspects of their work:

- Do all team members know how team progress/performance will be measured?
- Do all team members understand how success is defined for the team?
- Do all team members understand how team decisions are made?
- Do all team members know their respective responsibilities?
- Do all team members know the responsibilities of all other team members?
- Do all team members understand their authority within the team?

CONDUCTING A TEAM ASSESSMENT. Once the team in question is formed—construction team or project staff team—the project manager will soon begin to learn where things stand with each team member. But rather than rely on informal observations, it is important for project managers to formalize their assessments of team members can conduct the assessments in a structured way. An efficient and effective way to do this is to turn the criteria listed in the previous sections into an assessment instrument the project manager can use to record his observations of team members and their respective strengths and weaknesses from the perspective of teamwork.

Turning the criteria into an assessment instrument can be accomplished by adding a rating scale. A rating scale such as the following example that assigns numerical values to each possible response is recommended:

Completely true	6
Somewhat true	4
Somewhat false	2
Completely false	0

Once project managers have completed the assessment, they use the scores to determine the team's strengths and weaknesses. Any criterion that receives a score of less than "4" or "somewhat true" represents a weakness that should be singled out for improvement. Any criterion that receives a score that is higher than "4" represents a strength that can be exploited to maximize the team's performance. Figure 9.2 is an example of a teambuilding assessment instrument.

Developing a Teambuilding Plan

Teambuilding activities should be planned on the basis of the results from the assessment of strengths and weaknesses. For example, assume the assessment shows that the team does not understand the mission. Clearly, an important teambuilding activity will be to either rewrite the existing mission statement or do a more thorough job of explaining it. If the results of the assessment show that team members do not understand the schedule, an important teambuilding activity will be to clarify the schedule and all intermediate deadlines. Regardless of what the assessment reveals about the team, the results should be used as the basis for planning activities to correct weaknesses and exploit strengths.

Instructions

To the left of each item is a blank for recording your perception regarding that item. For each item, record your perception of how well it describes your team. Is the statement *Completely True (CT), Somewhat True (ST), Somewhat False (SF),* or *Completely False (CF)?* Use the following *numbers to* record *your* perception.

      CT = 6
      ST = 4
      SF = 2
      CF = 0

Direction and Understanding

_____ 1. The team has a clearly stated mission.

_____ 2. All team members understand the mission.

_____ 3. The team has a set of goals that translate the mission into more specific terms.

_____ 4. All team members understand the goals.

_____ 5. The team has a schedule for completing the project.

_____ 6. All team members know the schedule and intermediate goals within it.

Characteristics of Team Members

_____ 7. All team members are open and honest with each other.

_____ 8. All team members trust each other.

_____ 9. All team members put the team's mission and goals ahead of their personal agendas all of the time.

_____ 10. All team members are committed to accomplishing the team's mission.

_____ 11. All team members are willing to cooperate to accomplish the team's mission.

_____ 12. All team members will take the initiative to ensure the project is completed on time, within budget, and according to specifications.

_____ 13. All team members are patient with each other when working on solutions to problems.

_____ 14. All team members are resourceful in finding ways to accomplish the team's mission in spite of obstacles.

_____ 15. All team members are punctual for meetings, assignments, and deadlines.

_____ 16. All team members are tolerant of the individual differences of team members.

_____ 17. All team members are willing to persevere when the work becomes difficult.

_____ 18. All team members are mutually supportive.

_____ 19. All team members are comfortable expressing opinions, pointing out problems, and offering constructive criticism.

_____ 20. All team members support team decisions once they are made.

FIGURE 9.2   Sample assessment instrument.

Accountability

_____ 21. All team members know how team progress/performance will be measured.

_____ 22. All team members understand how team success is defined.

_____ 23. All team members understand how team decisions are made.

_____ 24. All team members know their respective responsibilities.

_____ 25. All team members know the responsibilities of all other team members.

_____ 26. All team members understand their authority within the team.

FIGURE 9.2 (continued)

Executing Teambuilding Activities

Teambuilding is an ongoing, never-ending process. The idea is to make a team better and better as time goes by. The project manager may need to work with individual team members away from the others. For example, if the other members of the team do not appear to trust one of the team members the project manager may need to pull that member of the team aside and work quietly to help correct the situation. In other instances, the entire team might be involved in a teambuilding activity. For example, when the team has just been formed and is having its first organizational meeting, presenting the members with a mission statement, goals, and a schedule can be a teambuilding activity because it can serve to get everyone on the same page from the outset. All subsequent team-building activities should also be based on the results of an assessment. Project managers should never make the mistake of assuming strengths and weakness of team members. Assess first and then base decisions concerning what training to provide on the results of the assessment.

Evaluating Teambuilding Activities

If teambuilding activities are effective, weak areas pointed out by the assessment will be strengthened. A simple way to evaluate growth and improvement in the team is for the project manager to periodically review the current behavior of team members against the original assessment. If these ongoing observations show that progress is being made, nothing more is required for the time being. If not, the project manager has additional work to do. If a given teambuilding activity appears to have been ineffective, the project manager should try something else. Involving team members in identifying ways to strengthen the team and improve its performance can itself be an effective teambuilding activity.

INITIATING THE TEAM'S WORK

Before beginning work on a project, it is a good idea for project managers to hold a team initiation meeting. The purpose of this meeting is to give the team a good start at working together to complete the project successfully. This is achieved by: (1) discussing the team's mission with team members; (2) confirming that all team members are committed to completing the project on time, within budget, and according to specifications; and (3) reviewing the work to be done, the schedule for the work, project milestones, and important deadlines.

The project initiation meeting is the same type of meeting a football coach has in the locker room just before the beginning of the game. If handled well, the meeting will ensure that all team members understand what is to be done, who is supposed to do what, and when assignments are to be completed. Last minute details are clarified during the initiation meeting, and last minute questions are answered in this meeting.

TEAMS SHOULD BE COACHED

Project managers soon learn that if team members are going to work together, they have to be coached. If the team is a construction team—architect, engineers, jobsite superintendent, and project manager—the project manager has line authority over only one member of the team: the jobsite superintendent. In other words, he or she is not the "boss" in the traditional sense of the concept. If the team in question is the project staff team, the project manager is the boss, but even in this situation coaching will be more effective than bossing.

Project managers need to understand the difference between bossing and coaching. Bossing, in the traditional sense, involves giving orders and evaluating performance. Bosses approach the job from a perspective that says, "I'm in charge—do what I say." Coaches, on the other hand, approach the job from a *follow-me* perspective. Their overriding mission for the team is to achieve the best possible outcome in completing the project in question. Consequently, coaches work with team members to influence, encourage, quip, and empower them to do their part in making the project a success. Construction project managers can become effective coaches by doing the following:

- Developing a team charter. A team charter contains the team's mission, goals—which are project milestones, and ground rules.
- Continuing team development and teambuilding activities throughout the duration of the project.
- Mentoring individual team members or providing mentors for them as appropriate.
- Promoting mutual respect between and among team members.
- Working to make human diversity within a team an asset.
- Setting a positive example of everything that is expected of team members.

Team Charter

It is not hard to imagine a baseball, football, basketball, soccer, or track coach calling his team together and saying, "This year we have one overriding purpose—to win the championship." In one simple statement, this coach has clearly and succinctly defined the team's mission—the first component of a team charter. From this statement, team members should immediately realize that everything they do will be aimed at winning the championship. The coach's statement of the mission was brief, to the point, specific, and easily understood. Project managers should be just as specific in explaining their team's mission, ground rules, and goals/milestones to team members. Writing a mission statement for a team was explained earlier in this chapter.

DEVELOPING THE TEAM CHARTER. A team charter is a document consisting of three major components: (1) mission statement for the team, (2) ground rules for team members, and (3) team goals/milestones (see Figure 9.3). The team charter is used to provide direction for

---

Team Charter
XYZ PROJECT TEAM

Mission Statement

Project XYZ is a seafood restaurant on Okaloosa Island. The mission of this team is to complete the project on time, within budget, and according to specifications.

Ground Rules

As we work together to accomplish the team's mission, members will be guided by the following ground rules:

- **Punctuality.** Team members will be punctual in arriving at work and in meeting deadlines.

- **Honesty.** Team members will be open, honest, and frank with each other at all times.

- **Dependability.** Team members will conduct themselves in ways that show they can be depended on.

- **Responsibility.** Team members will take responsibility, individually and as a group, for accomplishing the team's mission.

- **Mutual support.** Team members will work cooperatively in helping each other complete the project successfully.

- **Initiative.** Team members will take the initiative in finding ways to solve problems, overcome roadblocks, and keep the work of the team flowing.

- **Conflict management.** Team members will do what is necessary to prevent counter-productive conflict from undermining the team's work. Team members who disagree over ideas and recommendations will do so without becoming disagreeable.

Project Goals/Milestones

1. Complete site work by January 15.

2. Complete forming/pouring of the slab by January 22.

3. Complete framing of the walls by February 7.

4. Complete framing of the roof by February 9.

5. Complete rough electrical work by February 15.

6. Complete window and door installation by February 15.

7. Complete roof finish by February 22.

8. Complete installation of insulation by February 24.

9. Complete installation of drywall by March 3.

10. Complete installation of exterior siding by March 10.

11. Complete interior painting by March 17.

12. Complete painting of exterior trim by March 21.

13. Complete finished electrical work by March 30.

14. Closeout project by April 15.

FIGURE 9.3 Sample team charter.

team members and to make sure they understand their responsibilities for completing the project in question. The mission statement is developed by the project manager. The goals/milestones come from the construction project or the construction schedule. The ground rules are developed by the project manager and presented to the team or with the assistance of team members.

DEVELOPING THE MISSION STATEMENT.   Developing a team mission statement was covered earlier in this chapter. This is a reminder that a well-written team mission statement is brief but comprehensive and easy to understand, even by laypeople. The example in Figure 9.3 satisfies these criteria.

DEVELOPING THE GROUND RULES.   A project team's ground rules answer the following question: As we work together to complete this project, how are we to interact with each other? There are two ways to develop the ground rules, which are as follows: (1) the project manager develops them on the basis of past experience and his or her observations of the team members or (2) the project manager provides team members with a list of potential ground rules and asks them to choose the ones they think are most important. With this approach the project manager typically asks team members to choose their top eight, nine, or ten ground rules.

Developing the ground rules for a team presents the project manager with an interesting challenge. On the one hand, in order to achieve buy-in, it is important that all team members accept the ground rules as their own—rules they established themselves. On the other hand, if the team in question is a construction team, the owner, architect, and engineers may not want to participate in developing the ground rules. Consequently, it is important for the project manager to be able to *read* team members in this regard. The approach used should be the one the project manager intuitively feels will work best with the specific team in question. On one project, it might be best to simply provide the ground rules. On another project, the team members might be open to developing the ground rules. The challenge for the project manager is to accurately determine which approach is best for the team in question.

What follows is a list of possible ground rules for team charters. The following list can be used as a guide in developing ground rules for any kind of construction team:

- *Honesty.* Team members will be open and honest with each other at all times.
- *No personal agendas.* Team members will put the team's needs ahead of their personal agendas in all cases.
- *Dependability.* Team members will conduct themselves in ways that show they can be depended on.
- *Commitment.* Team members will be committed to completing the project on time, within budget, and according to specifications.
- *Responsibility.* Team members will take responsibility for their individual performance as well as the team's performance.
- *Mutual support.* Team members will be mutually supportive in carrying out their responsibilities in the team.
- *Initiative.* Team members will take the initiative in helping the team accomplish its mission.
- *Patience.* Team members will be patient with each other as they work together to solve the problems that can occur during construction projects.

- **Resourcefulness.** Team members will be resourceful, innovative, and creative in finding ways to get the project completed on time, within budget, and according to specifications in spite of obstacles.
- **Punctuality.** Team members will be punctual for team meetings and other project activities.
- **Tolerance.** Team members will be tolerant of individual differences in team members.
- **Perseverance.** Team members will persevere in getting the job done when difficulties arise and during times of adversity.
- **Conflict management.** Team members will express their opinions, make recommendations, point out problems, and offer constructive criticism in a tactful manner. When team members disagree, they will do so without being disagreeable. In addition, team members will solve differences among themselves in a responsible, professional manner that contributes to team morale and performance.
- **Decisions.** Team members will participate in the decision-making process by offering input before decisions are made. Once a team decision has been made, all members will support it fully and do their best to carry it out, even if they do not agree with it (unless, of course, there are ethical problems with the decision).

PUTTING TOGETHER THE FINAL TEAM CHARTER.   Once the team's mission statement has been finalized and the team's ground rules selected, the team charter can be finalized. It will contain the team's mission, ground rules, and goals/milestones. Although these components should be reviewed from time to time, typically they will not change during the course of the project. However, there may be occasions when contract changes will cause corresponding changes in the team's charter. In these cases, the team charter should be revised accordingly so that it always accords with the contract.

Coaching and Team Development

Project teams should be handled in a manner similar to athletic teams when it comes to team development. Coaches work constantly on developing the skills of individual team members and the team as a whole. Project managers should do the same. Team development activities should be continual, and they should go on for the duration of the project. Developing the skills of individual team members and building the team as a whole should be viewed by project managers a normal part of the job. If the team in question is the project staff team, formal teambuilding activities may be in order. If the team is the construction team, the project manager may have to spend one-on-one time with the owner, architect, or engineers to help them become good team players.

Coaching and Mentoring

Project managers need to be good coaches for several reasons. One of these reasons is that good coaches are good mentors. This means they establish nurturing, developmental relationships with team members. Developing the team-oriented capabilities of team members, improving the contributions individuals make to the team, and helping team members become effective team players are all mentoring activities. Project managers who are effective mentors help team members by:

- Helping them develop teamwork skills
- Making sure they understand all components in the team's charter

- Helping them learn to be good team players
- Helping them understand other people and their points of view
- Teaching them how to behave in unfamiliar settings or circumstances
- Giving them insight into differences among people

Coaching and Mutual Respect

It is important for team members to respect their coach, for the coach to respect team members, and for team members to respect each other. This applies to construction teams, project staff teams, and project managers as well. The following strategies will help project managers earn the respect of their team members and team members earn the respect of each other:

- ***Trust made tangible.*** This strategy applies to all of the various types of teams that construction project managers might lead. Trust is established by: (1) setting a positive example of being trustworthy, (2) honestly and openly sharing information, (3) explaining personal motives, (4) refusing to play favorites, (5) giving sincere recognition for a job well done, and (6) being consistent in applying discipline. Doing these things will build trust and mutual respect. The project manager must set the example of doing what is necessary to build trust, but may also have to conduct one-on-one conversations with the owner, architect, engineers, and jobsite superintendent to help them see the importance of trust and how to build it. With the project staff team, the project manager can make instruction on how to build trust part of team meetings and can even conduct meetings devoted to teambuilding.

- ***Appreciation of people as assets.*** This strategy is especially important for building the project staff into a strong team. It will also help project managers gain the support and cooperation of the members of the construction team. Even in this age of advanced technology, people are still a team's most valuable asset. To perform at their best, team members must be treated like assets that can appreciate in value if properly developed over time. Appreciation for people is shown by: (1) respecting their thoughts, feelings, values, and fears; (2) respecting their individual strengths and differences; (3) respecting their desire to be involved and to participate; (4) respecting their need to be winners; and (5) respecting their need to learn, grow, and develop. Organizations often claim that people are their most valuable asset, but too few actually follow through and treat people like valuable assets. Words are not enough. Project managers must work with team members in ways that continually increase their value to the team.

- ***Communication that is clear and candid.*** This strategy applies to all of the various types of teams construction project managers might lead. Communication can be made clear and candid if project managers do the following: (1) open their eyes and ears—observe and listen, (2) be tactfully candid, (3) give continual feedback and encourage team members to do the same, and (4) confront conflict in the team directly and immediately before it can fester and blow up. People in teams want to be informed and they want to know that the information they are given is the truth.

- ***Unequivocal ethical standards.*** This strategy applies to all of the various kinds of teams construction project managers might lead. Ethical standards can be made unequivocal by: (1) adopting the organization's code of ethics at the team level or, if the organization does not have one, working with the team to develop its own code of ethics, (2) identifying ethical conflicts or potential conflicts as early as possible and acting to resolve them, (3) recognizing and rewarding ethical behavior, (4) correcting unethical behavior immediately—never ignoring it, and (5) making all members of the team aware of the team's code of ethics. In addition to these strategies, project managers must set a consistent example of living up to the highest ethical standards themselves.

Coaching and Human Diversity

America is the most diverse country in the world and this diversity is reflected in the composition of construction teams. Diversity in all of its forms—racial, gender, cultural, political, religious, and intellectual—can be an asset to project teams or it can be a liability. The difference is in how project managers and team members handle diversity. For this reason, it is important that project managers invest the time and effort necessary to make diversity an asset in their project teams.

One of the most difficult project the author ever managed was one in which the owner was racially biased and the jobsite superintendent was Asian. At the beginning of the project, the owner refused to trust anything the superintendent said or did. Because the Asian superintendent had a limited command of the English language, the owner could not accept that he was a talented professional who knew what he was doing. It took the author many one-on-one conversations and guided tours of the jobsite to finally convince the owner that the jobsite superintendent knew what he was doing. By the end of the project, the owner had made great strides in overcoming his anti-Asian bias but it took some work on the author's part.

Most of the future growth in the labor force in the United States will consist of women, minorities, and immigrants. People in these groups will bring new ideas and new perspectives to their jobs, precisely what construction firms need if they are going to stay fresh, current, and competitive. Consequently, construction project managers will need to be effective in handling diversity in ways that make it an asset for their teams. This can be a challenge.

In spite of the progress that has been achieved in making the American workplace both diverse and harmonious, some people—consciously and unconsciously—still erect barriers between themselves and those who they view as being different. These barriers can quickly undermine the trust and cohesiveness on which teamwork is built, especially when the team has a diverse membership. To keep this from happening, project managers can apply the coaching strategies listed in Figure 9.4.

**Coaching Strategies For
HANDLING HUMAN DIVERSITY IN PROJECT TEAMS**

- Identify the needs of different groups within the team (e.g., racial, gender, age, cultural, national origin).

- Confront cultural clashes directly and immediately—do not allow them to simmer below the surface until they eventually blow up.

- Eliminate cases of institutionalized bias (e.g., too few female restrooms in an organization that once had a predominantly male workforce).

- Help people find common ground.

FIGURE 9.4 These strategies can be used to prevent diversity-related problems from undermining the effectiveness of project teams.

Identify the Specific Needs of Different Groups

Project managers should ask women, ethnic minorities, and older workers in their teams to describe any unique inhibitors they face in trying to do their jobs—things they see but the project manager might not. Then make sure that all team members understand these barriers, and are willing to work together as a team to mitigate, eliminate, or accommodate them.

Confront Cultural Clashes Directly and Immediately

When diversity-based conflict occurs in teams, project managers should confront it directly and immediately. This approach is particularly important when the conflict is based on such issues as race, culture, ethnicity, age, and/or gender. Because these issues are so deeply personal to individuals, they are potentially more volatile than everyday disagreements over work-related matters. Consequently, conflict that is based on or aggravated by human diversity should be dealt with quickly and effectively. Few things will polarize a team faster than diversity-related disagreements that are allowed to fester and grow. When diversity-related disagreements are allowed to persist, more of the team's productive energy will be devoted to conflict than to completing the project in question.

Identify and Eliminate Institutionalized Bias

Construction firms that have done things a certain way for a long time can suffer from institutionalized bias. This is a situation in which the bias—although not necessarily intended—results from a failure to match changes in the composition of the workforce with corresponding organizational changes. Consider the following example. A construction that has historically had a predominantly male workforce now has one in which there are a growing number of women. However, the organization's facility still has 10 rest rooms for men and only one for women, a circumstance left over from how things used to be. This is an example of institutionalized bias. Although there is no bias intended, it exists nonetheless.

Project teams can find themselves unintentionally slighting members simply out of habit, tradition, or unwitting circumstances. When the demographics of a team change but its habits, traditions, procedures, and work environment fail to accommodate the change, the result can be unintended discrimination. Eliminating institutionalized bias is important because failing to do so will undermine a project team's morale and performance. If the bias applies more broadly than to just one team, it will ultimately undermine organizational excellence.

An effective way to eliminate institutionalized bias is to circulate a blank notebook and ask team members to record—without attribution—instances and examples of it they have encountered. After the initial circulation, repeat the process periodically. The input collected will be helpful in identifying institutionalized bias that can then be eliminated. By collecting input directly from team members and acting on it promptly, project managers can ensure that discrimination that results from organizational inertia is not creating or perpetuating resentment among personnel who need to be focused on completing the construction projects they are assigned to.

Help People Find Common Ground

People are surprisingly similar—regardless of outward differences such as race, ethnicity, gender, and native language. The hopes, fears, desires, and ambitions of people of every race, gender, national origin, and culture are not that different. In fact, they are very similar. For example, when the author graduated from Marine Corps boot camp at Parris Island, South Carolina many years ago, the proud parents in attendance represented a collage of American culture.

There were Hispanic parents of young men from the inner cities of America's northeast. There were African-American parents of young men from the rural south. There were Caucasian parents from northern and southern suburbs. Some parents were well-off economically and some were obviously poor. Others were in the economic middle class. On a normal day, these parents would probably have had no contact with each other and nothing in common—at least outwardly. But on the day their sons became United States Marines, they had a shared pride that transcended differences of language, race, and economic status. Cultural differences did not matter to these proud parents because they had something in common that was more important than their differences.

Project managers can strengthen their teams by helping members find common ground, by helping them see that they are more alike than different. The one thing they all have in common is the desire to complete the construction project in question on time, within budget, and according to specifications. They have different reasons for wanting the project to be a success, but the fact that they share this mission gives them common ground nonetheless.

HANDLING CONFLICT IN TEAMS

In even the best teams there will be conflict. Even when all team members agree on a goal, they can still disagree on how best to accomplish it. Conflict is an ever-present reality in teams, a reality that can undermine effective teamwork unless it is confronted promptly and resolved in a positive manner. Members of project teams have sufficient energy to do their work or to engage in conflict, but not both. Conflict can sap the energy of a team, cause it to lose its focus on the project, and undermine its performance.

Preventing Problems That Can Cause Conflict

The best way to handle conflict is to prevent it. Project managers will never completely eliminate conflict, but they can limit the amount of it they have to deal with by preventing the problems that cause it. The following strategies will help project managers prevent conflict:

- ***Clarify assignments.*** It is important for people in teams to understand what work is assigned specifically to them. On a construction team, the architect should not attempt to do the work of the owner and engineers should not attempt to do the work of the architect. None of the team members should attempt to do the work of the project manager. Clarity of assignments is also important in the project staff team. If it is not clear who does what, two situations—both of them bad—can arise: (1) team members get into turf battles and (2) team members are confused concerning who is responsible for completing a given assignment. These two situations almost always lead to conflict.
- ***Clarify roles.*** Although it is important for members of teams to be mutually supportive in doing their work, it is also important for them to understand their specific and

primary roles in the team. For example, on a baseball team each player has a primary role: pitcher, catcher, first base, second base, third base, shortstop, right field, center field, and left field. Although the players in these positions are responsible for backing each other up and mutually supporting each other, each is primarily responsible for his specific position. This is how teams in construction are supposed to work. On the construction team, the architect is the architect and should not attempt to be the project manager. The owner is the owner and should not attempt to be the jobsite superintendent. In order to prevent confusion and conflict, project managers must clarify the roles of all team members so that conflict does not arise over the issue of territoriality. This means that project managers must be prepared to take team members aside—including the owner—and have one-on-one discussions about roles. The author once had to deal with an owner who had just enough knowledge of construction to be dangerous but not enough to be helpful. This owner, thinking he knew more than he did, got into the habit of visiting the jobsite and getting between the jobsite superintendent and the subcontractors who were doing their work. He would countermand assignments made by the jobsite superintendent and micromanage subcontractors who were trying to do their jobs. It took several one-on-one conversations and documentation of problems he had caused to convince the owner to play his role in the construction team and let others play theirs.

- ***Encourage team members to talk through their differences.*** People in teams will have differences of opinion and perspective. This is normal and natural. In fact, it is good as long as it is not allowed to become personal and counterproductive. However, differences of opinion and perspective can easily develop into conflict. To prevent this from happening, project managers should encourage their team members to talk through their differences. The better they communicate, the less likely it is that differences will escalate into conflict. Occasionally, project manager will need to serve as mediators and facilitators to help team members talk through their differences.

- ***Encourage team members to learn how to disagree without being disagreeable.*** Frank and open discussions and even debates concerning project-related issues are good things provided team members can disagree without being disagreeable. Consequently, project managers should help team members learn that there is nothing wrong with disagreeing with each other as long as they do not become disagreeable in the process. Disagreement can be a good thing if handled well. The opinion of one team member might sharpen the opinion of another. The point of view of one team member might change the point of view of another. But these things will happen only if team members can disagree without taking things personally and becoming disagreeable. The author once had to deal with an architect who was hypersensitive about his designs. If any member of the construction team—including the owner who was paying all the bills—suggested even the slightest change, he would become defensive and even belligerent. It took a number of one-on-one meetings with the architect to convince him that the opinions of others on the team were not attacks on him or his design—just suggestions on how to make the project stronger.

- ***Handle conflict promptly.*** When project managers see conflict brewing between team members, they should deal with it right away. Putting off dealing with simmering conflict is a sure way to guarantee it will escalate and eventually blow up. The time to intercede and deal with conflict between and among team members is the moment it is noticed.

**Potential Human Responses To
CONFLICT IN TEAMS**

ESCAPE RESPONSES

- *Denial.* Pretending there is no conflict.

- *Flight.* Running away from conflict either physically or mentally.

- *Suicide.* Giving in to being overwhelmed by conflict.

ATTACK RESPONSES

- *Litigation/formal grievances.* Becoming so mired in conflict that one team member sues another.

- *Assault.* When conflict becomes violent and one team member attacks another.

- *Murder.* The ultimate violent response to conflict.

RESOLUTION RESPONSES

- *Overlook.* Ignore the negative behavior of another team member and move on because the source of the dispute is not important enough to warrant the conflict.

- *Reconcile.* Acknowledge the conflict but forgive, forget, and move on.

- *Negotiate.* The issue in dispute is important so the parties involved try to find a compromise solution that suits both through discussion and mutually-respectful give and take.

- *Mediate.* The issue in question is important and negotiations have not worked. Consequently, the Project manager is asked to "referee."

- *Arbitrate.* When mediation does not work and the project manager must serve as a "judge" and decide the issue.

FIGURE 9.5 Project managers should be aware of the various ways their team members might respond to conflict and act accordingly.

Potential Human Responses to Conflict in Teams

Conflict in project teams is, if not inevitable, highly likely. Consequently, it is important for project managers to understand the various ways in which people respond to conflict and how to ensure that their responses are team-positive and resolution-oriented. The various responses to conflict fall into one of the following categories (see Figure 9.5):

- *Escape responses.* Escape responses represent one extreme of the continuum of possible responses to conflict. Escape responses are negative because they hurt the team and the person in question. Escape responses include denial, flight, and suicide. While it is not likely that construction project managers will have to deal with suicides on the part of team members, one should not assume. Consequently, it is best for project managers to be aware of all escape responses to conflict. People who respond to conflict by trying to escape it in any of these ways are so averse to conflict that they will go to extremes to avoid it. All escape responses are harmful. Consequently, project managers should be vigilant in observing team members and in acting quickly if they notice any who might take an escape response to conflict.

- **Attack responses.** Attack responses represent the other extreme of the response continuum. They are negative in that they hurt the team, the victims, and the person who perpetrates the attack. Attack responses include litigation/formal grievances, assault, and murder. Murder and assault are not common in construction project teams, but litigation and formal grievances are. One team member filing a lawsuit or, more likely, a formal grievance against another can tear a team apart. The author was once involved in a construction project in which the owner, architect, engineers, and construction firm all filed lawsuits against each other over who was responsible for the failures that led to liquidated damages being assessed. Only assault or murder could have been more detrimental to team cohesion. When any of the attack responses to conflict happen, team members take sides or go into defensive mode. When team members take sides, the conflict spreads and becomes more intense. When they go into defensive mode, they focus only on the conflict and its potential outcomes and not on the project. Consequently, project managers must stay closely connected to team members so they can anticipate the potential for an attack response and move quickly to prevent it.

- **Resolution responses.** Resolution responses are positive in that they can lead to a resolution that is good for all stakeholders. Although all of the responses explained in this paragraph are resolution responses, it is important to understand that they are not all equally positive. Consequently, the various resolution responses are listed in order of preference: overlook, reconcile, negotiate, mediate, or arbitrate. Often the best way to resolve conflict is for those involved to simply *overlook* what brought it on in the first place and move on. This occurs when both parties realize that fighting over a certain issue is going to get them nowhere, so they agree to forget it and move on. When this does not work, *reconciliation* is the next option. Reconciliation is the forgive-and-forget response. This means that those involved shake hands, apologize, and agree to refocus on the project's mission. When this option does not work the parties in question can *negotiate* some type of mutually agreeable resolution. In this case, neither party gets everything desired, but each gives enough that both parties are willing to accept the resolution and move on. *Mediation* is called for when the other responses have not worked and a mediator must be brought in to facilitate discussions and an agreement. Mediation is a legally recognized way to settle disputes in construction projects and mediators receive special training to prepare them for this task. A mediator serves as a referee and tries to guide the conflicting parties to a resolution that is in the best interests of the team and the project. When this does not work, the only remaining positive response is *arbitration.* With this option, the arbitrator serves as a decision maker. Arbitrators are empowered to decide how a dispute will be settled and may use their authority to impose a resolution.

CONSTRUCTION PROJECT MANAGEMENT SCENARIO 9.2

All this diversity is causing conflict

Melinda Morris has been a project manager for less than a month and things are not going well. The team members she selected for her project staff for the XYZ Project are the best the construction firm has to offer in each of their respective positions. Consequently, one would think the project staff team for the XYZ Project would be effective and smooth-running. It's not. In fact, diversity-related disagreements have turned the team into a group of warring cliques that devote more time to disagreements than to the team's work. Morris is at a loss concerning what to do. When a colleague asked her how things were going with her team, Morris responded: "Not well. I have a very diverse team, and all of this diversity is causing conflict. I don't know what to do. If you have any suggestions, I would like to hear them."

Discussion Questions

In this scenario, Melinda Morris is a new project manager with a big problem. Her team has fallen into diversity-related squabbling and she does not know what to do about it. Have you ever been a member of a team in any setting where members did not get along because of diversity-related differences? If so, what problems did the disagreements in your team cause? If Melinda Morris asked for your advice concerning how to resolve the conflict in her team and how to handle the diversity issues so that problems do not recur, what would you tell her?

SUMMARY

Construction project managers either lead or serve on a variety of different kinds of teams over the course of a construction project. Prior to the beginning of construction, these teams include the construction cost estimating team, planning and scheduling team, risk management team, and procurement team. During construction, the project manager leads the construction team consisting of the owner, architect, engineers, jobsite superintendent, and project manager. He or she also leads the project staff team—which includes such positions as the jobsite superintendent, assistant superintendents, clerical/administrative staff, field engineers, and the safety director. Consequently, teambuilding and leading teams are important responsibilities of construction project managers.

Having a common purpose is essential to effective teamwork because the very basis of teamwork is a common purpose. Ensuring that all members of a team buy into its collective purpose is one of the most important duties and sometimes difficult challenges of the project manager. The first step in building a project team is developing its mission statement. The

core of a project team's mission is always to complete the project on time, within budget, and according to specifications. A complete mission statement contains three distinct elements: name of the project, description of the project, and a statement of purpose.

Fit is an important consideration when selecting the members of a team. People who do not work well with others, no matter how talented they may be in their respective professions, do not make good team members. Having team members who do not fit in can disrupt the team's work. Members who lack the characteristics of good team players can undermine the effectiveness of the whole team.

Project managers are responsible for serving as: the project team's connection with higher management, the official record keeper for the team, participants in team discussions (without dominating the discussions), implementer of team recommendations, and facilitator of positive working relationships among team members. Teambuilding proceeds in four steps: assess, plan, implement, and evaluate. For teams to be effective they must have a clear direction, members who are good team players, and accountability.

An assessment is conducted to identify strengths and weaknesses in a project team. The results of the assessment are used as the basis for developing a teambuilding plan. The plan is implemented and then given time to take effect. Finally, the team is evaluated to determine if implementing the teambuilding resulted in improvements. Teams should be coached rather than bossed. Project managers can become effective coaches by: developing a team charter for the project team (mission, ground rules, and goals/milestones), making teambuilding and team development ongoing activities, mentoring individual team members or providing mentors for them, promoting mutual respect among team members, working to make diversity an asset in the team, and setting a positive example of everything that is expected of team members.

Conflict will occur in even the best teams. Consequently, project managers must be prepared to deal with it. There are three categories of human responses to conflict: escape, attack, and resolution responses. Escape responses are denial, flight, and suicide. Attack responses are litigation/formal grievances, assault, and murder. Resolution responses are to overlook, reconcile, negotiate, mediate, or arbitrate. Project managers should work with team members to encourage resolution responses.

KEY TERMS AND CONCEPTS

Fit
Assess
Plan
Implement
Evaluate
Direction and understanding
Characteristics of team members
Accountability
Team assessment
Teambuilding plan
Executing teambuilding activities
Evaluating teambuilding activities

Team charter
Mission statement
Ground rules
Coaching and team development
Coaching and mentoring
Coaching and mutual respect
Coaching and human diversity
Institutionalized bias
Conflict in teams
Escape responses
Attack responses
Resolution responses

REVIEW QUESTIONS

1. What is the basis of effective teamwork?
2. What is the overall purpose of teams in a construction setting?
3. Explain the concept of fit and why it is important when choosing the members of teams.
4. Summarize the responsibilities of project managers in leading teams.
5. How can project managers develop positive working relationship in their teams?
6. Summarize the four steps for building effective teams.
7. What is meant by giving a team direction and understanding?
8. List five characteristics that are important for members of teams.
9. What is meant by accountability as it applies to teams?
10. Explain briefly how to assess the strengths and weaknesses of a team.
11. Explain briefly how to develop a teambuilding plan.

12. What is the difference between "bossing" and "coaching" a team?
13. What is a team charter?
14. Develop an example of a ground rule that might be part of a team charter.
15. Why is it important for project managers to confront cultural clashes in their teams directly and immediately?
16. What is institutionalized bias? Give an example.

17. List five techniques for preventing problems in teams that might lead to conflict.
18. List and explain the three categories of human responses to conflict.
19. List the various reconciliation responses to conflict.
20. Explain the difference between mediation and arbitration from the project manager's perspective.

APPLICATION ACTIVITIES

The following activities may be completed by individual students or by students working in groups:

1. Use the assessment instrument in Figure 9.2 to complete this activity. Choose two criteria from each section of the assessment instrument (e.g., two from "Direction and Understanding," two from "Characteristics of Team Members," and two from "Accountability"). Assume that these six criteria were rated very low during the team assessment. Develop a teambuilding plan for overcoming the team's weaknesses in these six areas.

2. Assume that you are a new project manager about to lead your first construction team. The team consists of the owner, architect, engineers, jobsite superintendent, and you. Only the jobsite superintendent reports to you. Develop a plan for: (1) holding your team members accountable for completing the project on time, within budget, and according to specifications and (2) preventing conflict within the team.

CHAPTER 10

Motivating Team Members

Construction project managers need to be good motivators. They must be able to motivate members of any kind of project team, especially the construction team and the project staff team. Members of the project staff team report to the project manager as their supervisor. This can have a positive effect on the project manager's ability to motivate them. Most members of the construction team—owner, architect, engineers—do not report to the project manager. This can make it more difficult for the project manager to motivate them—difficult but not impossible. The best tool project managers have for motivating the owner, architect, and engineers on the construction team is the *carrot* and the *stick*. The *carrot* is the mutual need of the owner, architect, and engineers to complete the project on time, within budget, and according to specifications. The *stick* is the negative consequences of failing to do so.

The motivational techniques presented in this chapter can be used most effectively with members of the project staff, cost estimating, planning/scheduling, risk management, and procurement teams. These are internal teams made up of personnel who either report to the project manager or, at least, work for the construction firm. Consequently, project managers are better able to create an environment and provide incentives that will motivate their team members.

Motivating others is an important skill for construction project managers because people can be surprisingly persistent and innovative at accomplishing things they are motivated to achieve. Motivating team members is about creating an environment and, in some cases, providing incentives that, together, encourage people to give their best on behalf of the mission. A highly motivated team member can be determined and persistent when it comes to pursuing the team's mission, even in the face of roadblocks, setbacks, and adversity. Highly motivated team members will typically find a way to get the job done, including working together in mutually supportive ways, even when they do not especially like each other. Consequently, motivating team members to adopt the team's mission of completing the project on time, within budget, and according to specifications as their own is a worthwhile endeavor for construction project managers.

The short-term goal of motivation is to encourage team members to do their best to help complete the projects they are part of on time, within budget, and according to specifications. The long-term goal of motivation is to help team members become self-motivated. Encouraging self-motivation is important because self-motivated team members are more valuable to project managers than those who have to be continually motivated and remotivated. Project managers can create a motivational environment and provide incentives, but in the long run team members have to motivate themselves. Fortunately, under the right circumstances and with proper leadership, most will motivate themselves. When team members become self-motivated, striving for successful completion of the project becomes normal behavior. When a team member is self-motivated, doing his or her best to help accomplish the team's mission becomes a personal goal. Self-motivated team members require little prodding, convincing, or supervision and can be depended on to do the right thing for the project when problems arise, and thus, decisions must be made.

MOTIVATION DEFINED

When it is said that people are motivated, it means that they are driven to do something. This drive can be internal, external, or a combination of both. The ideal state is for motivation to be internal. When construction project managers speak of motivation, they mean the drive in team members to strive for the successful completion of a project. A highly motivated member of a project team will consistently strive to do what is best for the project in all situations. Team members who do this are self-motivated. Hence, developing self-motivated team members should be a goal of construction project managers.

The goal of developing self-motivated team members can be accomplished by project managers who are willing to do the following: (1) provide a positive example consistently, (2) create an environment that encourages doing the right thing for the project, and (3) provide incentives that encourages a wholehearted commitment to completing the project on time, within budget, and according to specifications.

When the people being motivated are members of the project staff team, the performance appraisal form can be an excellent motivational tool for the project manager. Often, the performance appraisal form used by project managers to evaluate members of the project staff team will contain a criterion that asks how well an individual works without supervision. This is an important criterion because project staff teams need members who will do their best, not just when project managers are watching, but also when they are not. Team members who need close and constant attention can take up so much of a project manager's time and energy that there is little left for other responsibilities. When this happens, project managers are no longer leading, managing, guiding, or assisting. They are babysitting. Having to hover over team members to ensure that they are working is counterproductive and can undermine a project manager's effectiveness.

MOTIVATIONAL CONTEXT

When learning specific motivational techniques, it helps to view them in a context that allows for systematic study. Abraham Maslow established a workable context for motivational techniques with his well-known *Hierarchy of Human Needs*.[1] Maslow posited that people are motivated by basic human needs that can be categorized according to level—from the lowest level of need to the highest level. An important point to understand about motivating team members is that there are two sides to the equation. One side involves proactively doing things that motivate them. The other involves eliminating factors that undermine their motivation—factors that are demotivators.

When studying the motivational techniques explained in this chapter, remember that there will be times when the best way to motivate team members will be to remove demotivators. Demotivators are factors that rob team members of their drive to do what is necessary to complete the project on time, within budget, and according to specifications. All motivational techniques presented in this chapter are explained within the context of Maslow's hierarchy of needs. The levels of human need set forth by Maslow from lowest to highest are as follows:[2]

- Basis survival needs
- Safety and security needs
- Social needs
- Esteem needs
- Self-actualization needs

Maslow posited that in any situation people will focus on their lowest unmet needs. Although his work concerning human needs applies to life in the broader sense, it can be applied specifically to construction teams. Maslow's five levels of human needs provide a context for systematically learning techniques that can be used for externally motivating members of project teams. Team members who are externally motivated to give their best efforts to the project will experience the rewards that come when projects milestones are completed successfully. Once they experience the rewards that are the result of meeting project milestones, team members will be more likely to become self-motivated.

BASIC SURVIVAL NEEDS AND MOTIVATION

The basic survival needs of a human being include air, water, food, clothing, and shelter (see Figure 10.1). In Maslow's hierarchy, these are the lowest human needs. Unless these basic needs are satisfied, people will focus on little else but these basic unmet needs. For example, in the days before a natural disaster such as a hurricane, tornado, or earthquake, stakeholders in construction projects are focused on such concerns as meeting project milestones, staying

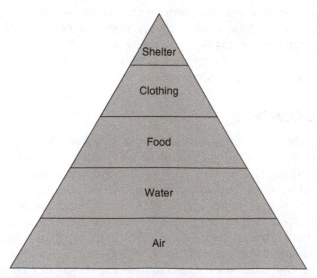

FIGURE 10.1   The most basic survival needs of humans.

within budget, and meeting specifications. However, when their basic survival needs are threatened by a natural disaster or any other factor, these work-related concerns suddenly go by the wayside. The focus of people in the aftermath of a natural disaster becomes basic survival needs, and what motivates them is anything that will contribute to meeting those needs.

At first glance, it might seem that basic survival needs have nothing to do with an individual's job performance, but a closer look will reveal direct ties between an individual's performance at work and the quality of the air, water, food, clothing, and shelter available to that individual and his family. This is because how well people perform their jobs has a lot to do with how secure their jobs are. There are two sides to the job security issue: (1) people who perform their jobs well are more likely to keep their jobs and (2) by performing their jobs well people help the construction firm to remain competitive which, in turn, means the firm will continue to have jobs to offer them.

Job security is important to people because without jobs they have no way to provide the basic needs for themselves and their families. This tie between job performance and basic survival needs can be used to help motivate members of the project staff team and the various preconstruction teams—cost estimating, planning/scheduling, risk management, and procurement. People who understand basic survival needs as part of Maslow's hierarchy are more likely to appreciate the significance of performing well on their jobs.

Motivational Technique Based on Basic Survival Needs

Project managers need to understand that the key to using basic survival needs to motivate members of the various types of internal teams on construction projects—cost estimation, planning/scheduling, risk management, procurement, and project staff—is found in helping them make the connection between their jobs and providing for these needs. More specifically, it involves helping members of teams understand the connection between performing well on a given project and their job security.

Some people take life's basic needs for granted. As long as they have a good job, they give little or no consideration to basic survival needs such as air, water, food, clothing, and shelter. But people who have experienced the debilitating effects of long-term unemployment have a different perspective. They know firsthand how being without a job can change their circumstances and threaten their basic survival needs. Consequently, using the basic survival needs to motivate works better during challenging economic times than during times of economic prosperity.

During the great recession that began in 2007, many people who had been corporate executives and highly paid professionals one day found themselves in the unemployment line the next. People who had never had to do without anything suddenly found themselves worried about securing the most basic of necessities. This connection between employment and life's basic needs can be used to motivate team members if it is used appropriately. Using basic survival needs to motivate is simply a matter of helping members of teams see, in a nonthreatening but factual manner, the connection between basic needs and their jobs. Team members who understand the connection are less likely to take their basic needs, their jobs, and their performance on projects for granted.

Talking about the connection between basic human needs and performing well on a given project during team meetings and in one-on-one conversations can enhance the value of having a job and performing well on projects in the eyes of team members. Further, it can eliminate the tendency of people to take life's basic needs and their jobs for granted. An effective and nonthreatening way to initiate conversations about the connection between performing well on a project and basic human needs is to explain the connection between performing well and getting future contracts. Just as general contractors make a determination concerning whether a subcontractor is a *responsible* subcontractor, owners and architects make the same determination about construction firms.

Whether or not a constructional firm is considered responsible by owners who need something built depends to a large extent on the performance of the firm's personnel and teams. The key to helping team members make the connection without feeling threatened is being tactful. Think of tact as driving in the nail without breaking the board. Of course, connecting basic survival needs to how well team members perform on a given project works better during difficult economic times than during times of prosperity. It also works better in construction firms whose personnel understand that the firm depends on repeat business from satisfied owners for its survival. Helping team members understand this is the project manager's responsibility.

CONSTRUCTION PROJECT MANAGEMENT SCENARIO 10.1

My team members just don't seem to get it

Max Renfroe was delighted when he was selected as the project manager for the huge Delta Project for his company, off-base housing for a large Air Force base. The Delta Project represented the largest contract his company had ever won. If his team did a good job, there was no question there would be future contracts from the Air Force at other bases, not to mention the boost it would give to his career. This was the good news. The bad news was that things weren't going very well with his project team. The best word Renfroe could come up with to describe the problem was "complacency." It was as if his team members took their jobs and the future of the company for granted. In discussing his concerns with another project manager in the firm, Renfroe said: "My team members just don't seem to get it. There is no urgency on their parts to show the Air Force that they can do a good job on the Delta Project." Renfroe feels like he is the only person on the team who cares about getting the project done on time, within budget, and according to specifications.

Discussion Questions

In this scenario, the project manager understands the need for the team to perform well but his team members do not seem to share his sense of urgency. Have you ever worked in a situation where the people involved seemed to take things for granted as Renfroe's team members apparently do? What can Renfroe do to motivate his team members to take their responsibilities more seriously?

SAFETY AND SECURITY NEEDS AND MOTIVATION

The next level of human needs in Maslow's hierarchy is safety and security. Safety and security needs include safety from physical harm and security from crime, health problems, and financial adversity. For most people, these needs relate directly to their employment. By working, people are able to earn the income necessary to provide a measure of safety and security for themselves and their families. The better the job and the higher the income, the

---

JOB SECURITY CHECKLIST

✓ When individuals in a construction team perform well, the construction firm is able to perform better.

✓ When a construction firm performs well, it is more competitive.

✓ When a construction firm is competitive, it is better able to provide job security for its personnel.

FIGURE 10.2 Individuals in construction firms contribute to their job security.

more the individual can invest in safety and security. A home in a safe neighborhood, a security system, financial investments, quality health care, and various types of insurance are all things people use to gain a measure of safety and security for themselves and their families. Of course, all of these things cost money.

Most people must work to earn the income necessary to provide a measure of safety and security for themselves and their families. Consequently, for working people job security is a major concern. The correlation is simple: no job—no security. This fact allows project managers to use job security as another tool for motivating their team members. The key to using safety and security needs to motivate team members is helping them understand the role their individual performance can play in securing their own job security.

In order to provide jobs for people, construction firms must be able to compete in their markets, and win the competition on a consistent basis. For some, the competition is local. For others it might be regional, national, or global. Regardless of the nature of the competition, all construction firms face the daily challenge of having to outperform the competition in order to survive. Employees in construction firms can contribute much to their own job security, a fact that can be used to motivate them.

People place a high value on job security because their job is the principal vehicle for satisfying their other security needs. This is why project managers must make sure that their team members understand how their individual performance affects the overall performance of the firm. The connection that members of teams need to understand is as follows: (1) the better they perform, the better the firm performs; (2) the better the firm performs, the more competitive it is; and (3) the more competitive the firm is, the more job security it can offer its employees (see Figure 10.2).

Project managers should never make the mistake of assuming that the connection between employee performance and employee job security is intuitively understood by their team members. This is a bad assumption. Not only do construction firms have employees who do not understand the connection, but such employees are surprisingly common. However, once employees understand the connection between their individual performance and their personal job security, the connection will be motivational. Consequently, explaining the connection between employee performance and job security can be an excellent motivational technique.

SOCIAL NEEDS AND MOTIVATION

Project managers should always remember the simple fact that humans are social beings. It is part of the nature of most people to want and need relationships with others. The social needs of people include having positive relationships with family members, friends, and colleagues as well as satisfying the natural human desire to belong, to be part of a group. An individual's job can go a long way toward meeting these needs. On the other hand, if a job fails to help meet these needs, the job can become 40 hours a week of high stress and drudgery. Many of the relationships in people's lives are tied to their jobs. In addition to providing for relationships, a job can also help satisfy an individual's desire to belong to a group.

Positive relationships with other team members and the project manager are important to people in construction firms because these relationships help satisfy their social needs. People who have negative relationships at work often experience low morale and high stress. Low morale and high stress, in turn, rob people of their motivation to strive for excellence. Positive relationships, on the other hand, can help motivate employees to perform at their best. Project managers who understand the importance of positive relationships with their team members can use relationship building as a motivating strategy.

Using Relationships to Motivate Employees

For project managers, relationship building has the following three components: (1) establishing positive working relationships with team members, (2) facilitating relationship building among team members, and (3) repairing damaged or broken relationships (see Figure 10.3). Project managers can establish positive working relationships with their team members by doing the following:

- Communicating with them often and well
- Being good listeners
- Being honest and trustworthy
- Encouraging input and feedback from team members
- Being fair and equitable
- Treating team members with respect
- Being a positive and consistent example of the behavior expected of team members
- Sharing the same challenges, burdens, and circumstances team members face
- Recognizing team members for doing a good job
- Caring about team members and the work to be done
- Being an advocate for team members with higher management
- Forthrightly apologizing when wrong

In addition to building positive relationships with their team members, project managers must be attentive to facilitating the establishment and maintenance of positive relationships among team members. This can require the application of both teambuilding and conflict management skills. Both of these topics are covered in detail elsewhere in this book. At this point,

---

Relationship Building Checklist
FOR PROJECT MANAGERS

✓ Establish positive working relationships with all members of the team.

✓ Help team members develop positive working relationships among themselves.

✓ Help repair damaged or broken relationships within the team.

FIGURE 10.3 Relationship building can be an effective motivational strategy for project managers.

what is important to understand is that positive peer relationships at work motivate while negative peer relationships demotivate. Creating opportunities for team members to get to know each other as people and interceding when necessary to resolve conflicts will help motivate team members while, at the same time, minimizing the demotivating effects of conflict.

Other relationship-oriented techniques that can be used to motivate team members grow out of the concept of *affiliation*. Some people are achievement-oriented, while others are affiliation-oriented. Affiliation-oriented people are motivated by feeling as if they belong to some type of defined group such as a team. This is one of the reasons college students join the various student organizations on campus. They are motivated by the positive association with others and, in the case of teams, by the esteem they receive from helping the team perform well. Affiliation-oriented people are natural team players. Every team member can and should learn to be a good team player, but affiliation-oriented people are natural team players—they do not have to learn the concept.

The needs of affiliation-oriented team members can be used to motivate them by applying the following techniques:

- Provide opportunities for social interaction with their teammates (e.g., after-hours social events).
- Ask affiliation-oriented employees to help other team members improve in specific areas.
- Ask affiliation-oriented employees to keep their fingers on the pulse of the team's morale and share any problems they think might be developing.

A word of caution is in order here. Affiliation-oriented team members will sometimes choose team harmony over team performance. Consequently, project managers should make a point of talking with affiliation-oriented team members one-on-one and reminding them that going along to get along will rarely improve team performance and that a happy team is not necessarily a productive team.

ESTEEM NEEDS AND MOTIVATION

People have an inherent need for self-esteem as well as for the esteem of others. Esteem relates to the self-respect, worth, and dignity people feel. People with self-esteem feel good about themselves. They have self-respect and self-worth. Typically, people who lack self-esteem also lack self-respect and self-worth. A lack of self-esteem can result in feelings of inadequacy, and such feelings can be powerful demotivators. Consequently, as with most motivational techniques, when using esteem needs to motivate team members it is important to remove demotivators.

In helping team members build self-esteem and win the esteem of their peers, a variety of techniques are available in the following areas: (1) achievement, (2) competition, (3) the potential for promotions, (4) incentives, and (5) legacy (see Figure 10.4). All of the techniques explained in the following paragraphs can help team members build self-esteem and win the esteem of peers.

Using Achievement Activities to Motivate

Achieving something important can give team members a sense of accomplishment. Accomplishing tasks that are both important and difficult will build self-esteem. Many people have

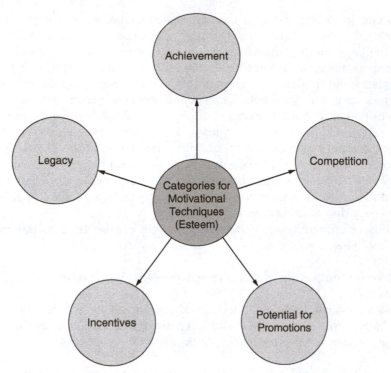

FIGURE 10.4 Esteem-based motivational techniques fall into these five categories.

an inherent need to achieve, although some do not. Because there are people who are not motivated by achievement, the first step in applying the techniques explained in this section is to identify those team members who are achievement-oriented. Such people are usually easy to recognize. They tend to be task- and goal-oriented. They typically need continual positive reinforcement and focus intently on evaluations of their performance. Achievement-oriented people like to collect physical evidence of their achievements, such as certificates, trophies, plaques, and other recognition memorabilia.

The message conveyed to project managers is that recognition is important to achievement-oriented people. Project managers who understand this can use the following techniques to motivate achievement-oriented employees:

- Tell them specifically what they can do to help accomplish the team's mission.
- Put them in charge of specific high-priority tasks. Let them know what needs to be done and give them a deadline. Then step back and give them room to work. Do not micromanage achievement-oriented personnel.
- Recognize them whenever they accomplish a goal or a specific assigned task. The recognition need be nothing more than a public pat on the back from time to time. The important aspect of this technique is the public aspect. Achievement-oriented team members need to be recognized publically. Public recognition in the presence of their peers and others whose opinions they value is part of what achievement-oriented team members seek.

A word of caution is in order when using achievement to motivate. Achievement-oriented team members who appear to be self-serving and more interested in personal glory than the team's best interests can damage the morale of other team members. Affiliation-oriented team members are especially sensitive to self-serving behavior in others. To get the most from achievement-oriented team members without damaging team harmony, project managers should confront the issue head on.

Talking with achievement- and affiliation-oriented team members one-on-one and as a group and explaining the contributions each can make to team performance is an effective approach. Explaining how achievement- and affiliation-oriented team members are both needed as well as how they can complement each other will help get the best from both groups. Reminding achievement-oriented team members to thank those who supported them whenever they are recognized will help maintain team harmony. It also helps if project managers can maintain a balance between achievement- and affiliation-oriented people on project teams to the extent this is possible.

Using Competition to Motivate

Most people like to win personally and to be affiliated with a winning team. Winning builds self-esteem. Watch children as they play games. They like to outperform other children who challenge them. A child's competitive spirit is nurtured through play and sometimes through participation in organized sports. Project managers can use the natural competitive spirit many people have to motivate team members, but caution is the order of the day when doing so. In order to have the desired effect, competition must be carefully planned, closely monitored, and scrupulously controlled.

Competition that is not properly planned, monitored, and controlled can go awry and result in a win-at-any-cost mentality becoming pervasive. When this happens, team members can lose sight of the real goal of doing what is necessary to complete the projects they are assigned to on time, within budget, and according to specifications. If winning at any cost becomes the goal instead of peak performance, undermining the work of other team members becomes an effective strategy. Out-of-control competition can lead to cheating and hard feelings among team members. If this is allowed to happen, the competition will do more harm than good.

The following tips will help project managers ensure that competition is properly controlled and that it contributes to improved performance:

- Involve the team in planning the competition, and explain to them that making sure the project is completed on time, within budget, and according to specifications is the purpose of the competition. Be frank in letting team members know that behavior that undermines another member's performance is unacceptable and will only hurt them and everyone else on the team.
- To the extent possible, plan the competition so that it is between teams rather than individuals within team. Competition between individuals can quickly become personal and counterproductive. Ideally, the competition would be against another team's record that is already established. For example, say that team XYZ completed a construction project two weeks ahead of schedule, 10 percent below budget, and in accordance with all applicable criteria in the specifications. Beating this record would be an excellent basis for the competition.

- Make sure the competition is as fair as it can possibly be. For example, do not ask a team of rookies to compete against a team of experienced members or the record of such a team.
- Be specific in selecting the basis of the competition—decide what will actually be measured. With teams, this will usually mean competing in such areas as beating deadlines and budget projections. Of course, meeting or exceeding specifications is the first and essential criterion. A team that beats a scheduled deadline but does not fulfill customer specifications has achieved nothing.

This section is not intended to argue against using competition to motivate. Rather, it is provided to remind project managers to take human nature into account when using competition. Once the competition has been planned, but before starting the competition, conduct a *roadblock analysis.* This is done by brainstorming with other project managers and team members about what roadblocks or unintended consequences might arise that could derail the competition and undermine performance. Once all potential roadblocks and unintended consequences have been identified, find ways to eliminate, mitigate, or control them before beginning the competition.

Using the Potential for Promotions to Motivate

People in construction firms are just like most others in that they want to advance their careers. Ambitious, hard-working people hope that over the course of their careers they will be promoted to increasingly higher levels and higher paying jobs. When this happens, it builds self-esteem and helps earn the esteem of colleagues. Consequently, opportunities for promotions can be used to motivate team members who are career-minded.

However, like most motivation strategies, promotions can have either a positive or negative effect depending on how they are handled. The two basic approaches to filling vacant positions in construction firms are promoting from within and hiring from outside the firm. Of the two, promoting from within is the approach most likely to be a motivator, provided the promotion process is handled properly. This can be important for project managers because they are sometimes asked to make promotion recommendations for personnel who have served on one of their teams.

The following rules of thumb will help ensure that when project managers make recommendations for filling vacant positions by promoting from within, the result is positive (Figure 10.5):

- ***Never promote solely on the basis of seniority.*** Experience is important, but it does not necessarily make an individual the best qualified person for an open position. Seniority is a legitimate factor to consider when making promotions, but it should not be the only factor. It is better to use seniority as a tie breaker when the other factors considered are equal. If an individual with seniority is promoted over a less senior but more qualified person, morale will suffer and the promotion process will be a demotivator.
- ***Do not promote on the basis of popularity.*** Personal popularity is no guarantee of effectiveness in a new and higher level position. It is not uncommon for a team member to be popular for reasons that have nothing to do with performance. Even if an

---

Checklist for
USING POTENTIAL PROMOTIONS TO MOTIVATE

✓ Do not promote solely on the basis of seniority.

✓ Do not promote on the basis of popularity.

✓ Do not promote on the basis of friendship.

✓ Do make performance the key factor in promotion decisions.

FIGURE 10.5 Project managers should base promotion recommendations primarily on performance.

individual is well-liked, he or she will still have to be able to do the new job and do it well. If the popular individual is not able to perform the new job effectively, he or she will not be popular long, and the credibility of the promotion process will suffer. If this happens, the promotion process will be a demotivator.

- ***Do not promote on the basis of friendship.*** Friendship should never be the reason behind an employee promotion. Promotions that are viewed by other team members as being influenced by friendship are doomed from the outset. Further, allowing friendship to influence promotions will not only serve to demotivate employees, but it will also undermine the promotion process.

To ensure that the possibility of being promoted is a motivator, project managers must tie promotions to performance. Team members must know that if they consistently meet or exceed performance expectations, they have a realistic chance of being promoted. If they believe the process is tainted by seniority, popularity, or friendship, the promotion process will be a demotivator.

Using Incentives to Motivate

Like most other organizations, one of the most common topics of informal conversation among people in construction firms is money or, more specifically, salaries and wages. There is more than just economics to an individual's paycheck. There is also the issue of esteem. Generally speaking, the more people earn the more they are esteemed by their peers and the greater their self-esteem. One can certainly debate the advisability of tying self-esteem to income, but there is no question that people do it.

One way for people in construction firms to increase their income is by earning incentive pay. For those motivated by money, incentive pay can be doubly motivating. First, there is the obvious motivation of receiving the extra money. But, this is not the only motivational benefit of properly managed incentive programs. Incentives also give people opportunities for achievement. Just the fact that people receive incentives over and above their normal pay—irrespective of the actual amount—can be a motivator because doing so represents an achievement. This is one of the main reasons that achievement-oriented people typically respond well to incentives.

In order to gain the motivational benefits of incentives, construction firms must plan and manage incentive programs carefully. Poorly planned and managed incentives can quickly become demotivators. Consequently, project managers' firms must be prepared to

---

Checklist for
MOTIVATING WITH INCENTIVES

✓ Define the objectives of the incentives.

✓ Set a positive example of the incentivized behaviors.

✓ Award incentives to teams rather than individuals.

✓ Involve team members to ensure that incentives are meaningful to them.

✓ Establish specific performance criteria.

✓ Communicate the objectives of the incentives to team members as well as the performance criteria.

FIGURE 10.6 Project managers can use these strategies to make incentives effective.

provide input that will help bolster the effectiveness of incentive programs. The following strategies can help make incentives programs effective (Figure 10.6):

- ***Define the objectives.*** The overall purpose of an incentive program is to motivate people to higher levels of performance. This should be understood by project managers as well as by all personnel who might have opportunities to earn the incentives. But just stating this overall purpose is not enough. Project managers must ensure that the firm takes the next step and defines specifically what is to be accomplished. Does the firm want to improve productivity? Quality? Safety? All of these? In addition to making people aware of the overall purpose of the incentive program, it is important to make them aware of the specific objectives.
- ***Set a positive example.*** By offering incentives, construction firms establish high performance expectations for members of project teams. It follows then that project managers must make a point of exemplifying the behaviors that are expected of their team members. Project managers should always remember that one of the most fundamental principles of leadership is to set a positive example. Project managers do this by exemplifying the behaviors they expect of employees.
- ***Award incentives to teams whenever possible.*** Awarding incentives to teams can be more effective than awarding them to individuals. Project managers should always be prepared to make this case with higher management. Construction work is done in teams. In all of the different kinds of construction teams, people depend on each other to get the project completed according to specifications, within budget, and on time. Because of this interdependence, team members who have contributed to the team's success might, understandably, resent just one member receiving incentives. If this happens, the incentive program can backfire and do more harm than good.
- ***Make incentives meaningful.*** For an incentive program to be effective, the incentives offered must be meaningful to potential recipients. Giving team members rewards they do not value will not produce the desired results. This is a critical point that project managers must be prepared to make when involved in designing incentive programs for construction firms. An effective way for ensuring that incentives are meaningful to people in construction firms is to involve them in developing the list of

incentives that will be made available. The people who are to be motivated by incentives know better than anyone else what types of incentives will motivate them and what types will not. In other words, if you want to know what types of incentives will motivate team members, ask them. Then use this feedback to develop the list of actual incentives that will be made available.

- **Establish specific criteria.** If the purpose of the incentives is to encourage peak performance—beating scheduled deadlines and budget projections, productivity, quality, safety—there must be specific criteria for measuring performance. On what basis will incentives be awarded? Specific criteria define the levels of performance that are to be rewarded. For example, assume that one of the areas to be measured is the project's budget. The incentive goal relating to budget is to complete the project in question at least 10 percent below budget. Project managers might ask that incentive dollars be earned for savings that exceed 10 percent on the basis of a certain amount per percentage point. It is important for team members to know specifically how performance improvements will be measured. This means establishing benchmarks against which performance will be measured and establishing intermediate as well as overall performance targets.

- **Communicate, communicate, communicate.** Team members must be completely informed about the incentive program if the incentives are going to motivate. Team members should know the purpose of the incentive program, its specific objectives, the performance criteria, when incentives will be awarded and how often, and anything else that will help maintain the programs' effectiveness and credibility. With incentive programs, there should be no surprises and no confusion concerning the details.

Using the Legacy Question to Motivate

People in construction firms go through stages in their careers. In each stage, they tend to focus on different concerns. For example, when young people first begin their careers, money is one of the most important of their work-related concerns. This is the *income phase* of their career. People in this phase are concerned primarily with income because they are at the bottom of the pay scale in their respective positions and are struggling to get established financially. People who have just begun their careers are often shocked to learn how much it costs to buy or rent a home and pay for utilities, groceries, and upkeep. This is why they tend to focus so intently on income and why they are typically good candidates for being motivated by monetary incentives.

Once people reach the point in their careers where they are past the initial shock of how much it costs to live from day to day, they begin to focus on whether or not they like their jobs. This is the *personal satisfaction* phase of their career. People in this phase want their job to provide them with a sense of personal satisfaction and enjoyment. Of course, all jobs have their good days and bad days, but people in this stage want to like their jobs, at least generally speaking, after balancing the good days with the bad ones. In this phase, they are concerned less with money because they are earning a sufficient income and more with whether or not they enjoy their jobs. For example, there are many people who hate their jobs even though they make plenty of money. On the other hand, there are people who love their jobs in spite of earning less than they would like in those jobs.

In the final phase of their careers, people who are relatively satisfied with their income and generally like their jobs begin to focus on the *legacy question*. The legacy question grows out of a need people have to know that their work matters—that their work is important, that it has meaning, and that it allows them to make a difference. For most people, knowing that their work is important builds self-esteem. Knowing that others think their work is important adds to the esteem. When they believe that their work matters, people are more likely to believe that they matter. Philosophers and theologians will argue that people have value irrespective of their work—and they certainly do. However, there is no getting around the fact that people tend to tie their self-worth to the relative worth of their jobs. People with this perspective consider their life's work a major part of their legacy.

This human desire to matter and to leave a worthwhile legacy can be used to motivate team members who are in the *legacy phase* of their careers. Legacy-minded people will work harder and smarter when they believe their work matters. Consequently, project managers can motivate legacy-minded team members by helping them see the importance of their work. This is not a difficult task for project managers in construction firms. Construction professionals, technicians, and support staff build things that improve the lives of individuals as well as the quality of life in communities. Whether they work on residential, commercial, industrial, or infrastructure projects, people in construction make a significant difference. The author still finds it rewarding to drive around the community and see parking decks, malls, office building, stadiums, retail stores, and other projects he managed. Project managers can use this legacy-related fact to help motivate team members.

SELF-ACTUALIZATION NEEDS AND MOTIVATION

Self-actualization refers to the human need to achieve one's full potential. In order to achieve self-actualization, people must first satisfactorily meet all of their other needs: basic survival, safety/security, social, and esteem. In reality, few people ever reach the level of self-actualization. However, with self-actualization the pursuit may be more important than the accomplishment, at least from the perspective of motivation. The U.S. Army was appealing to the human need for self-actualization when its recruiting slogan was: "Be all that you can be."

The key to using self-actualization to motivate is the concept of potential. Project managers in construction firms can use the concept of potential to motivate team members who want to climb the career ladder as well as those who want to broaden their career horizons. Broadening one's career horizons in the current context refers to learning new career skills or even a new job through cross-training. The key to using potential to motivate is to tie both concepts—advancement and expansion—to performance. Those who consistently perform at peak levels must be the ones who advance the fastest in the firm and who are given opportunities to expand their horizons by learning new job skills. Project managers use potential to motivate by letting their team members know that the team members they will recommend for advancement and broadening opportunities are those that exceed performance expectations.

The fact that advancement and expansion should be tied to performance would seem to go without saying, and it should. However, there are construction firms that tie advancement and expansion to nonperformance factors such as seniority. Further, there are cases in which advancement and expansion decisions are influenced by such factors as friendship, who knows whom, and favoritism. Self-actualization needs can be used to motivate only if the potential for advancement or career broadening is tied directly to performance.

DEVELOPING PERSONAL MOTIVATION PLANS

There is no one-size-fits-all strategy project managers can use to motivate their team members. Because motivation is based on appealing to individual needs—needs that vary from person to person—project managers must be prepared to personalize their motivational strategies. What will motivate a given team member depends on where that person is in Maslow's Hierarchy of Needs as well as other factors specific to the individual. This is why project managers must get to know their team members well enough to understand where they fit into Maslow's hierarchy and to be aware of other motivation-related factors affecting them at any given time.

An effective approach for personalizing motivation strategies is to develop Personal Motivation Plans (PMPs) for individual team members. A PMP is a brief plan containing strategies for motivating a project team member that takes into account that individual's specific human needs. If the individual in question is achievement-oriented, his or her PMP should be based on meeting those kinds of needs. If the individual is concerned about his legacy, the strategies in the PMP should be based on meeting legacy needs. The key is to personalize the strategies in the PMP for the individual in question rather than applying the same one-size-fits-all strategies for everyone on the team.

ADDITIONAL MOTIVATION STRATEGIES

In addition to the motivational strategies explained earlier in this chapter, there are several strategies that have proven to be effective in helping motivate members of project teams. Project managers may wish to try the following strategies for motivating their team members:

- *Explain the benefits of the project in their terms.* With this strategy, project managers explain why it will be good for the team and its members to complete the project on time, within budget, and according to specifications. If doing so might result in additional contracts, higher profits, or any other benefit, explain how these things will also benefit the team members (e.g., better job security, potential for incentive bonuses, possible raises, and promotions in the future).
- *Eliminate the fear factor.* When first beginning a project, team members may feel overwhelmed. The project might appear to involve more work than they can possibly get done. When this is the case, the fear factor must be eliminated. This can be done by making the project appear more feasible by: (1) breaking it into smaller subprojects; (2) sharing the schedule with team members and discussing it; (3) allowing team members to voice their concerns, ask questions, and make suggestions; and (4) calmly expressing confidence in the team's ability to get the project done on time, within budget, and according to specifications.
- *Give continual feedback.* Team members tend to focus on their individual assignments when working on a project. They do not always see the big picture and how the overall project is progressing. Giving them continual feedback on how the overall project is progressing can reassure team members and help them see that the team can actually complete the project on time, within budget, and according to specifications.
- *Recognize the work of team members continually in real time.* Project team members are just like anyone else in that they need to know how they are doing and how they need to be encouraged. Both of these needs can be satisfied by providing

continual informal recognition to team members and the team as a whole. For example, sending e-mail notes of appreciation to individual team members and the team as a whole, giving public pats on the back to team members, and nominating team members for various types of rewards and bonuses are all effective techniques for encouraging team members. The key is to do these things in real time rather than waiting until the project is completed.

A prerequisite to motivating people in teams is to develop an understanding of human needs. People are motivated by actions that meet their specific individual needs. Although the specific needs of individuals differ, there are generic needs common to most people. Maslow summarized and categorized these generic human needs in his hierarchy of needs. Project managers who learn to tie the needs Maslow identified to motivational strategies can become effective at motivating their team members to perform at their best.

CONSTRUCTION PROJECT MANAGEMENT SCENARIO 10.2

I don't know how to motivate these people

Marie played organized sports from the time she was a child right through college. In fact, she was an All-American volleyball player and team captain at Northwest Florida Institute of Technology where she earned her construction management degree. Consequently, Marie knew plenty about motivating team members. At least she thought she did, but lately she has begun to have doubts. The motivational techniques that always worked so well with her volleyball teammates do not seem to be working with the members of her project team.

The XYZ Project is her first as a project manager, and Marie is determined to make a good impression on her company's vice president. But the motivational techniques she is using do not seem to be having the desired effect. The various members of her project staff team just do not respond the way her volleyball teammates from college did. Marie is at a loss concerning how to get her team motivated.

Discussion Questions

In this scenario, Marie is using the one-size-fits-all approach in trying to motivate her team members and it is not working. Have you ever been involved with a team in which the coach or team leader used the same motivational techniques for everyone? If so, describe how that worked out. If Marie asked for your advice concerning how to motivate her team members, what would you tell her?

SUMMARY

When it is said that people are motivated it means they are driven to do something. Motivation can be internal, external, or a combination of both. When project managers speak of motivation, they mean the drive that team members have to do their best to ensure that projects are completed on time, within budget, and according to specifications. Ideally, members of teams will be or become self-motivated. A good context for motivational techniques is Abraham

Maslow's Hierarchy of Human Needs. The levels of human need set forth by Maslow from lowest to highest are basic survival, safety and security, social, esteem, and self-actualization needs.

Basic survival needs include air, water, food, clothing, and shelter. These needs can be used to motivate team members by tying them to their jobs and how well they perform them. Safety and security needs include safety from physical harm and security from crime, health problems, and financial adversity. Again, these needs can be used to motivate team members by tying them to their job security and job performance. Social needs include having positive relationships with family members, friends, and colleagues. By facilitating positive relationships among team members and by repairing damaged relationships in the team, project managers can use the social needs to motivate.

People have an inherent need for self-esteem and for the esteem of others. Esteem relates to the self-respect, worth, and dignity people feel. Project managers can use achievement, competition, the potential for promotions, incentives, and the legacy question to build self-worth and self-esteem in team members. This, in turn, will motivate them. Self-actualization refers to the human need to achieve one's full potential. "Potential" is the key to using self-actualization needs to motivate team members. When team members believe that they have the potential to move up in terms of salary, benefits, positions, and perquisites based on performance, they will be motivated by the fact. However, if they think that moving up in the organization is based on nonperformance factors such as seniority and friendship, the result will be demotivation and a loss of morale.

There is no one-size-fits-all strategy project managers can use to motivate team members. Because motivation is based on appealing to individual needs—needs that vary from person to person—project managers must be prepared to personalize their motivation strategies. An effective approach for personalizing motivation strategies is to develop Personal Motivation Plans (PMPs) for team members. A PMP is a brief plan containing strategies for motivating an individual team member and it takes into account that individual's specific human needs.

KEY TERMS AND CONCEPTS

Motivation
Motivational context
Hierarchy of human needs
Basic survival needs
Safety and security needs
Social needs
Esteem needs
Self-actualization needs
Affiliation

Achievement
Competition
Potential for promotions
Incentives
Legacy
Seniority
Popularity
Friendship
Personal Motivation Plans (PMP)

REVIEW QUESTIONS

1. Define the term "motivation."
2. What is meant by the term "motivational context."
3. Explain Maslow's Hierarchy of Human Needs.
4. Explain one motivational strategy from each category of need in Maslow's hierarchy.
5. For project managers, what three components comprise relationship building in a project team?
6. List three strategies project managers can use for establishing positive working relationships with their team members.

7. Explain the concept of "affiliation" and how it can be used to motivate certain members of teams.
8. Explain how achievement can be used as a motivational technique.
9. Explain how competition can be used as a motivational technique.
10. How can project managers keep competition from getting out of hand and undermining team performance?
11. Explain how the potential for promotions can be used as a motivational technique.
12. Explain how offering incentives can be used as a motivational technique.
13. Explain how the legacy question can be used as a motivational technique.
14. Explain three rules of thumb that will help ensure that when project managers make recommendations for filling vacant positions by promoting from within, the result is positive.
15. Explain how organizations can ensure that performance incentives will actually be effective at motivating project team members.
16. Explain how the concept of self-actualization can be used to motivate project team members.
17. Explain what a Personal Motivation Plan is, why it is necessary, and how to develop one.

APPLICATION ACTIVITIES

The following activities may be completed by individual students or by students working in groups:

1. Assume that you are a new project manager and you want to motivate your team members to perform at their best. Develop one strategy for each of the categories of human need in Maslow's hierarchy. Tell specifically how you will use each specific category of need to motivate team members.

2. Think of things that would motivate you personally to perform at your best as a member of a team. Use this information to develop a PMP for yourself. If this activity is done in groups, pair off and have one student serve as the project manager and one as the team member for whom the PMP is being developed.

ENDNOTES

1. Saul McLeod, "Maslow's Hierarchy of Needs." Retrieved from www.simplypsychology.org/Maslow.html on July 26, 2013.

2. Ibid.

CHAPTER 11

Communicating, Influencing, and Negotiating

Effective communication is an essential skill for construction project managers. Construction project involves a lot of different entities, all of which have their own concerns, goals, agendas, responsibilities, egos, and attitudes. In other words, there are a lot of moving parts in a construction project. To keep all of the parts moving in the right direction and working together, effective communication is essential. Communication skills help project managers gain a commitment from team members; understand the views, perspectives, and messages of others; clearly, accurately, and succinctly convey work instructions to team members; offer constructive criticism in a positive, helpful manner; convince others of the veracity of their ideas; promote teamwork, cooperation, and collaboration; enhance interpersonal relationships; ensure that team members understand the big picture and where they fit into it; and prevent and resolve conflict.

Good communication is critical to the success of construction projects. In fact, if a team can be viewed as a machine, communication is the oil that keeps it running smoothly. Of all the skills needed by project managers, effective communication may be the most important. It is at least one of the most important. This chapter explains how to develop the communication skills needed to help effectively lead project teams.

BEGIN WITH A COMMUNICATION PLAN

Project managers are responsible for keeping a number of different stakeholders informed about their projects. Some of the stakeholders are part of the project manager's construction firm while others are from outside the firm. Internal stakeholders include the construction firm's executive management team and members of various project teams including the cost estimation, planning/scheduling, risk management, procurement, and project staff teams. External stakeholders include the owner, architect, engineers, trades people, subcontractors, inspectors, and code enforcement officials. All of these stakeholders will want to be fully informed about the specific aspects of the project that apply to them, how the project is progressing, what problems have been encountered, what is being done to overcome problems, and so on.

It is incumbent on project managers to keep all members of the project team fully informed of progress, problems, budget issues, and anything else they need to know to be fully informed, contributing members of their respective teams and to do their parts to ensure that the project is completed on time, within budget, and according to specifications. An ill-informed team member is not equipped with the information needed to do the best possible job in this regard.

Since there are so many different stakeholders who will want to stay informed about projects, it is important for project managers to develop a communication plan so that communication is well-organized and systematic. A communication plan for a given project contains the following information:

- A list of stakeholders—internal and external
- The types of communication to be provided for each stakeholder (written report, team meeting with verbal communication, stand-and-deliver presentations to selected groups, e-mail updates, etc.)
- Frequency of communications with each stakeholder
- Standard content to be contained in each type of communication
- Who is responsible for collecting and providing the information for each type of communication and when it should be provided (e.g., members of the project staff team are typically tasked with preparing draft reports of various kinds for the project manager to distribute)

Higher management and other stakeholder groups within the organization will probably receive periodic written reports. The same is true of the owner, although project managers can expect to occasionally be asked to make stand-and-deliver presentations to higher management and the owner. Team members will typically receive verbal updates presented during team meetings along with a written summary of the updates. The types of communication that will be expected of project managers will be either specified for them or negotiated during the initiation phase of the project.

COMMUNICATION SKILLS CAN BE LEARNED

Communication is a human process. Hence, it is an imperfect process. Doing it well takes work. This is because the quality of communication is affected by so many different factors. These factors include speaking ability; hearing ability; language barriers; differing perceptions and meanings based on age, gender, race, nationality, and culture; attitudes; nonverbal cues; writing ability; and the level of trust between senders and receivers to name just a few. Because of these and other factors, communicating effectively can be difficult. Regardless of the difficulty, good communication skills are essential for project managers. Fortunately, most people can learn to be effective communicators.

With sufficient training and persistent practice, most people—regardless of their innate capabilities—can learn to communicate well. Project managers must be good communicators. Consequently, those who want to be project managers should strive to develop their communications skills. A project manager who lacks communication skills is like a carpenter who lacks a hammer.

CONSTRUCTION PROJECT MANAGEMENT SCENARIO 11.1

I'm just not a good communicator

Nancy Powers is becoming increasingly frustrated. She worked long and hard to become a project manager, and was pleased and proud when she was promoted. But now less than six months later Powers is beginning to think she made a mistake. Her problem is communication. Powers does not seem to be able to communicate clearly with her team members. Too many mistakes are being made in her team as the result of poor communication. Even when she is sure that her team members heard what she told them, they do not seem to get things right. There always seems to be confusion over who is supposed to do what and when. Yesterday, her supervisor asked why there seemed to be so many problems with her current project. A frustrated Powers responded: "I am just not a good communicator. I will never be a good communicator."

Discussion Questions

In this scenario, Nancy Powers is having problems moving the work of her project team forward because of poor communication. Have you ever known someone whose poor communication skills inhibited progress by creating confusion or other problems? If so, relate your experience. If you were Nancy Powers' supervisor, how would you respond to her statement that she "…will never be an effective project manager"?

COMMUNICATION DEFINED

Project managers who want to be good communicators should never confuse *telling* with *communicating*. Unfortunately, some do. When a problem develops these confused project managers are likely to protest, "I don't understand why he didn't get this done. I told him what to do." In addition, some project managers confuse *hearing* with *listening*. When there is a problem, these confused project managers are likely to say, "That isn't what I told her to do. I know she heard me. She was standing right next to me!"

In these examples, the project managers did not understand the concept of communication. Project managers should always remember that what they say is not necessarily what the other person hears, and what the other person hears is not necessarily what they intended to say. Often, the missing ingredient when attempts at communication go awry is comprehension. Communication may involve telling, but it is not *just* telling. It may involve hearing, but it is not *just* hearing. The following definition clarifies the concept of communication:

> Communication is the transfer of information that is received and fully understood from one source to another.

A message can be sent by one person and received by another, but until the message is fully understood communication has not occurred. This qualifier applies to spoken, written, and nonverbal communication.

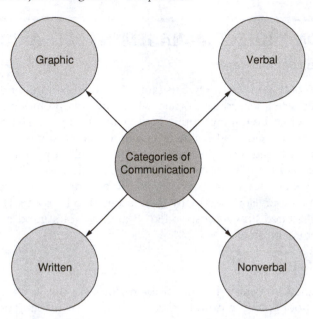

FIGURE 11.1 Project managers will use all four categories of communication.

COMMUNICATION IS A PROCESS

Communication is a process. As such it has several components: *sender, receiver, method, medium*, and the *message* itself. The sender is the originator or source of the message. When a project manager conveys work instructions to a team member, he or she is a sender. The receiver is the person or group for whom the message is intended. When team members receive instructions from a project manager they are receivers. The message is the information that is to be conveyed, understood, and acted on. In the two previous examples, the work instructions were the message. The medium is the vehicle by which the message is carried (e.g., telephone, e-mail, social networking software). The method is the type of communication chosen for conveying the message.

There are four basic categories of communication methods: *verbal, nonverbal, written,* and *graphic* (see Figure 11.1). Verbal communication includes face-to-face conversation, telephone conversation, speeches, public announcements, press conferences, and other venues for conveying the spoken word. Nonverbal communication includes gestures, facial expressions, voice tone, body poses, gestures, and proximity. Written communication includes letters, memorandums, billboards, bulletin boards, manuals, books, and all of the various electronic means of conveying the written word. Graphic communication involves using pictures and nonalphanumeric symbols to convey a message.

Technological developments have significantly enhanced the ability of people to send and receive information, but not necessarily their ability to communicate. Technological development in the broad field of communication include the Internet, e-mail, social networking, word processing, satellites, telephones, cellular phones and an ever-growing variety of other handheld devices, answering machines, facsimile machines, and pocket-sized dictation devices. Project managers should make a point of becoming skilled at using the various technological aids that are readily available to enhance communication in their project teams.

NOT ALL COMMUNICATION IS EFFECTIVE

When the information conveyed from one source to another is received and understood, communication has occurred. However, understanding alone does not guarantee effective communication. *Effective communication* occurs when the information received and understood is accepted and acted on in the desired manner. For example, a project manager might ask a team member to retrieve some old files from the company's warehouse. The team member in question verifies that he received and understood the message. However, rather than make the short walk to the company's warehouse, he decides to put the task off. He is right in the middle of another assignment and does not want to put it aside until the task is complete. By the time he is ready to make the trip to the warehouse, it is time to go to lunch. He decides to retrieve the file needed by the project manager after lunch.

Unfortunately, by the time he returns from lunch he has forgotten all about the file in the warehouse. When the project manager drops by his office to pick up the file—a file he needs for a meeting that begins in just five minutes—the embarrassed team member has to admit that he does not have it. There had been communication when the project manager gave the team member his assignment, but it was not effective communication because it did not result in the desired action.

Effective communication is a higher level of communication because it requires not just understanding but acceptance and action by the receiver. The acceptance aspect of effective communication requires influence, persuasion, and monitoring. Since acceptance of the message is essential to effective communication, project managers need to know how to gain acceptance of the messages they communicate.

The first step toward ensuring acceptance of messages is to gain credibility with those who receive the messages. Project managers who have credibility with employees will find it easier to also have influence with them. The more influential project managers are, the more likely it is that their messages will be accepted and acted on in the desired manner. When credibility and influence are lacking, receivers tend to question or even doubt the veracity of the message. They may not voice their doubts, but they will make them known by their hesitance to accept the message and their corresponding reluctance to respond to it.

Persuasion can be an important factor in gaining acceptance of messages. Project managers can be more persuasive by explaining: (1) the *why* behind their messages, (2) the benefits of accepting their messages and acting on them in the desired manner, and (3) the consequences of failing to accept their messages and act on them in the desired manner. This is why a dictatorial approach to communication that says "do what I say and don't ask any questions" does not work very well. Of course, in crisis situations there may not be time to explain. When this is the case, acceptance of the message can still be gained provided the sender has gained credibility with the receiver(s). This is another reason why credibility is essential to effective communication—project managers will not always have time to explain. However, when there is time to do so, explaining the reasons for as well as the benefits and consequences of compliance will help gain acceptance of the message.

Finally, monitoring is an effective way to ensure acceptance of the message and that the desired action is taken. For example, in the earlier example in which the individual failed to retrieve the file for his project manager, the embarrassed team member would probably

---

Checklist of
FACTORS THAT CAN INHIBIT COMMUNICATION

✓ Differences in meaning

✓ Insufficient trust

✓ Information overload

✓ Interference

✓ Condescending tones

✓ Listening problems

✓ Premature judgments

✓ Inaccurate assumptions

✓ Technological glitches

FIGURE 11.2   Project managers must learn how to overcome these inhibitors.

have acted differently if his supervisor had monitored him. Instead, the project manager simply told the individual to retrieve the file he needed. He did not call or e-mail in the interim to ensure that the task had been completed. This is an example of both poor communication and poor leadership.

FACTORS THAT CAN INHIBIT COMMUNICATION

There are several factors that can inhibit communication. Project managers should be familiar with these inhibitors and understand how to overcome them. Factors that can inhibit communication include the following (Figure 11.2):

- *Differences in meaning.* Differences in meaning are inevitable in communication because people have different backgrounds, experience, and levels of education. In a country as diverse as the United States, project teams are likely to mirror that diversity (e.g., different races, cultures, and nationalities). Because of this diversity, the words, gestures, and facial expressions used by people can have altogether different meanings. To overcome this inhibitor, project managers must invest the time necessary to get to know their team members and learn what they mean by what they say.
- *Insufficient trust.* Few factors can inhibit communication more than insufficient trust. If team members do not trust their project manager, they are not likely to believe what she tells them. If they do not believe what she tells them, they are not likely to accept it and act on it in the desired manner. Team members who do not trust their project manager often question the motives behind her messages. They will tend to concentrate on reading between the lines and looking for a "hidden agenda." In fact, they might focus so intently on reading between the lines that they miss the real message. Project managers should understand this and, as a result, strive to build trust with their team members.
- *Information overload.* Because advances in communication technology have enabled and encouraged the rapid and continual proliferation of information, members of project teams can find themselves dealing with more information than they can process

effectively. This is known as *information overload*, and it can easily cause a breakdown in communication. Project managers can protect their team members from information overload by screening, organizing, summarizing, and simplifying the information conveyed to them. For example, project managers should never take the reports they receive from higher management, the architect, engineers, or inspectors and just hand them over to their team members. This concept is known as an *information dump*. Instead, project managers should take the time to extract or at least highlight the information that is pertinent so that their team members do not get bogged down wading through superfluous information.

- *Interference.* Interference is any external distraction that inhibits effective communication. It might be something as simple as background noise or as complex as atmospheric interference. Regardless of its nature, interference can distort or even completely block out communication. Consequently, project managers should be attentive to the setting and the environment when trying to communicate with team members. The author once had to move an entire audience of 100 people when giving a speech in a resort on the Gulf of Mexico. The beautiful emerald waters of the Gulf were not the problem. Rather, a contractor was doing renovations to the resort that hosted the meeting and one of the workers was using a jack hammer. Sometimes, to eliminate interference it is necessary to change the setting.

- *Condescending tones.* Communication problems created by condescension result from the tone rather than the content of the message. People do not like to be talked down to. If team members sense that the project manager is talking down to them, they might respond by tuning out. Worse yet, they might express their resentment by intentionally ignoring the message.

- *Listening problems.* Listening problems are one of the most common inhibitors of effective communication. They can result from the sender not listening to the receiver and vice versa. To be good communicators, project managers must be good listeners. This topic is important enough to warrant a section of its own later in this chapter.

- *Premature judgments.* Premature judgments by the sender or the receiver can inhibit effective communication. This inhibitor exacerbates listening problems because as soon as people make a premature judgment they stop listening. One cannot make premature judgments and maintain an open mind, and an open mind is essential to effective communication. Therefore, it is important for project managers to listen nonjudgmentally and avoid making premature judgments when receiving a message.

- *Inaccurate assumptions.* Perceptions are influenced by assumptions. Consequently, inaccurate assumptions can lead to inaccurate perceptions. Here is an example. Janice will go to great lengths to avoid participating in presentations her project team has to make to the owner or higher management. Having never worked with Janice before, her project manager assumes that she is lazy. As a result, whenever Janice makes suggestions in team meetings, the project manager either ignores her input or simply tunes out. The project manager is making an inaccurate assumption about Janice. She is actually a highly motivated team member who works hard to help the team. Her reluctance to help make presentations is the result of fear not laziness. Janice is mortified at the thought of getting up in front of an audience, but she is embarrassed to admit it. Because of an inaccurate assumption, Janice's project manager is missing out on the suggestions of a highly motivated team member. In addition, his misperception points to a need for trust building. Perhaps if Janice trusted her project manager more, she would be less embarrassed to discuss her fear of public speaking.

- ***Technological glitches.*** Software bugs, computer viruses, dead batteries, power outages, holes in coverage, and software conversion problems are just a few of the technological glitches that can interfere with communication. The more dependent project teams become on technology for conveying messages, the more often these glitches will interfere with and inhibit effective communication.

LISTENING WELL IMPROVES COMMUNICATION

Hearing is a physiological process, but listening is not. A person with highly sensitive hearing can be a poor listener. Conversely, a person with impaired hearing can be an excellent listener. Hearing is the physiological process of receiving sound waves, but listening is about perception.

Understanding the following definition of listening will help project managers to be better listeners:

> Listening is receiving a message, correctly decoding it, and accurately perceiving what is meant by it.

Notice that this definition contains three critical elements—all of which must be present: (1) receiving the message, (2) correctly decoding the message, and (3) accurately perceiving what is meant by the message. If even one of these elements is missing, there will not be effective listening.

Challenging Construction Project

BURJ KHALIFA

The Burj Khalifa, a mixed-use skyscraper in Dubai, is the latest in a long line of structures holding the record for being the world's tallest structure. It was preceded by the Warsaw Radio Mast, KVLY-TV Mast, CN Tower, Taipei 101, Willis Tower, and a number of other structures dating all the way back to the Empire State Building. As of this writing Burj Khalifa is the tallest building ever built. The skyscraper is 2,717 feet tall, has 163 inhabitable floors, 46 maintenance levels, and two parking garages in the basement, and occupies 3,331,100 square feet of floor area. The structure cost $1.5 billion. The general contractor for Burj Khalifa was Samsung C&T of South Korea.

Among the records originally held by Burj Khalifa on the day it opened are the following: 1) tallest skyscraper to top of spire, 2) tallest structure ever built, 3) tallest freestanding structure, 4) building with the most floors, 5) building with the highest occupied floor in the world (160th), 6) fastest elevators in the world (40 miles per hour), 7) highest observation deck in the world (124th floor), 8) highest vertical pumping of concrete (1,988 feet), 9) highest swimming pool in the world (76th floor), and 10) highest restaurant in the world (122nd floor).

The structural system for Burj Khalifa is reinforced concrete. Construction of the building required 431,600 cubic yards of concrete, 55,000 tons of steel reinforcing bar, and 22 million man-hours of labor. The building sits on a foundation of 192 concrete piles—each of which is more than a yard in diameter and 164 feet long. Because of the enormous pressure of the weight of the building and the intense heat of the desert environment, the concrete for the building had to be poured at night when it was cooler and ice had to be added to the concrete mix. The cooler mix cured more evenly with less chance of cracks.

Source: Based on The Burj Khalifa. Glass, Steel and Stone. www.BurjKhalifa.ae/

Inhibitors of
EFFECTIVE LISTENING

- Lack of concentration

- Preconceived notions

- Thinking ahead

- Interruptions

- Tuning out

FIGURE 11.3 Project managers must avoid these inhibitors.

Inhibitors of Effective Listening

Communication will not occur when the receiver hears but does not accurately perceive the message. Several inhibitors can cause this to happen. The most common inhibitors of effective listening include (Figure 11.3):

- Lack of concentration
- Preconceived notions
- Thinking ahead
- Interruptions
- Tuning out

To perceive a message accurately, it is necessary to concentrate on what is being said, and how it is being said—verbally and nonverbally. Nonverbal communication is explained in the next section. This section focuses on listening to verbal messages. *Concentration* requires that extraneous distractions be either eliminated or mentally shut out. When project managers concentrate, they clear their minds of everything but the message being conveyed and focus on the team member who is sending the message.

Preconceived notions also inhibit listening because they can cause people to make premature judgments. Making premature judgments shuts down listening. Leaders in organizations should practice being patient and listening attentively. People who prematurely jump ahead to where they think the conversation is going may get there only to find that the speaker was going somewhere else. *Thinking ahead* is typically a response to being impatient or in a hurry. Project managers are typically in a hurry. Project deadlines always loom large, and they are unforgiving. Consequently, project managers can be forgiven for getting in a rush and thinking ahead when team members try to convey a message. However, project managers need to understand that it takes less time to hear someone out than it does to start over after jumping ahead to a preconceived conclusion. The time-saving approach is to listen attentively and get the message right the first time.

Interrupting can be especially harmful in that it can inhibit effective listening and frustrate the speaker. Consequently, it is doubly bad to interrupt someone who is speaking to you. If clarification is needed during a conversation, project managers should make a mental note of it and wait for the speaker to reach a stopping point. Mental notes are preferable to written notes. The act of writing can, itself, distract the speaker or cause the listener to miss the point. If it is necessary to make written notes, project managers should keep them short. They should avoid the temptation to interrupt and should not allow cellular phones or other people to interrupt.

Tuning out inhibits effective listening. A person who has tuned out will typically appear distracted or have a far-away look on her face. However, some people become skilled at using body language to make it appear they are listening when in fact they are not. Project managers should avoid the temptation to tune out during conversations with team members and to engage in nonverbal ploys to make it appear they are listening when they are not. An astute team member might ask the project manager to repeat what was said. At any point during a conversation, the project manager should be able to paraphrase and repeat back to a team member what she has said.

Strategies for Promoting Effective Listening

Project managers who are trying to become effective listeners can improve their listening skills by applying the following strategies:

- Use the five-minute rule
- Remove all distractions
- Put the speaker at ease
- Look directly at the speaker
- Concentrate on what is being said
- Watch for nonverbal cues
- Take note of the speaker's tone
- Be patient and wait
- Ask clarifying questions
- Paraphrase and repeat what the speaker has said
- Control emotions

THE FIVE-MINUTE RULE   The *five-minute rule* is really a self-defense mechanism for project managers. Consider the case of Luke. As a newly minted manager, he wanted to maintain an open-door policy for team members and an open ear for their problems, concerns, complaints, and recommendations. Having come up through the ranks, Luke knew firsthand how it was to work for a project manager who was not accessible. Consequently, Luke was determined to be just the opposite.

On the other hand, listening to the problems, concerns, complaints, and recommendations of team members can be time consuming. Like all project managers, Luke had other duties that needed his attention. Before long his open-door policy had him spending most of the time in the workday doing nothing but listening to the input of his team members. Luke was making an "A" in listening, but an "F" in attending to his other duties.

Luke's open-door policy was popular with his team members and did produce some positive results beyond just the morale boost it gave them. On the other hand, he often found himself in the office late at night trying to finish his other duties that were interrupted by drop-in visits from his team members. Clearly, Luke needed to find a way to retain his open-door policy without allowing team members to monopolize all of his time. The answer that eventually solved his problem was what the author calls the *five-minute rule*.

The five-minute rule works like this. Project managers let their team members know that—within reason—they can have five minutes on a drop-in basis any time they have a complaint, recommendation, or any other type of input to offer. However, the time for these drop-in visits will be limited to five minutes. Lest the reader think this policy is too restrictive, five minutes is actually plenty of time provided the team member has thoroughly considered what she wants to say. Preparation is the key. Spending time listening to a team member who rambles on because of poor preparation is wasting time. In addition to saving time, the five-minute rule helps team members learn how to organize their arguments and prepare brief, succinct, but comprehensive explanations that get right to the point without wasting time. This is a skill that will serve them and the team well.

The allotted five minutes are not to be used for thinking out load or brainstorming. There is a time and place for these things, but it is not during five-minute rule sessions. During the allotted five minutes, the team member is expected to explain his problem and offer a recommended solution that is realistic. Recommending poorly conceived solutions is a major faux pas on the team member's part. Proposing a $1000 solution to a $100 problem is not acceptable. Team members who ask for five minutes are expected to have already conducted a cost-benefit analysis for the solution they plan to propose or recommendation they plan to make.

The cost-benefit analysis might amount to just carefully thinking through the recommendation to be made, but even this will help team members realize that some solutions are better than others. Nothing is free. There is a cost associated with everything. Consequently, team members who make recommendations should: (1) be aware of the costs associated with their recommendations and (2) make sure the potential benefits of their recommendations outweigh the costs. The cost-benefit analysis requirement of the five-minute rule can prevent time from being wasted considering unrealistic solutions.

Not all issues in teams can be properly dealt with in five minutes. Issues that are too complex to fit into the five-minute format should be handled in the normal manner (i.e., the team member makes an appointment and asks for as much time as is needed). The five-minute rule is a strategy for facilitating effective listening while allowing project managers to maintain an open-door policy. It is not intended as a replacement for traditional problem-solving methods such as brainstorming, focus groups, or team meetings.

OTHER LISTENING-IMPROVEMENT STRATEGIES AND THE FIVE-MINUTE RULE To gain the most from five-minute sessions with team members, project managers should apply the other listening-improvement strategies explained earlier. The strategy of removing distractions and giving full attention to the speaker is important. Anyone who has ever tried to talk with someone who was distracted by other concerns will understand why. Removing distractions typically involves such things as turning off cellular phones, putting a temporary hold on landline calls, allowing no other visitors to drop in, and getting away from distracting paperwork on the desk top.

An easy way to get away from distracting clutter without having to clean off the desk is to have two chairs in front of or away from the desk. Project managers should not try to sit at their desks and listen to team members. Trying to concentrate on what someone is saying while paperwork on the desk beckons takes more self-discipline than most people have.

Before asking the speaker to begin, put her at ease—particularly if you sense nervousness or discomfort. Asking about something unrelated to the job such as children, grandchildren, ball games, or hobbies will usually do the trick. Then, once the speaker begins, look directly at him and concentrate on what is being said. Project managers should never waste a moment of their team members' time by being inattentive. Rather, they should concentrate on what is being said and learn to listen not just with their ears, but also with their eyes. In other words, project managers should learn to watch for nonverbal cues. Nonverbal communication is explained in the next section.

Project managers should learn to avoid interrupting or pushing the team member along. One of the keys to effective listening is to be patient and wait. When there are hesitant pauses in a team member's explanation it can mean that he or she is trying to decide: (1) how to say what is really on his or her mind or (2) if he or she is really going to say what is on his or her mind. If project managers interrupt or try to prompt a hesitant speaker, they risk missing out on the real reason he or she asked for five minutes in the first place—especially if what the team member has to say is sensitive or embarrassing. Better for project managers to give hesitant team members a positive, affirming facial expression and then be patient and wait.

Once a team member has stated his or her case, the project manager should ask clarifying questions to gain a more complete and accurate understanding. Once the project manager has a complete and accurate understanding, an effective strategy is to paraphrase what the team member has said and repeat it back to him or her. Paraphrasing can be beneficial in two ways. First, it shows the team member that the project manager listened. Second, if the message has been misperceived the team member can clarify further. Paraphrasing can prevent a situation in which the project manager wastes time trying to solve the wrong problem or something that is not a problem.

The final strategy—control emotions—is critical. A good rule of thumb for project managers to remember about communication is this: When dealing with team members, losing one's temper will undermine communication and trust. One of the differences between being a project manager and a member of a project team is that there are higher expectations for the project manager. When construction professionals step up from the ranks to leadership positions, it is not just their pay that increases, but it is also their level of responsibility and behavioral expectations that increase.

Project managers who lose their tempers when team members bring them unwelcome information soon find themselves without messengers or messages. This is one of the worst things that can happen to project managers because the more unwelcome the message the more they need to hear it. They may not want to hear it, but they need to hear it and the sooner the better. Bad news that goes unattended has a way of turning into even worse news.

NONVERBAL COMMUNICATION

Nonverbal messages represent one of the least understood but most powerful modes of communication. Nonverbal messages can reveal more than verbal messages for those who are attentive enough to observe them. Nonverbal communication is sometimes called *body language*, an only partially accurate characterization. Nonverbal communication does include body language, but body language is only part of the concept. There are actually three components of nonverbal communication: body factors, voice factors, and proximity factors (Figure 11.4):

Body Factors

An individual's posture, facial expressions, gestures, and dress—in other words his or her body language—can convey a variety of messages. Even such factors as makeup or the lack of it, well-groomed or messy hair, and clean or scruffy shoes can convey a message. Project

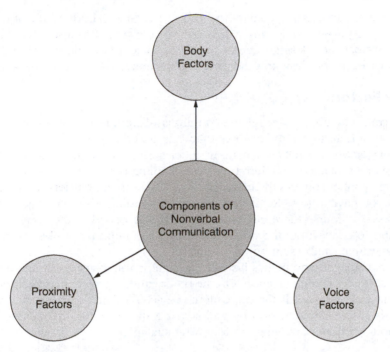

FIGURE 11.4 Nonverbal communication is more than just body language.

managers who learn to be attentive to these body factors can improve their communication skills markedly. The key is to understand that nonverbal messages should agree with, support, and enhance verbal messages. For example, when someone says "yes" with a smile and a nod of the head, the verbal message—yes—and the nonverbal—a smile and a nod—all agree. However, if a person says "yes" but frowns and shakes his head from side to side in the universal symbol for "no" there is disagreement between the verbal and nonverbal messages. In this example, the disparity between the verbal and the nonverbal is obvious. However, this is not always the case. Differences between the verbal and nonverbal can be subtle and require careful attention to perceive.

The key to understanding nonverbal messages lies in the concept of consistency. In a conversation with another individual, are the spoken messages and the corresponding nonverbal messages consistent with each other? They should be. In a conversation, if nonverbal messages do not seem to match the verbal message, something is probably wrong and it is a good idea to find out what it is. An effective way to deal with inconsistency between verbal and nonverbal messages is to tactfully but frankly confront it. A simple statement such as, "Andrew your words say that you put the XYZ File back in the drawer yesterday, but your body language says you didn't." Such a statement can help project managers get to the truth in conversations with team members.

Voice Factors

Voice factors are also important elements of nonverbal communication. In addition to listening to the words team members speak, it is important to listen for voice factors such as volume, tone, pitch, and rate of speech. These factors can reveal feelings of anger, fear, impatience, uncertainty, interest, acceptance, confidence, and so on. As with body factors it is important to look for consistency when comparing words and voice factors. It is also advisable to look for groups of nonverbal cues. A single cue taken out of context has little meaning. But as one of a group of cues, each appearing to validate it, a given nonverbal cue can take on significance.

Proximity Factors

Proximity factors range from the relative positions of people in conversations to how an individual's office is arranged, the color of the walls, and the types of decorations displayed. A project manager who sits next to a team member during a conversation conveys a different message than one separated from him by a desk. Coming out from behind a desk and sitting next to a team member conveys the message that, "There are no barriers between us—I want to hear what you have to say." Remaining behind the desk sends a message of distance and standoffishness. Of course there are times when this is precisely the message the project manager wants to convey, but the point here is that it is important to be aware of the nonverbal messages that can be sent by proximity.

A project manager who makes his office a comfortable place to sit and talk is sending a message that invites communication. On the other hand, a project manager who maintains a cold, impersonal office sends the opposite message. To send the nonverbal message that team members are welcome to stop by and take advantage of the five-minute rule, project managers should consider applying the following strategies:

- Have comfortable chairs available for team members.
- Arrange chairs so as to be able to sit beside team members rather than behind a desk.
- Choose neutral colors for the walls of the jobsite office rather than harsh, stark, overly bright, or busy colors.
- If possible, have refreshments such as water, coffee, tea, and soda available for team members.

Some people like to turn their offices into to shrines displaying their achievements. In offices that are shrines, visitors will find trophies, plaques, photographs taken with important people, award certificates, and various other career mementoes displayed prominently. There is nothing wrong with a project manager having a "love me" wall in his office, but the concept can be overdone. To make a positive impression on team members, evidence of career and personal achievements can serve a valuable purpose. However, when trying to encourage team members to open up and reveal their concerns, issues, and problems, it is helpful to have a more inviting place to meet—one that is comfortable and inviting. A good rule of thumb for decorating your office is this: three walls for visitors and one for the occupant.

VERBAL COMMUNICATION

Effective verbal communication ranks close in importance to effective listening. Even in the age of technology, talking is still by far the most frequently used method of communication. This is why project managers should strive to continually improve their verbal communication skills. Being attentive to the following factors will help project managers improve the quality of their verbal communication (Figure 11.5):

**Factors That Affect
VERBAL COMMUNICATION**

- Interest

- Attitude

- Flexibility

- Tact

- Courtesy

FIGURE 11.5 Project managers can use these factors to improve verbal communication.

- ***Interest.*** When speaking with team members, project managers should show interest in their topic—show that they are sincerely interested in communicating the message in question. Project managers should demonstrate interest in the team members—the receivers of the message—as well. It is a good idea to look listeners in the eye, or if in a group, spread eye contact evenly among all receivers. Project managers who sound bored, ambivalent, or indifferent concerning their own message cannot expect receivers of the message to be enthusiastic about it.
- ***Attitude.*** Maintaining a positive, friendly attitude enhances verbal communication. This is because people are more open to listening to someone who is friendly and positive. A caustic, superior, condescending, disinterested, or argumentative attitude will shut down or, at least, inhibit communication. To increase the likelihood that their messages will be received in a welcome or at least open-mined manner, project managers should make an effort to be positive and friendly.
- ***Flexibility.*** Project managers who are dogmatic and dictatorial in their verbal communication increase the likelihood that their messages will be rejected by receivers. Flexibility and a willingness to hear other points of view will usually improve the chances of having a message received in a positive manner. For example, if during a team meeting the project manager presents a case for solving a problem, she should let team members know that their views and opinions will be heard and appreciated. Even if no one has an alternative idea to propose, the fact that the project manager is open and flexible enough to ask will improve communication. In fact, an effective communication tactic is to ask team members for their ideas first. When the project manager presents his or her ideas first, some team members may be reluctant to appear to disagree.
- ***Tact.*** Tact is an important factor in verbal communication, particularly when delivering a sensitive, potentially controversial, or unwelcome message. Using tact can be thought of as hammering in the nail without breaking the board. The key to tactful verbal communication is to think before speaking. Tact does not mean being less than forthright. Rather, it means finding a way to candidly say what has to be said without adding insult to injury.
- ***Courtesy.*** Being courteous means showing appropriate concern for the needs and feelings of the receiver. Calling a meeting as team members are leaving for home on a Friday evening is inconsiderate and will inhibit communication. Courtesy also dictates that project managers avoid monopolizing conversations. When communicating verbally, they should give receivers ample opportunities to ask questions for clarification

---

**STRATEGIES FOR ENHANCING THE EFFECTIVENESS
OF CORRECTIVE FEEDBACK**

- Be positive

- Be prepared

- Be realistic

FIGURE 11.6 Corrective feedback can backfire unless it is given properly.

and to state their own points of view. Project managers are wise to remember that one-sided conversations are not conversations at all—they are broadcasts.

COMMUNICATING CORRECTIVE FEEDBACK

Project managers occasionally need to give corrective feedback to team members. In fact, this is an important responsibility for project managers who are trying to complete their projects on time, within budget, and according to specifications. Corrective feedback is given to help team members individually and help the whole team perform better. Effectively given corrective feedback will do this. But in order to be effective, corrective feedback must be received in a positive manner by those at whom it is directed. This means it must be communicated properly and effectively. The following guidelines will help project managers enhance the effectiveness of their corrective feedback (Figure 11.6):

- ***Be positive.*** To actually improve performance, corrective feedback must be accepted and acted on by the team member. This is more likely to happen if it is delivered in a positive and tactful manner. Corrective feedback that is delivered in a less than tactful manner may cause the receiver to become defensive. If this happens, one is more likely to get excuses than improved performance. Project managers should give the team member being corrected the necessary feedback, but avoid focusing only on the negative. They should try to find something positive to say. For example, assume a team member has arrived late for work twice in one week. One approach would be to confront the tardy team member and say, "I'm glad you could show up today. I certainly hope this job isn't interfering with your social life." The tardy team member would certainly get the message about coming to work on time, but he might also be offended by the sarcasm. Also, if there is some legitimate reason for the tardiness, he might resent the assumption that his social life is the cause. A better approach would be to say, "Is everything OK? I noticed that you have been late twice this week. Is there anything I can help you with?" This approach lets the team member know that his tardiness has been noticed, but without boxing him into a corner where his only option is to become defensive. Another positive approach would be to say, "Mark I am really proud of how you helped solve that problem we had last week. Now let's talk about how you can do an equally good job of getting to work on time." The latter two examples let the individual know that his tardiness is unacceptable but without creating resentment or defensiveness. Project managers in these situations should remember that the goal is to correct and improve, not to punish.

- *Be prepared.* Before giving corrective feedback, project managers should do their homework. They should avoid kneejerk reactions and gather the facts. Then they should give corrective feedback based on the facts. It is always best to give specific examples of the poor performance or unacceptable behavior that needs to be improved. A normal human response is for team members to view corrective criticism as just plain criticism and become defensive. Return to the earlier example of the team member who was late for work two days in one week. If this team member becomes defensive he might try to deny being late. However, if he knows that the project manager has the facts, this is less likely to happen. A project manager who is poorly prepared might say, "Mark, you have been late a couple of times this week." The vagueness of this statement might encourage Mark to challenge it. However, if the project manager says, "Mark, you were 30 minutes late on Monday and 20 minutes late on Tuesday," he will know better than to challenge the statement.

- *Be realistic.* Project managers who are trying to improve the performance of team members can be forgiven for wanting to see a change for the better right now if not sooner. This is understandable since they are always are under pressure to meet deadlines, stay within budget, and comply with specifications. Unfortunately, developmental activities seldom produce immediate results. Improving performance can take time, and often does. Consequently, when giving corrective feedback for improving performance it is important to be realistic. First, project managers should use their experience to determine how much improvement is realistic to expect over a given period of time. They should then be guided by this knowledge in setting improvement goals. Second, project managers should make sure the performance they are trying to improve is within the control of the team member. Project managers should never make the mistake of expecting a team member to correct something over which he or she has no control. An effective approach is to explain the situation to the team member, ask for his or her feedback, and listen carefully. If there is an organizational inhibitor standing in the way of the desired improvement, removing that inhibitor is the project manager's job.

WRITTEN COMMUNICATION

Project managers can expect to be required to prepare written reports of the team's progress for different stakeholders. The following techniques will help project managers prepare written reports that are concise but comprehensive, readable, and understandable:

- *Begin by identifying the audience(s) for the report.* A report is written for a specific audience. That audience might be the construction firm's higher management, the owner and architect, the project staff team, or local government officials. The audience determines the approach that should be used for preparing the report, the language that can be used, the format, and the point of view.

- *Choose a format that will allow the report to be concise but comprehensive.* The key to making a report comprehensive enough to cover all necessary information, but concise enough to be read, understood, and remembered is to choose the right format. A good format for project reports presents its information in categories such as the following: (1) results to date (how much of the project work has been competed at the time the report is written), (2) comparison of the schedule versus actual progress,

(3) comparison of budget projections to actual expenditures, and (4) explanation of problems and proposed solutions.

- ***Use graphics where appropriate.*** It is true that a picture can be worth a thousand words, especially if it is a well-developed graphic. If information can be displayed on a simple graph or chart rather than using several written paragraphs of text, project managers should do so.
- ***Use language that is appropriate for the audience.*** Determine how much technical terminology can be used in a report by considering the makeup of the audience. Project reports are provided to different stakeholders. Consequently, they should be written in language that can be understood by the stakeholder in question. The same report modified for the audience in question can be an effective way to solve the language problem.
- ***Highlight actions that need to be taken by readers of the report.*** If a project report contains actions that must be taken by the stakeholder reading it, these actions should be highlighted to call the stakeholder's attention to them. Two effective ways to accomplish this include: (1) by having the actions in question typed in boldface in the report or (2) by adding an "actions needed" component at the end of the report that contains the list of required actions and who is responsible for each.
- ***Keep reports as short as possible.*** As a rule, people are inhibited by lengthy reports. Add to this that they are busy and the need to keep reports short becomes apparent. People are more likely to read short reports than long. Historians claim that during World War II, President Franklin D. Roosevelt would not read a report of more than one page. Project reports will not always fit into a one-page format, but they should be kept as short as possible.

Communication is an imperfect, but essential process. Without effective communication, members of project teams cannot do their part to complete projects on time, within budget, and according to specifications. Consequently, investing the time and effort necessary to become an effective communicator is a worthwhile endeavor for project managers. Few things will serve project managers better than learning to: (1) be effective listeners, (2) communicate well verbally, (3) understand and make use of nonverbal communication, (4) provide constructive criticism that is helpful and tactful, and (5) develop written project reports that are concise, comprehensive, readable, and understandable.

INFLUENCING AND NEGOTIATING IN PROJECT MANAGEMENT

Two important ways that project managers use their communications skills are in influencing others and in negotiating on behalf of their project. The best project managers are influential and they are good negotiators. They have to be. Every project has a number of different stakeholders, all of whom see the project from their own self-interested perspectives. The project manager needs the cooperation and support of all these stakeholders. Getting cooperation and support from stakeholders can require influence and the application of negotiating skills.

To influence stakeholders, project managers apply some of the various people skills explained in Part Two of this book (i.e., leading by example, communicating effectively, managing time well, managing adversity, and managing diversity). Stakeholders can also be influenced through effective negotiations. Negotiating is a skill that must be learned and can be learned. Like any skill, the more it is practiced the better one becomes at it. The word "become" is important here because being good negotiators is not something that people *are,* it's something they *become* through hard work and persistent practice.

NEGOTIATION DEFINED

It is important for students and professionals who want to be project managers to understand what *negotiation* means before beginning to develop negotiating skills. This is important because people tend to view the concept as Party A cleverly out maneuvering Party B in ways that give him all the value while leaving the hapless Party B empty-handed and wondering what happened. This view of negotiating is both inaccurate and shortsighted in any setting, but it is especially so in project management. A better way to define negotiation is as follows:

> Two or more parties working together to reach a mutually beneficial agreement that serves them well in the present and leaves the door open for mutually beneficial agreements between them in the future.

This is a simple definition, but it is loaded with meaning. The words "working together" are used to make it clear that the two sides in a business negotiation should not be enemies, adversaries, or even opponents. The reason for this is found in the "mutually beneficial" element of the definition. For example, a project manager for a large construction firm might need to negotiate with one of his colleagues to *borrow* a certain highly skilled employee for one of his jobs. To make the agreement mutually beneficial, he offers to help his colleague solve a problem he is currently facing. In this way, the eventual agreement is mutually beneficial and it leaves the door open to future mutually beneficial agreements. Construction professionals who trick or coerce others into one-sided agreements put future agreements at risk. Nobody wants to deal with a dishonest or one-sided negotiator.

An agreement negotiated today should lay the groundwork for future agreements with the same party in the future. Mutually supportive relationships and repeat business are critical to construction professionals who operate in a competitive environment. Negotiating lopsided agreements is a sure way to undermine future agreements with the *victim* of the one-sided negotiation. Consequently, negotiating for a lopsided result can mean that a construction professional "wins" in the short term, but loses in the long run. When a negotiation is concluded, there should be no victims.

CHARACTERISTICS OF EFFECTIVE NEGOTIATORS

Good negotiators are not slick dealing, pushy, domineering types who overpower their witless opponents; nor are they indecisive, submissive, unimaginative types afraid to advocate for their point of view. Rather, good negotiators are typically patient, fair, well informed, cooperative, innovative, imaginative, and intuitive. These are the characteristics that are most likely to lead to agreements that are fair, balanced, and mutually beneficial.

Construction professionals who develop these characteristics will typically display the following behaviors during negotiations:

- Quick to see through the fog of debate to the heart of a matter.
- Solve problems in real time before they can derail the process.
- Think clearly and quickly under pressure.

- Depersonalize comments that are made and see through emotionally charged language to the issues in question.
- Listen carefully and with patience.
- Approach the process with an open-mind knowing there is always more than one way to achieve a desired result.
- Develop alternative options and solutions quickly and on the spot.
- Watch for verbal and nonverbal cues and use them to assess people.
- Think critically (i.e., recognize assumptions, rationalizations, justifications, and biased information that are presented as facts).
- Take the long-term/repeat business perspective.
- Give themselves and the party they are negotiating with room to maneuver (i.e., they avoid boxing themselves and others in).
- Maintain a sense of humor and a positive attitude throughout the negotiating process.
- Consider issues from the other side's point of view as well as their own.
- Understand that timing is important to success in negotiating.
- Prepare, prepare, prepare.

PREPARATION AND SUCCESSFUL NEGOTIATIONS

The behaviors explained in the previous section are all important to successful negotiations. But none is more important than the last one—preparation. Going into a negotiation unprepared will almost guarantee a bad result. The following questions will help ensure that construction professionals are well prepared for negotiations with customers, subcontractors, colleagues, higher management, and government officials who control the planning, zoning, and permitting processes:

- What do we want out of this negotiation? What does the other party want?
- What are we willing to give up in the negotiation in order to get what we want? What is the other party willing to give up?
- What is at risk here for us and for the other party? In other words, if we cannot reach an agreement, what do we lose and what does the other party lose?
- How much do we know about the other side and their needs? How much do they know about us?
- Do we have any "hot-button" issues? Does the other party have "hot-button" issues?
- What don't we know about the other party, and who can help us learn what we don't know?
- Are there factors that might affect the outcome that we or the other party have no control over? What are those factors?
- What is our bottom-line—at what point do we just walk away? What is the other party's bottom-line?
- What are the easy issues we can use to generate early agreement? What are the other party's easy issues?

Good preparation is essential to a successful negotiation. Before beginning a negotiation—whether with customers, subcontractors, or internal personnel—project managers must know what they want and what they do not want, what they can accept and what they cannot accept, what they are willing to give up in order to get what they want and what they are not willing to give up. It is also important for construction professionals to understand that there is more than one way to get what they want.

The story of Sherry illustrates this last point. Sherry was a project manager for a large construction firm. As the result of a divorce, she was a single mother. Child care costs were making it difficult for Sherry to make ends meet. Consequently, she arranged a meeting with her vice president to ask for a raise. The vice president respected Sherry, admired her work, and was sympathetic to her plight. Sherry was an excellent project manager, and the vice president wanted to help her. Unfortunately, his hands were tied when it came to giving Sherry a raise.

The company they worked for had a salary schedule that paid project managers based on their level of education, years of experience, and performance. Sherry's performance was excellent. Consequently, she would certainly receive a raise, but not for about a year. At the time in question, Sherry had just been promoted to a new level, meaning she had recently gotten a raise. The company's policy was that an employee had to perform well at a given level for at least one full year before receiving the next raise.

The vice president told Sherry that if she could make it for another 11 months, she was virtually guaranteed the raise she wanted. However, Sherry was adamant that she needed another raise immediately. In an attempt to solve the problem without violating company policy on raises and promotions, the vice president offered Sherry a counter-proposal. The company had a new program that would help pay the child care costs of employees who qualified for it. If Sherry would accept this program instead of a raise, her child care costs would be reduced by an amount equal to the 75 percent of the raise she was demanding. But Sherry was so intensely focused on the raise that she failed to see compensation for child care costs as an equally viable solution. Out of anger and frustration, Sherry quit.

Her resignation was a problem for the company. Sherry was managing an important project at the time she quit. But the resignation turned out to be an even bigger problem for Sherry. Having given up her only source of income and needing a job quickly, Sherry was compelled to accept a lower-paying job. She had received better job offers, but all of them would have required relocating to another state—something she could not do. Sherry needed a local job because her divorce decree required that she live within 75 miles of her ex-husband to accommodate his visitation rights.

This is an example of why it is so important for project managers to understand what they really need out of a negotiation. Sherry thought she needed a raise when, in fact, what she really needed was financial relieve. As it turned out, accepting the offer of a company-provided child care subsidy would have benefited Sherry almost as much as the raise she demanded after tax considerations were factored in.

CONDUCTING NEGOTIATIONS

Once construction professionals have prepared to negotiate, there are strategies they can use that help the process work better. These strategies fall into the following broad categories: (1) Negotiate in stages and consider timing, location, and image, (2) create favorable momentum, and (3) observe certain behaviors during negotiations. Strategies in each of these categories are presented in this section.

Negotiate in Stages

People who are unskilled in negotiating typically want to jump right in, make an agreement, and quickly close the deal. Closing is certainly an important part of the process, but it is the third stage in a process that has three distinct stages. A good rule of thumb is to go through each stage, even if a given stage takes only a short time. This is because the first stage sets up the second and the second sets up the third. Project managers who skip a stage are likely to find themselves going backwards in the process just when they thought they were done.

The author created the *bridge analogy* to describe the stages in a negotiation. At the beginning of the process, a river runs between the two parties to a negotiation that keeps them apart. The project manager is on one bank and the other party is on the other bank. In order to get together, they must build a bridge. Stage 1 in the process involves building the foundation. Stage 2 involves building the skeletal structure for the bridge. Stage 3 involves adding the finishing touches. Once Stage 3 has been completed, the project manager and the other party can come together at the center of the bridge. A factor that leads to negotiations more frequently than any other is change orders. Project managers often have to negotiate with the owner, architect, and subcontractors over change orders.

Based on this analogy, the three stages in the negotiating process can be labeled as follows: Stage 1—Building the foundation; Stage 2—Erecting the structure; and Stage 3—Completing the bridge. What occurs in each of these stages is explained in the following paragraphs:

STAGE 1—BUILDING THE FOUNDATION   Building the foundation involves laying the groundwork for the negotiating process. This is the stage in which the project manager convinces the other party that he needs to hear what the project manager has to say. In other words, project managers state their case explaining how a successful negotiation will have mutual benefits. If this stage cannot be successfully completed, there is no reason to go to the next. Both parties must agree that there is a need for negotiating before proceeding with the remainder of the process. For this reason, it is important for project managers to practice explaining the need for a negotiation from the perspective of how the other party will benefit.

STAGE 2—ERECTING THE STRUCTURE   Once it is apparent that the other party understands the need to negotiate, the project manager can get into the specifics. The specifics include coming to an understanding of the expectations of both parties, an explanation of how both parties will benefit from what is being proposed, and specifics concerning the needs and concerns of both sides.

When buying a car, this is the point in the process when the buyer explains the details of what she wants (e.g., color, features, size) and the seller explains what is available. This is not the step in which price is discussed. When both parties have agreed on expectations, needs, concerns, and potential benefits, then they can move to the final stage.

STAGE 3—COMPLETING THE BRIDGE   At this point both parties have agreed that they would like to come to an agreement and both their needs and expectations have been explained. The final stage—the one most often associated with negotiating—involves bargaining over the specifics of the agreement such as price, deadlines, specifications, and other negotiable factors. This step will be much easier to conduct if a thorough job has been done in stages 1 and 2.

Create Favorable Momentum

Baseball teams try to score at least one run in their first time at bat. Football teams try to score a touchdown on their first possession. Tennis players try to win the first game or break the other player's serve during the first game. In all of these examples, the goal is to create favorable momentum. Momentum is the impetus or tendency of something to go in a certain direction. If project managers can get negotiations going in the right direction from the outset, they will tend to keep going in the right direction the rest of the way. For project managers, it is wise to create favorable momentum from the outset.

Additional Strategies for Effective Negotiating

Once a negotiation has commenced and is proceeding in the right direction, it is important to keep it going in the right direction. Momentum gained can be quickly lost if participants fail to do what is necessary to keep the ball rolling. The following strategies can be used during negotiations to keep things moving in the right direction:

- ***Think critically.*** Do not confuse facts with opinions or issues with positions. Tactfully insist on facts to back opinions and be quick to point out that issues are not positions. Issues can be resolved. Positions are negotiated.
- ***Listen to what is said and what is not said.*** Unfortunately, the other party in a negotiation might not begin the process as a partner. Consequently, project managers must be prepared to bring the other party along and turn that individual or team into a partner. To do this, project managers should listen attentively not just to what is said, but also to what is not said during negotiations. If the other party appears to be holding back, withholding information, or putting a certain "spin" on proposals and counter proposals, this could be evidence of a hidden agenda. Do not be afraid to tactfully say, "Something seems to be missing in the discussion. Can you clarify?" It might take a while to earn sufficient trust to convince the other party to drop his guard and be a partner. But if he is negotiating in good faith, he will eventually come around. If the other party is not negotiating in good faith, continuing the negotiation will be counterproductive.
- ***Keep the other party's needs and hopes in mind.*** When preparing for a negotiation, some determinations are made concerning what the other party wants to achieve. Project managers should keep the other party's needs and hopes in mind during the process. Before making a proposal or a counter proposal, ask how it might affect the other party's needs and hopes. Can the proposal be made in a way that will achieve the goals or a sufficient enough portion of them for both parties? Making a proposal that stomps on the needs and hopes of either party is not a good negotiating strategy.
- ***Be patient.*** Do not rush negotiations. Rushing will create suspicion in the mind of the other party. Be patient. Give things time to develop. Remember, this principle can be applied only if the negotiation has been scheduled early enough that there is no need to rush. Consequently, keep scheduling in mind when arranging the timing of negotiations.
- ***Ignore personal comments.*** Negotiations can become heated. Occasionally the other party might make a comment that is offensive. When this happens, project managers should simply ignore the negative comments and refuse to take them personally. Project managers should practice being objective and refusing to take things personally. It is likely in these cases that the other party is just a poor negotiator and does not know how to make proposals or counter proposals without getting personal. However, it could be that the other party is purposefully trying to agitate as a way to gain an advantage. Stay calm, depersonalize, and stay focused. Negotiators who use personal remarks to gain advantage are trying to break the other team's focus. Once they see that their tactic is not working, they will drop it.

- **Leave room to maneuver.** Avoid stating bottom line positions. Stating bottom line positions can paint a negotiating team into a corner, and make its members look foolish if, after stating a bottom line position, they are forced to change positions. There are many different ways to solve the same problem or meet the same need. Although project managers attempt to anticipate these various ways to meet needs, it is not possible to anticipate all problems and corresponding solutions. Consequently, it is wise to remain open to a better idea the other party might propose.

AFTER AGREEING—FOLLOW THROUGH

The negotiation process does not end once an agreement is made. As soon as one negotiation is concluded, it is time to begin paving the way for future agreements with the same party. Remember, negotiations have more than one purpose. The first, of course, is to arrive at an agreement concerning the issue in question at the moment. The second is to pave the way for future agreements with the other party. For example, assume that a project manager found it necessary to negotiate with a colleague to have a certain employee serve on his project team. This is a situation that is likely to arise repeatedly over time. Consequently, it is important to leave the door open for future negotiations with this colleague. The best way to leave the door open for future negotiations, assuming a win-win agreement has been concluded, is to follow through and do what was agreed to. The best negotiators refuse to make promises they cannot keep and always follow through on the promises they make.

Whatever a project manager promises to do during a negotiation must be done both properly and on time. If unanticipated problems arise, the project manager should make the other party aware of them immediately. The trust and credibility developed during the negotiation will now be either reinforced or lost based on how well the project manager performs and keeps his promises. Stay in touch with the other party, keep them informed, solicit feedback, and always be available and responsive to them. The better the project manager performs on the current agreement, the easier the negotiations will be for future agreements. Following through on agreements that were negotiated is important regardless of whether the negotiations are with the owner, architect, engineers, subcontractors, or internal colleagues.

CONSTRUCTION PROJECT MANAGEMENT SCENARIO 11.2

Your problem is that you have no tact

John has a strong grasp of the process aspects of project management. He is an excellent scheduler and does a good job of monitoring jobsite work. But when it comes to the people aspects, he has some glaring weaknesses. One of them is in providing constructive criticism to team members. John can certainly criticize, but his criticism is seldom constructive. In fact, it is often downright offensive—something John seems oblivious to. As a result, John's team members often resent his attempts to improve their performance or correct their behavior. The resentment manifests itself in a variety of different ways, but all of them are negative.

In an attempt to learn why his team members always seem to react so negatively to his "constructive criticism," John approached a fellow project manager, one who had formerly been a team member on one of John's teams. His colleague was blunt: "John, I would rather take a beating than get constructive criticism from you. Your problem is that you have no tact." After learning how it felt to receive some of his own brand of constructive criticism, John asked his colleague how he could improve.

Discussion Questions

In this scenario, John is oblivious to the damage done by his tactless attempts at constructive criticism. Have you ever worked with a person who was tactless when dealing with others? If so, how was this person received by others? If you were John's colleague in this scenario, what advice would you give him concerning how to improve?

SUMMARY

Because project managers are responsible for keeping several different constituent groups up-to-date concerning progress and problems, it is important for them to develop a communication plan for each project they manage. A communication plan identifies the constituent groups that need to be kept informed, the types of communication that will be used to keep them informed, the frequency of communication, the content of each type of communication, and the individual responsible for collecting and providing the information for each type of communication.

Communication is the transfer of information that is received and fully understood from one source to another. Communication is a process that has the following components: sender, receiver, method, medium, and the message. There are four basic types of communication: verbal, nonverbal, written, and graphic. Effective communication occurs when the information that is received and understood is accepted and acted on in the desired manner. Communication can be inhibited by a number of factors, including differences in meaning, insufficient trust, information overload, interference, condescending tones, listening problems, premature judgments, inaccurate assumptions, and technological glitches.

Listening means receiving a message, correctly decoding it, and accurately perceiving what is meant by it. Inhibitors of effective listening include a lack of concentration, preconceived notions, thinking ahead, interruptions, and tuning out. The five-minute rule allows project managers to maintain an open-door policy for listening to their team member's complaints, suggestions, recommendations, and problems. It means that, within reason, team members can have five minutes of the project manager's time at any time to discuss a problem. However, during the five minutes the team member must convey the information about the problem and recommend a viable solution.

Nonverbal communication consists of body factors, voice factors, and proximity factors. The key to understanding nonverbal communication is to look for agreement or disagreement between what is said verbally and what is said nonverbally. To improve the quality of their verbal communication, project managers can show interest in their topic, maintain a positive attitude, be flexible in making their points, use tact in delivering the message, and be courteous. When communicating corrective feedback, project managers should be positive, prepared, and realistic.

Written communication can be improved by: (1) identifying the audience before writing reports, (2) choosing a format that will allow the report to be concise but comprehensive, (3) using graphics wherever it is appropriate, (4) using language that is appropriate for the audience, and (5) highlighting actions that need to be taken by readers of the report.

Project managers must be good negotia-tors. Many of the factors necessary for bringing a construction project in on time, within budget, and according to specifications are out of the project manager's control. This means that some of these factors must be negotiated. Important considerations when negotiating include the fol-lowing: (1) Negotiate in stages and consider tim-ing, location, and image, (2) create favorable momentum, and (3) observe certain behaviors during negotiations..

During the negotiating process, project managers should think critically, keep the other party's hopes and needs in mind, be patient, ignore personal comments, and leave room to maneuver. After an agreement has been reached, project managers should follow through and do what they agreed to do.

KEY TERMS AND CONCEPTS

Communication plan
Communication
Sender
Receiver
Method
Medium
Message
Effective communication
Differences in meaning
Insufficient trust
Information overload
Interference
Condescending tones
Listening problems
Premature judgments
Inaccurate assumptions
Technological glitches
Listening
Lack of concentration
Preconceived notions
Thinking ahead
Interruptions
Tuning out

Five-minute rule
Nonverbal communication
Body factors
Voice factors
Proximity factors
Verbal communication
Interest
Attitude
Flexibility
Tact
Courtesy
Corrective feedback
Written communication
Negotiate in stages
Bridge analogy
Building the foundation
Building the structure
Completing the bridge
Think critically
Be patient
Ignore personal comments
Follow through

REVIEW QUESTIONS

1. What is a communication plan, why is one needed, and what does one contain?
2. Define the term "communication."
3. What are the components of the communication process?
4. List the four basic categories of communication methods.

5. Distinguish between communication and effec-tive communication.
6. List and briefly explain the factors that can inhibit communication.
7. Define the term "listening."
8. List and briefly explain the inhibitors of effective listening.

9. Explain the five-minute rule. Why is it needed and how does it work?
10. Explain briefly the three components of nonverbal communication.
11. List and briefly explain the factors that will help project managers improve their verbal communication.
12. How can project managers ensure that their corrective feedback does not make team members defensive?
13. Briefly explain how project managers can ensure that their written reports are concise, comprehensive, readable, and understandable.
14. Explain the stages in the negotiating process.
15. Summarize the bridge analogy and explain how it applies to negotiating.
16. Explain how the following factors can affect a negotiation: timing, location, and image.
17. What is momentum and how can a project manager create it during the negotiating process?
18. List and explain five strategies that can be helpful during the negotiating process.
19. How should a project manager handle the following situation during a negotiation: The other party begins to make negative personal comments?
20. Why is it important to follow through after negotiating an agreement?

APPLICATION ACTIVITIES

The following activities may be completed by individual students or by students working in groups:

1. Assume that you are a new project manager. You will be required to keep higher management, your project staff team members, the owner, and the architect fully informed about progress and problems. Develop a communication plan for your project.
2. For a week, observe people who are having conversations. Write down the various forms of nonverbal communication exhibited by people in these conversations. Make note of how people use nonverbal cues to emphasize points, convey agreement, convey disagreement, and so on.
3. Assume that you are a project manager who has to give corrective feedback to a team member who is coming to work late on a regular basis. Write down what you will say to this team member and how you will say it. If working in a group, pair off and complete this activity verbally.
4. Assume that you are a project manager for a major addition to a regional airport. After the contract has been signed and the project is underway, the owner calls and says he needs to negotiate some changes to the contract including a new deadline for completion. You have been tasked with putting together your construction firm's negotiating team. The members of your team all have the necessary discipline-specific knowledge to participate, but they have never been involved in a negotiation. Therefore, your first task is to develop a brief seminar to teach them how to play a positive role during the negotiation. Develop that seminar.

CHAPTER 12

Managing Time in Project Teams

Time is always of the essence in construction projects. This is why project managers put so much effort into ensuring that time is carefully controlled throughout the duration of a project. Chapter Four explained how to develop a schedule for a construction project. Chapter Seven explained how to monitor and control the schedule at the jobsite. This chapter explains how project managers can manage their own time. This is important because project managers who cannot manage their own time cannot manage the project's time. In fact, project managers who fail to effectively manage their time and that of their teams are not likely to complete projects on time, within budget, and according to specifications.

Time management affects all three of these areas. Poor time management can cause a number of problems, including wasted time, added stress, lost credibility, missed deadlines, poor follow through on commitments, inattention to detail, ineffective execution, and poor stewardship. Further, project managers who are poor time managers will not be able to help their team members be good time managers. Clearly, effective time management is an important skill for project managers.

When a project team becomes rushed because of poorly managed time, its members often respond by cutting corners on their work. When the time of project teams is not managed well, its members cannot meet their deadlines. Finally, when project teams become rushed because of poorly managed time, additional resources in the form of time, personnel, and technology have to be added to the team in order to get the project completed on schedule. Resources cost money. Hence, adding them to a project as the result of poorly managed time just adds to the cost of a project.

POOR TIME MANAGEMENT AND TEAM PERFORMANCE

The following scenario shows how poor time management can undermine the performance of a construction team or a project staff team. David is a project manager for his construction firm. Although he has excellent credentials, David's projects never seem to meet their deadlines or stay within budget. The main reason for the mediocre performance on his projects is poor time management. As is often the case when the leader of a team is a poor time manager, David's poor time management snowballs making it difficult for other team members to manage their time well and causing various other problems.

Consider just the example of David's project staff team. Because he is always running late, David never gets to spend much one-on-one time with its members. He claims to have an open-door policy, but in reality David is always so busy trying to catch up that he has no time to listen when his team members have problems, recommendations, or concerns. As a result, he typically just ignores their requests for face-to-face meetings. His most frequent response when team members need a few minutes of his time is: "I'll get with you later. I don't have time right now."

Unfortunately for David, complaints that are ignored often become problems that cannot be ignored. The longer they are ignored, the bigger the problems become. David and his team suffer from the snowball effect all the time, which means that the construction firm suffers from it too. Poor time management is a major inhibitor when it comes to completing projects on time, within budget, and according to specifications.

COMMON TIME MANAGEMENT PROBLEMS AND THEIR SOLUTIONS

Project managers in construction firms often face a common situation. They prepare their "to-do" lists for the day, but never even get to the first item on the list. Instead, they spend their day dealing with one crisis after another. The concept is known as *putting out fires*. By the time they have put out all the fires, the work day is over and their to-do list remains untouched. For many project managers, this type of day is not just common, it's typical. The reasons some project managers spend too much time putting out fires and too little time working on planned activities vary, but most of them are predictable.

The most common causes of time management problems for project managers are unexpected crises, telephone calls, taking on too much—failing to say "no," unscheduled visitors, poor delegation, disorganization, and inefficient meetings (see Figure 12.1). Project managers must learn how to prevent these common time wasters and how to overcome them when they cannot be prevented.

Checklist of
COMMON CAUSES OF TIME MANAGEMENT PROBLEMS

✓ Unexpected crises

✓ Telephone calls

✓ Taking on too much—failing to say "no"

✓ Unscheduled visitors

✓ Poor delegation

✓ Disorganization

✓ Inefficient meetings

FIGURE 12.1 Project managers must learn how to overcome these time wasters.

STRATEGIES FOR MINIMIZING TIME DEVOTED TO PUTTING OUT FIRES

- End each day by planning the next

- Schedule loosely

- Avoid being an amateur psychologist

- Remember that most tasks take longer to complete than planned

---

FIGURE 12.2   Putting out fires is a common time waster for project managers.

Reducing the Number of Crisis Situations

Crisis situations are a part of the job for project managers in construction. Bad weather threatens, critical materials are not delivered on time, an important subcontractor goes bankrupt, a unionized workforce calls for a strike. These types of crisis situations happen all the time on construction projects. Even the best project managers will never completely eliminate crisis situations—there are just too many causal factors that cannot be controlled. However, there is a correlation between planning and the number of crisis situations that must be dealt with. This correlation can be stated as follows: *the better the planning the fewer the crises.* This is the good news. The bad news is that even with good planning, crisis situations will still occasionally occur.

A crisis, by definition, is a situation that must be dealt with right away. Consequently, when crises occur they take precedence over other obligations. This is why it is important to limit the number of crises and, in turn, the amount of time devoted to dealing with them. The following strategies can help project managers minimize the amount of time they must devote to putting out fires (Figure 12.2):

- ***End each day by planning the next.*** One of the best ways to prevent crises is through good planning. By planning, project managers can take control of their days rather than letting their days control them. Part of being an effective project manager— as was learned in Part One of this book—is developing a plan and a schedule for completing the project in question on time. This is the big picture plan. Wise project managers will also develop small picture plans. The process of developing small picture plans can be as simple as developing a "to-do" list for tomorrow before leaving the office today or a schedule for the upcoming week. For daily to-do lists, an effective approach is to list everything that is planned for tomorrow down one side of a sheet of paper in time order (e.g., 8:00 A.M.—Task 1, 9:30 A.M.—Task 2). On the other side of the sheet of paper list the same tasks but in priority order. The latter step is done to accommodate the fact that even with the best planning, emergencies can still occur that throw the project manager's planned schedule into disarray. When this happens, it is important to know which tasks on the daily "to-do" list are the most important since the emergency might leave only limited time in the day to do other things. Spending the last 15 minutes of each day planning for the next will not eliminate emergencies, but it will benefit project managers in two important ways by: (1) allowing them to adjust more efficiently and effectively to the emergencies and (2) preventing the creation of emergencies that occur because the project manager forgot something he was supposed to do on a given day.

- **Schedule loosely.** Anyone who has been to a doctor's office is familiar with the concept of overbooking. Physicians tend to overbook. They schedule appointments so tightly that more often than not by mid-morning their work is hopelessly backed up. Too often, the doctor's appointment log is more of a dream sheet than a realistic schedule. Physicians who schedule too tightly are guilty of ignoring the management adage that *most things take longer than you think they will.* All it takes is an unexpected turn of events with just one patient, and the rest of the day is thrown off schedule. Project managers are often like physicians in that they schedule more appointments and activities in a day than they can realistically handle—a practice that creates crises. The solution to this problem is to schedule loosely. This means scheduling more time than you think appointments will require and, then, trying to complete the appointments on schedule. If an appointment should take 15 minutes, schedule 20 or even 30 minutes for it. Also, if someone says, "I just need five minutes of your time," do not take this literally. Owners and architects rarely take just five minutes. Schedule 10 or 15 minutes. Scheduling loosely also means leaving catch-up time in between appointments. Often, there is follow-up work that needs to be done at the conclusion of an appointment. Leave sufficient time between appointments to get this work done immediately. Follow-up work that is left undone will add up throughout the day and become tomorrow's crisis.

- **Avoid being an amateur psychologist.** Project managers must be open to listening when employees have ideas, concerns, recommendations, or complaints relating to their jobs. Allowing project managers to maintain an open-door policy is the purpose of the five-minute rule that was explained in Chapter Eleven. However, it is not uncommon for members of project teams to ask for time to discuss personal problems. Few things can rob project managers of productive time faster than the personal problems of their team members. Project managers who allow themselves to get bogged down in the personal problems of team members typically end up devoting too much of their time to personal counseling and too little to project management. The predictable result of this imbalance is that project works falls behind schedule and crises begin to occur. Consequently, when team members bring their personal problems to project managers, the appropriate response is for the project manager to make a helpful referral. Making a referral is appropriate, but it is as far as the project manager should go in such cases, especially during work hours. Project managers should avoid the temptation to play amateur psychologist when team members want to discuss personal problems. There are several good reasons for this and some of them are listed as follows: (1) project managers are not qualified to give advice on personal problems, (2) providing personal counseling is not in the project manager's job description, (3) if the project manager's advice exacerbates the problem in question, the team member will blame him, and (4) playing amateur psychologist is a time-consuming exercise, and time is a precious asset to project managers. Problems that relate to the job should receive the full and immediate attention of project managers, but the personal problems of team members should be referred to human resource professionals who are better equipped to deal with them.

- **Remember that most tasks take longer to complete than planned.** This is a good rule of thumb to follow. No matter what has been planned, experience shows it will probably take longer than planned. Consequently, it is wise to build a little extra time into your schedule. For example, if a task should take 30 minutes, allow 45 minutes, and then try to finish in 30 minutes. In this way there will be extra time, if it is needed, without having to rush. If the task is completed on time, the extra time gained can be put to good use returning telephone calls or getting a head start on other obligations.

CONSTRUCTION PROJECT MANAGEMENT SCENARIO 12.1

All I ever do is put out fires

Sherry Jackson is ready to pull her hair out. She has an excellent plan for completing the Tele-Tec Project on time, within budget, and according to specifications. She also has an excellent project staff team to work with. The problem is that she never seems to be able to get around to executing the plan or to get her team members engaged. Instead, she spends all of her time putting out fires. Jackson has all the qualifications to become a good project manager except one: She is a poor time manager. She does not control her days. Rather, events in the day control her. In frustration, Jackson told a colleague: "I can't get this project going because all I ever do is put out fires."

Discussion Questions:

In this scenario, Sherry Jackson is frustrated because she never seems to be able to things moving in the right direction. When it comes to project management, Jackson may be her own worst enemy because she is a poor time manager. Have you ever worked with someone in any setting who was a poor time manager? How did poor time management affect this individual's work? What advice would you give Jackson about how to reduce the number of crises she has to deal with everyday?

Making the Telephone a More Efficient Tool

The telephone can be a time saving device for project managers, but it can also be a time waster. Spending time on hold, listening to irritating messages on answering machines, playing telephone tag, and engaging in dialogue unrelated to the purpose of the call are all common time wasters for project managers.

Cellular phones and other handheld communication devices with their almost infinite list of applications tended to compound the problem. On the other hand, if used wisely handheld communication devices can be time savers rather than time wasters. The following strategies will help project managers minimize the amount of time they waste on the telephone (Figure 12.3):

- ***Use e-mail instead of the telephone whenever appropriate.*** One of the best ways to avoid wasting time on hold, playing telephone tag, or listening to recorded messages is to use e-mail instead of the telephone whenever appropriate. Of course, e-mail is not always an appropriate option. However, when it is project managers can simply type a brief message, click "Send," and move on to their next task. With e-mail

---

STRATEGIES FOR MORE EFFICIENT USE
OF THE TELEPHONE

- Use e-mail instead of the telephone whenever appropriate.

- Categorize calls as important, routine, and unimportant.

- Use a cellular phone to return calls between meetings and during breaks.

- Block out time on the calendar for returning telephone calls.

- Limit unrelated dialogue during telephone calls.

FIGURE 12.3   The telephone can be a time saver or time waster.

there is no pressing one for this option or two for that option, no talking to answering machines, and no being put on hold. In addition, people are often more prompt about returning e-mail messages than they are about returning telephone calls or responding to telephone messages.

- ***Categorize calls as important, routine, and unimportant.*** Project managers will find that time invested in helping administrative assistants learn to distinguish between important and unimportant telephone calls and between important and routine calls will be time well spent. One of the ways to do this is for project managers to provide administrative assistants with a list of people they always want to talk to. Within reason, these calls should always go through. On others, the administrative assistant can take a message, let the caller leave a recorded message, or even suggest sending an e-mail message. If administrative assistants are going to take written telephone messages, train them to be comprehensive and detailed. In addition to who called and when, a good telephone message will contain the caller's telephone number, reason for calling, and a good time to return the call. This final item—a good time to return the call—is important because it will help project managers avoid the time wasted playing telephone tag.

- ***Use a cellular phone to return calls between meetings and during breaks.*** On the one hand, cellular phones can be obnoxiously intrusive. Being interrupted by the untimely ringing of a cellular phone during a conversation, in a restaurant, during a meeting, in a movie, or even during a funeral has become a common annoyance. On the other hand, cellular phones can help project managers turn time that might otherwise be wasted into productive time. Project managers can save valuable time by taking telephone messages to meetings and using a cellular phone to return them during breaks and between meetings. Project managers can also use cellular phones to return calls from their cars, provided they have the appropriate "hands-off" technology or that their car is off the road and parked. Causing an accident by trying to send text messages while driving with one's knees will ultimately cost project managers more time than the texting would have saved them.

- ***Block out time on the calendar for returning telephone calls.*** Telephone tag is one the project manager's most persistent and frustrating time wasters. Assume that a project manager really needs to talk with someone and e-mail is not an appropriate communication medium. She places the call, but the person in question is not available. She leaves a message. The individual in question calls back, but now the project manager is tied up in a meeting. The caller leaves a message. The project manager calls back, but just misses him. This frustrating situation is known as "telephone tag," and it repeats itself many times every day in construction firms. To minimize the amount of time wasted playing telephone tag, project managers can block out times on their calendars for returning calls and let the times be known to callers who leave messages. An effective approach is to schedule three blocks of time each day: one in mid-morning, one in mid-afternoon, and one near the end of the day. Project managers should block these times out on their calendars as if they are appointments. Then they should make sure that the administrative assistants who take their messages let callers know that these are the times during which they typically return calls. The times should also be included in recorded telephone answering messages if possible. In this way, if callers really need to talk to the project manager, they will make a point of being available during one of these times. Another version of this time-saving strategy is to send an e-mail or a text message that asks either: (1) What is a good time for me to call you today? or (2) Will you call me at (name a time) today?

- ***Limit unrelated dialogue during telephone calls.*** One of the reasons telephone calls are such time wasters is the human propensity to engage in dialogue about matters unrelated to the purpose of the call. Project managers can save a surprising amount of time on the telephone by simply getting to the point, and by tactfully nudging callers to do the same. There is certainly nothing wrong with a few appropriate comments on the latest ball game, movie, or news item, but the amount of time devoted to unrelated issues should be kept to a minimum. Project managers should practice being focused, staying on task, and tactfully nudging callers do the same.

Cutting Back on Nonessential Responsibilities

Taking the initiative and seeking responsibility are traits of effective project managers. There is both good news and bad news in this fact. The good news is that taking the initiative and seeking responsibility are traits that will help construction professionals become good project managers. The bad news is that these traits can result in project managers taking on too many responsibilities. When this happens they end up suffering the effects of overload. For people who take the initiative and seek responsibility, it is easy to fall into the trap of taking on too much. This is a common problem among project managers. When this happens, the following strategies can help get things back in balance:

- ***Make a list of all current and pending obligations.*** Make a list of all current and pending obligations and commitments. For each entry on the list, ask the following questions: (1) What will happen if this is not done? and (2) Is there someone else who could do this? Invariably there will be things on the list that really do not have to be done. These items sounded like good ideas originally, but in retrospect they do not really need to be done or, at least, not now. In addition, going through this exercise will often reveal that there are items on the list that could and should be delegated. Asking the two questions suggested will usually reveal ways to pare down the list. Once the list has been pared down to the items that must done, the remaining items should be prioritized. This will ensure that project managers are putting their efforts into the most important items on the list.
- ***Take stock of after-work activities.*** Taking the initiative and seeking responsibility are two traits common to people who become project managers. These traits apply as

much after work as they do during the work day. Consequently, project managers often get involved in professional organizations, civic clubs, and other outside activities. Participation in organizations and activities outside of work is an excellent way to grow as a leader. However, it is easy to fall into the trap of taking on too many outside responsibilities. Balance is the key. For project managers, outside activities are like food. A certain amount of the right types are essential, but too much—even of the right types—can be harmful. When project managers find themselves overextended and need to cut back on their obligations, outside activities can be a good place to begin. Project managers who overburden themselves with too many outside activities run the risk of distracting themselves from obligations at work.

Limiting Unscheduled Visitors

It is important for project managers to maintain an open-door policy for their team members. On the other hand, unscheduled visitors can take up a lot of time. Consequently, project managers must find the right balance between maintaining an open-door policy and requiring visitors to have appointments. Helping project managers strike the proper balance is the purpose of the five-minute rule that was explained in Chapter Eleven. To review, with this rule a member of the project staff team or other internal teams can have five minutes of the project manager's time at any time unless there are circumstances that make this impossible. Within the allotted five minutes team members are expected to state their problem or concern and recommend a well-considered solution. If the team member has thoroughly prepared before asking for a five-minute audience, five minutes is typically more than enough time.

If an issue cannot be handled in five minutes, team members should be required to make an appointment so the necessary time can be blocked out on the project manager's calendar. The five-minute rule, together with requiring appointments for issues that will take more than five minutes, will help alleviate the problem of people just dropping in unannounced to chat. Unfortunately, these two strategies will not eliminate the problem completely. After all, some of the people who just drop in to chat might be superiors in the organization or the owner. Consequently, it is important for project managers to understand how to minimize the amount of time taken up by drop-in visitors. The following strategies will help (Figure 12.4):

- ***Do not allow drop-in visitors during peak times.*** Some days are busier than others and some times of the day are busier than other times. During these peak times, it is best to ask drop-in visitors to come back at another time when they can receive undivided attention, unless the following situations occur: (1) they are bringing critical

---

STRATEGIES FOR REDUCING TIME TAKEN BY UNSCHEDULED VISITORS

- Do not allow drop-in visitors during peak times.
- Train administrative support personnel to come to the rescue.
- Remain standing.

FIGURE 12.4 Drop-in visitors can waste much of the project manager's time.

information that needs to be conveyed immediately, (2) they are warning of an emergency, or (3) they are someone who must be seen right away no matter what else is on the schedule.

- ***Train administrative support personnel to come to the rescue.*** Project managers can minimize the intrusions of drop-in visitors by working out an arrangement with administrative support personnel to come to the rescue after a prearranged amount of time, for example, five minutes. It works like this. Whenever a drop-in visitor has been in the office for five minutes or so, the secretary buzzes or looks in and says "it's time to place that important call." This will tactfully let the drop-in visitor know that he needs to make an exit.

- ***Remain standing.*** One way to convey the message that "I am busy" without having to actually say it is to remain standing when an unannounced visitor walks into the office. Once visitors sit down and get comfortable, it can be difficult to uproot them. Continuing to stand, will tactfully convey the message that, "I can give you a few minutes, but only a few."

Overcoming Poor Delegation of Work

Poor delegation is one the easiest time-wasting traps to fall into. Some project managers find it difficult to let go of work they are accustomed to doing themselves. For example, it can be difficult to be a top-performing jobsite superintendent one day and a project manager the next. Yet, that is often how it happens. In addition, some project managers find it difficult to delegate work to team members who do it differently than they would. These two phenomena can result in poor delegation, a major time waster. Tasks that are not project management tasks—planning, scheduling, budgeting, controlling, teambuilding, motivating, leading, and so on—should be delegated. If team members cannot perform the tasks delegated to them satisfactorily, the problem is one of training or staffing not delegation. Training and staffing problems cannot be solved by refusing to delegate work. Project managers who must do work that team members should do simply because they refuse to delegate will not be effective. If a team member who should be doing certain work cannot do it properly, he should be trained, mentored, or replaced.

Overcoming Personal Disorganization

Project managers can waste a lot of time rummaging through disorganized stacks of paperwork looking for the folder, form, drawing, or document needed. The author once worked with an engineer who had the unfortunate habit of never returning files, documents, or drawings to the appropriate place when he was done with them. Wherever this engineer happened to be when he finished with paperwork is where he left it. As a result, this otherwise talented professional could be counted on to waste valuable time looking for "missing" paperwork. He eventually earned a reputation for being habitually disorganized—a reputation that hurt his career. Because getting people and work organized is an important part of what project managers do, this talented engineer never became a project manager. Project managers can use the following strategies to help decrease the amount of time they waste because of personal disorganization (Figure 12.5):

- ***Periodically clean off desks and workstations.*** This strategy sounds so simple that one might be tempted to ignore it. But before doing so, project managers should

---

**STRATEGIES FOR
GETTING ORGANIZED**

- Periodically clean off desks and workstations.

- Restack your work in priority order.

- Categorize work folders (e.g., read folder, correspondence folder, signature folder).

FIGURE 12.5 A lack of organization will cause project managers to waste time.

look at their work areas, check their in-baskets, and go through their stacks of pending work. The result of this examination of the work area is often that old and outdated paperwork is found cluttering up desks and taking up space in in-boxes. Finding irrelevant material in pending-work stacks is also common. Consequently, project managers should periodically go through everything on their desks and in their work areas and get rid of anything that is no longer pertinent. When a project manager decides to get organized, one of her best organizational tools will be a large trash can.

- ***Restack your work in priority order.*** Go through the in-box or stack of pending work and reorganize everything in order of priority. Paperwork is often stacked in the order it comes in, especially when project managers are in a hurry and do not have time to organize it. Because this can happen so frequently, it is a good idea to occasionally stop working long enough to go through the work stack and reorganize everything by priority. It's an even better idea to screen work as it comes in, placing all work in priority order from the outset.

- ***Categorize work folders.*** Organize paperwork in folders by category. Have a *Read Folder* for paperwork that should be read, but requires no writing or other action. Have a *Correspondence Folder* which contains correspondence that requires some action. Have a *Signature Folder* for paperwork that requires a signature (e.g., correspondence). Organizing work in this way will save time and increase efficiency.

Making Meetings More Efficient

In spite of their value in bringing people together for the purpose of conveying information, brainstorming, planning, and discussing issues, meetings can be major time wasters. Meetings waste time because some of them are not necessary in the first place while those that are necessary, in many cases, are inefficiently run. Project managers can minimize the amount of time wasted in meetings by: (1) making sure that all regularly scheduled meetings are actually necessary and (2) ensuring that necessary meetings are run as efficiently as possible. Strategies that will help minimize the time wasted by meetings include the following (Figure 12.6):

- ***Be aware of the causes of wasted time in meetings.*** Much of the time spent in meetings is wasted. The principal reasons for this are poor preparation, the human need for social interaction, time of participants spent in conversation unrelated to the meeting's purpose, interruptions, getting side-tracked on unrelated issues, no agenda, and no prior distribution of backup materials. In addition to these time wasters, there is also the *comfort factor*. Coffee, soda, tea, water, snacks, and social interaction can create a comfortable environment that people are reluctant to leave. Project managers

---

**STRATEGIES FOR
MAKING MEETINGS EFFICIENT**

• Be aware of the causes of wasted time in meetings.

• Determine if regularly scheduled meetings are really necessary.

• Hold impromptu meetings while standing up.

• Complete the necessary preparations before meetings.

• Begin meetings on time, stick to the agenda, and take minutes.

• Follow-up meetings promptly.

FIGURE 12.6   Inefficient and unnecessary meetings are a major time waster.

should make a point of eliminating these time wasters from internal meetings, especially those meetings they chair. These comfort items should still be made available when meeting with the owner and architect. Minimizing the comfort factor, minimizing unrelated conversation, preparing thoroughly, and getting organized will help enhance the efficiency of meetings.

• ***Determine if regularly scheduled meetings are really necessary.*** Most organizations have regularly scheduled weekly, biweekly, and monthly meetings of various groups. Project managers often have regularly scheduled meetings of their teams. When these meetings were established they probably had a definite purpose, and that purpose was probably valid. In fact, it might still be valid. However, it is not uncommon to find people in organizations meeting only because they have always met on a given day at a given time. Sometimes, meetings are perpetuated out of habit rather than need. If project managers call or attend regularly scheduled meetings, they should ask the following questions about each meeting: (1) What is the purpose of the meeting? (2) Is the meeting really necessary? (3) Can the meeting be scheduled less often, and (4) Can the purpose of the meeting be satisfied some other way (e-mail updates, written reports, webcasts, etc.)?

• ***Make impromptu meetings stand-up meetings.*** Impromptu meetings that should last no more than 10 minutes can be kept on schedule by holding them in standing positions. These are typically meetings without an agenda called on the spur of the moment to quickly convey information to a select group or to get input from that group. These types of meetings are best held in a location other than a conference room. When meetings are held in a conference room, it is difficult to keep participants from pulling up chairs and getting comfortable. Holding a meeting standing up conveys the message that "this is going to be a brief meeting" without the project manager having to actually say so.

• ***Complete the necessary preparations before meetings.*** For sit-down, scheduled meetings, project managers should have an agenda that contains the following information: purpose of the meeting, starting and ending time, list of agenda items with the person responsible for each item noted, and a projected amount of time to be devoted to each item. Set a deadline for submitting agenda items and stick to it. Require all backup material to be provided at the same time as the corresponding agenda items.

Distribute the agenda, backup material, and the minutes of the last meeting at least a full day before the meeting. Distributing meeting materials too far in advance is not advisable because participants will tend to put them aside and forget about them. In addition, distributing materials too far in advance limits the time participants have to submit agenda items and backup material. On the other hand, if project managers wait until during the meeting to distribute the agenda and backup materials they will waste time handing them out and waiting while participants read them. Better to ask participants to read the agenda and backup materials before the meeting so that they come prepared to participate in an efficient and effective manner. This is why the agenda and backup materials are distributed in advance. In other words, project managers should come to meetings well prepared and insist that other participants do the same.

- ***Begin meetings on time, stick to the agenda, and take minutes.*** An important rule of thumb for project managers who run meetings is: begin meetings on time. Waiting for latecomers to arrive only reinforces tardiness. Project managers who run meetings should remember the management adage that you get what you reinforce. Reinforce tardiness and you will get tardiness. If participants know that meetings are going to start on time, most will eventually discipline themselves to arrive on time. Project managers should assign someone to take minutes or bring an administrative assistant to meetings who can take minutes. In the minutes, all action and follow-up items should be typed in bold face so they stand out from the routine material. Make the minutes of the last meeting the first item on the agenda of the current meeting. In this way, the first action taken in each meeting will be following-up on assignments and commitments (those items that appear in boldface in the minutes) made during the last meeting. In running meetings, project managers should stay focused and insist that participants stick to the agenda and stay on task. The last agenda item should be "Around-the-Table Comments." This final item gives participants an opportunity to bring up issues that are not on the agenda without getting the meeting sidetracked before all agenda items have been discussed. Around-the-table comments should be limited to minor informational items that do not warrant a place on the agenda, require no discussion, and will take only a little time—matters such as announcements. Project managers who allow participants, during the around-the-table portion of the meeting, to discuss issues that should have been put on the agenda just encourage participants to ignore the agenda preparation process. When this happens, project managers lose control of their meetings and find themselves wasting time. Project managers should also ask participants to turn off cellular phones and other handheld communication devices during meetings so they can focus on the purpose of the meeting. Interruptions from cellular phones can be a major distraction and time waster in meetings.

- ***Follow-up meetings promptly.*** Project managers should ensure that the minutes of meetings are distributed as soon as possible following meetings—ideally on the same day the meeting occurred. Once the minutes have been distributed, the next step is to allow an appropriate amount of time for participants to act. Once team members have had sufficient time to get started on actions items from the meeting, project managers should begin the follow-up process. It is unwise to wait until the next meeting to follow-up on progress made by team members in completing action items from the previous meetings.

The strategies recommended in this chapter will help project managers take control of one of their most valuable assets—time. The same strategies will also help project managers teach team members how to manage their time efficiently. When project managers and their team members manage time well, project teams are more likely to complete projects on time, within budget, and according to specifications.

CONSTRUCTION PROJECT MANAGEMENT SCENARIO 12.2

His meetings waste too much time

Edward Andrews, Vice President for Future Tech Construction, Inc., is concerned about one of his project managers—Mack Day. It seems that nobody wants to serve on project teams led by Day. No one has said anything specific, but Andrews can tell that the company's personnel are reluctant to work with Day. To get to the bottom of the situation, Andrews asked one of his most trusted colleagues to stop by his office for a brief meeting. When the meeting convened, Andrews came right to the point. "I don't understand why nobody seems to want to work with Mack Day. I need to know why. What's going on?" The response Andrews got from his colleague was frank and straightforward.

"Mack Day is world-class time waster. He is disorganized, refuses to delegate, micromanages his team members, and his meetings waste too much time. He will answer his cell phone in the middle of a conversation, get sidetracked on unrelated issues, and is constantly frantically running around trying to find paperwork he has misplaced. I don't see how Mack manages to get dressed in the morning. I wouldn't be surprised if he shows up one day without his shoes."

Discussion Questions:

This case illustrates one of the problems that can occur when project managers are poor time managers: Nobody wants to work with them. Have you ever known a person who was so disorganized that nobody wanted to work with him or her? If so, describe the situation. If Mack Day was your colleague and he asked for advice about how to be a better time manager, what would you tell him?

SUMMARY

Project managers who fail to manage their time and that of the team effectively are not likely to complete projects on time, within budget, and according to specifications. Poor time management can cause a number of problems, including wasted time, added stress, lost credibility, missed appointments, poor follow through, inattention to detail, ineffective execution, and poor stewardship.

Crisis situations rob project managers of valuable time. Crisis situations may be reduced if project managers will end each day by planning the next, scheduling loosely, refusing to get bogged down in the personal problems of team members, and remembering that most tasks take longer to complete than planned. The telephone can be a time waster unless project managers learn to use e-mail instead of the telephone when appropriate; categorize calls as important, routine, and unimportant; use a cellular phone to return calls between meetings and during breaks; block out time on the calendar for returning calls; and limit unrelated dialogue during calls.

Project managers can create problems for themselves by taking on too much. When this happens, the key is to cut out all nonessential responsibilities. This can be done by making a list of all current and pending obligations and deciding which items on the list are not essential. It will also help to take stock of after-work activities to determine if they are essential. Unscheduled visitors can rob project managers of valuable time. The interruptions of unscheduled visitors can be limited by refusing to meet with them during peak work times (unless the visitor is a superior, the owner, or a team member with an emergency). It will also help to train administrative assistants to come to the rescue and to remain standing when unscheduled visitors pop in.

Poor delegation and personal disorganization can rob project managers of valuable time. Project managers who fail to delegate because they have no confidence in team members should either arrange training for these team members or replace them. Personal disorganization can be overcome by cleaning out one's desk every six months, stacking the work in one's in-basket in order of priority, and using categorized work files: read folder, signature folder, and correspondence folder.

One of the most effective ways for project managers to save time is to eliminate unnecessary meetings and to make necessary meetings more efficient. Strategies that will help project managers minimize the amount of time wasted by meetings include being aware of the causes of wasted time in meetings, determining if regularly scheduled meetings are actually necessary and if they can be scheduled less frequently, holding impromptu stand-up meetings, preparing well for meetings and requiring participants to prepare, beginning meetings on time and sticking to the agenda, and following up promptly after meetings.

KEY TERMS AND CONCEPTS

Time management
Poor time management
Crisis situations
Schedule loosely
Unrelated dialogue
Nonessential responsibilities
Pending obligations
After-work activities

Unscheduled visitors
Drop-in visitors
Remain standing
Poor delegation
Personal disorganization
Unnecessary meetings
Impromptu meetings
Stick to the agenda

REVIEW QUESTIONS

1. How can project managers reduce the amount of time in their days wasted by crisis situations?
2. Explain the concept of scheduling loosely.
3. Why should project managers avoid the temptation to play amateur psychologist with their team members?
4. How can project managers make the telephone a more efficient work tool?
5. What is the problem with unrelated dialogue during telephone calls?
6. When project managers begin to feel overloaded, how should they approach the problem?

7. How can project managers limit the time wasted by unscheduled visitors?
8. If you knew a project manager who wasted time because of personal disorganization, what advice would you give him or her about how to get organized?
9. Describe how you would go about reducing the time wasted in an organization by meetings.
10. Why is it important to begin meetings on time?

APPLICATION ACTIVITIES

The following activities may be completed by individual students or by students working in groups.

1. Make a list of upcoming activities in your life and estimate how long it will take you to complete each activity. Record how long each activity actually takes. Discuss your list and the results of your experiment with other students.

2. Record how much time you spend on the telephone for a week. During each call, estimate how much time was devoted to the subject of the call and how much was unrelated. Record your results and discuss them with other students.

3. Make a list of all of your current and pending obligations and responsibilities. Are there any on the list that could be eliminated without causing problems. Discuss your list and the results with other students.

4. Identify a project manager in an organization who is willing to be interviewed. Ask this individual to list all regularly scheduled meetings he or she must attend and how often. Then ask if the meetings are essential and if they could meet less regularly without causing problems. Discuss the results of your interview with other students.

Managing Change in Construction Projects

Change is as much a fact of life in construction firms as it is in any other kind of organization. Change in construction projects occurs primarily for two reasons: (1) in response to change orders and (2) in response to change initiatives undertaken to improve performance, productivity, quality, safety, or some other critical factor. Change orders are an ever-present part of construction projects and usually result from a change instigated by the owner, architect, or engineers. They can also be initiated to compensate for errors made in the design or quality of the construction work.

Change initiatives are sometimes undertaken by construction firms to improve in a specific area. For example, the author can remember a construction project in which the construction firm decided to require its framing crews to make the transition from hammers to nail guns. The idea was to improve productivity. The initiative was not popular at first, but it quickly caught on and had the desired effect. Now, of course, nail guns have all but replaced hammers in framing crews.

To manage change effectively project managers must be prepared to deal with an unchanging fact of life. People get comfortable with how things are and, as a result, are prone to resist change. Consequently, construction project managers need to be prepared to deal with a concept called comfort-induced inertia. Construction firms, project teams, subcontractors, tradespeople, and individual workers can all suffer from comfort-induced inertia.

Students might remember inertia as a concept from physics in which a body at rest will tend to stay at rest until sufficient force is applied to move it. People working on construction projects can be like the body at rest in physics in that their tendency is to maintain the *status quo* until sufficient *force* is applied to break the inertia. "Force" in this case refers to leadership provided by the project manager. The concept of comfort-induced inertia, as it applies to project teams, can be summarized as follows: A project team's members will prefer to stay in their comfort zones and maintain the *status quo* until sufficient leadership is applied to convince them to embrace a given change.

CHANGE MANAGEMENT MODEL FOR PROJECT MANAGERS

In a highly competitive environment, construction firms and, in turn, project managers must be attuned to the need for finding ways to continually improve performance, productivity, quality, safety, and all of the other success factors associated with construction projects. They must also be prepared to implement change orders on the projects they manage. Managing change effectively is an important responsibility of project managers. Therefore, project managers who find themselves in the position of being responsible for implementing a significant change or even just a part of one will find the change management model in Figure 13.1 helpful. This model can be used regardless of the cause of the change in question.

This model is designed to be used to implement changes that are caused by both change orders and improvement initiatives. It works best on major changes that affect factors that are important to stakeholders inside and outside of the construction firm such as time, money, quality, safety, long-standing procedures, and even personal preferences. Each step in the model is important. Consequently, no step should be skipped.

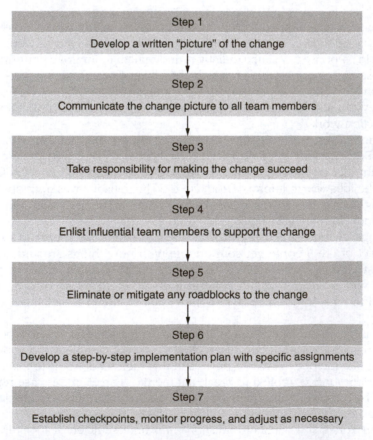

FIGURE 13.1 This model will allow change to be made systematically.

CONSTRUCTION PROJECT MANAGEMENT SCENARIO 13.1

Mary changes just for the sake of change

When Sandra Barker told her supervisor she would rather not serve on a project team managed by Mary Andrews, the supervisor was shocked. "I thought you liked Mary. In fact, I thought she was your best friend." "She is. That's why I know her so well, and knowing her so well is why I do not want to be on her project team." Barker went on to explain that Andrews become bored easily and that as a result she likes to change things. "She changes the way she drives to work every week, changes the arrangement of the furniture in her apartment at least once a month, and changes clothes at lunch. In fact, Mary changes just for the sake of change." Barker claimed that the last time she served on a team led by Mary Andrews her fellow team members nearly revolted over the constant and needless changes.

Discussion Questions

In this scenario, Mary Andrews seems to make changes that affect her team members for no reason other than boredom with standard operating procedures or established best practices. Have you ever known someone who liked to make changes just for the sake of change? If so, did this individual's restlessness cause problems? What advice might Sandra Barker give her friend, Mary Andrews, about when and why to make changes in a project team?

Step 1: Develop a Written "Picture" of the Change

What people fear about change is the unknown. The unknown typically makes people uncomfortable. This is one of the main reasons they tend to resist change. Even when people do not like the conditions in which they work, they still find a measure of comfort in the concept of familiarity. Even in undesirable conditions people gravitate toward the known out of a fear of the unknown. Because of their comfort with the familiar, members of project teams often adopt an attitude toward change that is best summarized by the old maxim, *the devil you know is better than one you don't.* Consequently, when making changes it is important for project managers to begin by eliminating the fear of the unknown. Thus, eliminating fear is the first step in the change management model explained in this chapter.

When a change must be made, project managers should remember that their team members know how things are now. What they do not know is how things will be after the proposed change. When a major change order is in the works or a major change initiative is being discussed, rumors will begin to circulate unless the project manager takes the initiative to get out in front of them. To eliminate fear of the unknown, project managers should replace the unknown with the known. The unknown can be transformed into the known by developing a compelling, informative change picture. A change picture is a comprehensive but brief written explanation of how things will be after the change.

The key to making a change picture compelling is for project managers to view the change from the perspective of those affected by it—their team members—and develop the change picture accordingly. People take change personally and react to it on an emotional level. Consequently, the first thing they will want to know about the proposed change is how it will affect them. Will the change mean more work or less? Will the change require workers to do things they do not yet know how to do? Will the change require workers to scrap their comfortable procedures and learn new ones? Will the change threaten or enhance job security? Project team members will have these types of questions, even if they do not verbalize them. Whether the news is good or bad, a good change picture will answer these types of questions as well as others by providing the following information: what, when, where, who, why, and how (five Ws and one H). Even if the news is will not be welcomed by stakeholders, it is better for them to know the facts than to depend on rumors that, by their very nature, tend to be exaggerated. A good change picture explains the nature of the change (what), when the change will occur, where the change will occur, who the change will affect, why the change is being made, and specifically how it will affect team members. This final consideration—how the change will affect team members—is important because it is the aspect of the change team members will be the most concerned about. A good change picture will explain the "how" question from the personal perspective of team members. For example, if there will be changes to deadlines, more work added to the project, cutbacks that could result in layoffs, changes in the normal work hours, altered working conditions, or anything else that will affect team members, the change picture should explain the change in terms that are open, honest, candid, and easily understood. Project managers who keep their team members guessing about how a change will affect them run the risk of losing their team members' trust and damaging their morale. It cannot be stressed enough that facts about changes—no matter how unwelcome the changes—are better than rumors.

Challenging Construction Project

GOLDEN GATE BRIDGE

When it was completed in 1937, the Golden Gate Bridge connecting San Francisco and Marin County, California was the longest suspension bridge in the world. Since this time, its span has been eclipsed by eight other bridges but it is still considered one of the world's most important construction projects. In fact, the American Society of Civil Engineers has declared the Golden Gate Bridge to be one of the modern wonders of the world.

Construction of the Golden Gate Bridge began in January 1933 and was completed in April 1937. Its design is suspension with truss and arch causeways. Total length is 8,981 feet, width is 90 feet, and height is 746 feet. The cost of construction was $35 million in 1930s dollars. Everyday almost 120,000 cars travel over the bridge. The Golden Gate Bridge is to the west coast of America what the Statue of Liberty is to the east coast.

The project manager for construction of the Golden Gate Bridge was Joseph Strauss. One of the groundbreaking construction innovations that came out of this project was the moveable safety netting that was strung under the bridges it was built. During the course of the project the lives of nineteen workers were saved when they fell into the net. Sadly, eleven other workers perished during construction—ten of them when the safety net failed because of a scaffold that had fallen.

Source: Based on Golden Gate Transportation District. http://www.goldengate.org

Step 2: Communicate the Change Picture to All Team Members

Once the change picture has been written it must be communicated to all team members. Unless the team is scattered among different locations and connected only electronically and by telephone, the change picture should be provided to team members in face-to-face meetings that allow for maximum give and take. However, no matter what communication methods are used, it is imperative that there be a convenient feedback mechanism so that stakeholders can ask questions, express concerns, point out problems, or just vent. This is why face-to-face meetings are the best approach for communicating the change picture.

An important part of communicating the change picture to employees involves conducting what the author calls a *roadblock analysis*. The roadblock analysis involves asking team members to identify any and all roadblocks that might derail the change process. A roadblock analysis can be accomplished by e-mail or even telephone, but the best approach is face-to-face. By asking for the help of team members who are closer to the situation than they are, project managers can identify roadblocks that might impede or even undermine the change implementation process. Once they have been identified, roadblocks can be removed or, at least, minimized before they derail the implementation. Removing or minimizing roadblocks is explained in a later step. This step involves only identifying them.

Step 3: Take Responsibility for Making the Change Succeed

Change in construction projects does not just happen—it requires commitment, persistence, and a lot of work, both mental and physical, from team members. More than anything, it requires team members to step forward and take responsibility for doing their part to make the change succeed. Often, the first response of team members to change is to resist it. Some team members will react to change by openly resisting it, but many will opt for a more passive form of resistance.

One of the more common forms of passive resistance is the wait-and-see attitude. Team members who adopt a wait-and-see attitude do not necessarily work against the change. On the other hand, they do not work for it either. Wait-and-see team members might say all the right things about supporting the change, but in reality they put no effort into making it succeed. This is the "passive" aspect of their resistance. Passive resisters often try to play on both sides of the fence until they know whether the change is going to succeed or fail. They like to be able to say "I told you so" if the change initiative fails and "I supported it from the beginning" if the initiative succeeds.

The wait-and-see crowd can be even more detrimental to change initiatives than those who openly oppose them. This is because making changes in the middle of a project is like pushing a boulder up a hill. The team needs all of its members pushing together in a coordinated and concerted effort. Those who openly oppose the change can be neutralized by isolating them from the implementation process. But those who act like they are helping when they are not can undermine the process. When trying to push a boulder up a hill, a lot of people are needed who will push with all their might.

Team members who act like they are pushing when they really aren't make the task more difficult for those who are. This is why wait-and-see team members do not have to openly work against changes they oppose. All they have to do is sit back and let them fail. However, when the team members who do their part manage to get the boulder over the crest of the hill the task is easier from that point on. When this happens, those who contributed nothing to getting the boulder up the hill typically join the parade of team members following it down the other side, all the while acting as if they had pushed with all their might. On the other hand, if the boulder bogs down and fails to make it up the hill, members of the wait-and-see crowd will be quick to join the naysayers who openly opposed the change.

Changes in construction projects—whether they are caused by change orders or change initiatives—will succeed only if the team members responsible for implementing them are willing to: (1) take responsibility for pushing the boulder up the hill and (2) commit to doing what is necessary to push the boulder over the crest. Once the organizational inertia has been broken and the boulder is moving uphill, responsible team members cannot rest until it is over the crest and the momentum is on the side of success.

Step 4: Enlist Influential Team Members to Support the Change

In every team, some members are more influential than others. The source of their influence might be seniority, talent, popularity, strength of personality, a combination of these, or a variety of other factors. Regardless of why they are influential, other team members look to these special few for direction and approval. These influential team members can contribute greatly to the success or failure of a change. This step is more important when implementing improvement initiatives than it is when implementing change orders, since there is an inherent urgency to change orders that is backed up by the construction contract. This contractual motivation is missing when implementing change initiatives. However, this step should not be skipped, even when the change in question is caused by a change order.

It is always best to enlist the support of influential team members when implementing changes. For example, assume that a major change order has been approved that is going to require several crews to rip out work they have already completed and redo it. It will be important to enlist the support of the leaders of these crews. Nobody likes to redo completed work, but the crews will do what has to be done if they see that their crew leaders support the change. Enlisting influential team members can be a challenge, especially if they are opposed to the change or fall into the wait-and-see category. Consequently, before attempting to enlist influential team members it is important for project managers to have face-to-face, one-on-one conversations with them.

During these conversations, project managers should determine where the influential team member in question stands concerning the change. Is he for it, against it, or just waiting to see what will happen? If he is against the change, does he intend to throw up roadblocks or just sit back and to see what will happen? If influential personnel are against the change or are in wait-and-see mode, project managers have two options: (1) isolate them from the implementation process to limit their negative influence or (2) enlist them in the effort by giving them responsibility for some aspect of the implementation plan.

When an influential team member is enlisted in this manner, it is important to apply appropriate reinforcement methods. Typically, this means applying both the carrot and the stick. The *carrot* consists of incentives the individual will receive when the change succeeds. The *stick* consists of the negative consequences that will occur if it does not. Assurance of full cooperation should be gained before putting an enlisted team member to work on behalf of the change. To do otherwise might just guarantee failure of the change initiative.

Step 5: Eliminate or Mitigate any Roadblocks to the Change

Part of an earlier step in this model involved conducting a roadblock analysis to identify any obstacles that might impede a successful implementation of the change initiative. In this step, the obstacles identified are either eliminated or minimized. In order to explain the change management model in a step-by-step manner, it is necessary to put this step here. However, in reality, removing or minimizing roadblocks—this step—begins as soon as they are identified. This step can be carried out in parallel with those that have already been explained.

More often than not, the obstacles identified are internal impediments decision makers have not thought of. For example, the author once worked on a project that was designed with prestressed concrete columns and beams, which was good because the general contractor had in-depth experience in erecting prestressed concrete buildings. When the price of concrete underwent a sudden and steep increase, the architect proposed that the material for the columns and beams be changed to steel. It was early enough in the process that the change would create no schedule problems and the cost differential was easily worked out. However, during the roadblock analysis the general contractor determined that there was a major roadblock. Its crews did not have any experience in erecting structural steel—they were prestressed concrete experts.

This was a serious roadblock, but not an impossible situation. The general contractor decided that it would have to procure a subcontractor with the appropriate expertise to erect the steel columns and beams. A change order was worked out between the general contractor and the architect that, in the end, was still less costly than using the prestressed columns and beams would have been.

Step 6: Develop a Step-by-Step Implementation Plan with Specific Assignments

Once the roadblocks to a successful implementation have been removed or minimized, a step-by-step implementation plan is developed. The plan lists every action step that must be taken to get the change initiative successfully implemented. Every action that is to be taken is then assigned to a specific individual—not a group or a team, but an individual. This is important. Although changes in construction projects often require an entire crew or crews to be involved in their implementation, the crew leader or leaders should be given responsibility for the assigned tasks. The crew leader, then, is held accountable for ensuring that the appropriate team member does what is necessary. Avoiding confusion concerning who is supposed to do what is essential to the successful implementation of a change initiatives and change orders.

Step 7: Establish Checkpoints, Monitor Progress, and Adjust as Necessary

One of the reasons changes in construction projects fail is that project managers ignore the need to monitor progress after implementation has begun. Assigning responsibility for specific tasks is important, but doing so is no guarantee of a successful implementation. To ensure a successful implementation, project managers must stay on top of the implementation. This means establishing a deadline for completion of each action step that is assigned to a team member, establishing incremental progress points that precede these deadlines, monitoring regularly to ensure that satisfactory progress is being made, and making adjustments when unanticipated problems occur.

When implementing a change initiative, it is almost guaranteed that problems will arise that even the roadblock analysis did not anticipate. When this happens, project managers need to know about it so they can take immediate corrective action. Taking corrective action involves adjusting as necessary to keep the implementation on track and the momentum on the side of a successful implementation.

The model for implementing changes in construction projects presented in this chapter will help project managers become effective change agents and change managers. Project managers who are good change managers will, in turn, be better able to ensure that their projects are completed on time, within budget, and according to specifications.

CONSTRUCTION PROJECT MANAGEMENT SCENARIO 13.2

I know how to manage projects, but I need to know how to manage change

Juan Padea has been a project manager for more than three years. During this time, his teams have always performed well. In fact, Padea is known throughout his firm for bringing projects in on time, under budget, and according to specifications. But Padea's current project is presenting him with a challenge he has not had to deal with on previous projects: a major change resulting from a contract amendment requested by the owner. The owner has decided he wants two additional floors added to the hotel that Padea's firm is building.

The owner is willing to pay generously for the change order. This is the good news. The bad news is that he wants to retain the original date for project completion. Word has spread about the contract amendment and Padea's team members are nervous. His team members want to know what the change will mean to them. In fact, Padea's crews are spending so much time worrying about the impending change order that they are not attending to the work at hand. Suddenly, the project is falling behind schedule even without the two new floors that will have to be added. When a colleague asked Padea how he planned to handle the change order, he replied: "I know how to manage projects, but I need to know how to manage change."

Discussion Questions

In this scenario, Juan Padea faces a problem that often challenges the leadership skills of project managers: having to make changes in the middle of a project that is well underway. Have you ever started on a project and had to make a major change in midstream? If so, how did you respond? Was the situation frustrating? If Juan Padea asked you for advice concerning how to manage the change his team is facing, what would you tell him?

SUMMARY

Change is a fact of life in construction firms. This is why the change order process exists. Even without change orders, continual improvement means continual change because doing something better than it is currently being done necessarily means doing it differently. To manage change effectively, project managers must be prepared to deal with a phenomenon known as *comfort-induced inertia*: the tendency of people to maintain the *status quo* until sufficient leadership is applied to change it.

An effective change management model has the following steps: (1) develop a written "picture" of the change initiative, (2) communicate the change picture to all team members and conduct a roadblock analysis, (3) take responsibility for making the change initiative succeed, (4) enlist influential team members to support the change initiative, (5) eliminate or minimize any road-blocks to the change initiative, (6) develop a step-by-step implementation plan with specific assignments, and (7) establish checkpoints, monitor progress, and adjust as necessary.

KEY TERMS AND CONCEPTS

Comfort-induced inertia

Change management model

Change picture

Roadblock analysis

Change initiative

Influential team members

Carrot

Stick

Implementation plan

Monitor progress

REVIEW QUESTIONS

1. Why is it important for construction project managers to be effective change managers?
2. Explain the concept of organizational inertia. How can it be overcome?
3. Explain what is meant by developing a written "picture" of a change initiative.
4. What information should be contained in a change picture?
5. Ideally, how should a project manager go about communicating a change picture to team members?
6. What is a roadblock analysis?
7. Explain how team members can passively resist a change initiative.
8. How can a project manager enlist influential team members to ensure that a change initiative succeeds?
9. Explain the carrot-and-stick approach for dealing with influential team members project managers want to enlist on the side of a change order or initiative.
10. Think of an example of a change that you have made or would like to make. Now identify any roadblocks that might keep you from making the change. How would you eliminate or, at least, minimize the roadblocks?
11. What is the significance of making specific assignments when implementing a change order or initiative?
12. Explain why it is important to establish checkpoints and monitor progress when implementing change initiatives.

APPLICATION ACTIVITIES

The following activities may be completed by individual students or by students working in groups.

1. Identify a construction firm in your region that will cooperate in completing this project. Ask the firm to see a major change order that occurred during a construction project. Then ask for an explanation of how the change order was handled. How could the seven-step model for implementing change presented in this chapter have been used to implement the change order?

2. Assume that you work for a large engineering/construction firm that plans to purchase new CAD software. The new software is much more advanced than the current CAD software used by your firm, but its operational procedures are much different. The software will also require a hardware upgrade. Using the change management model explained in this chapter, explain what will need to be done in each step of the model. Do not forget to conduct a roadblock analysis.

CHAPTER 14

Managing Diversity in Teams

Americans can trace their lineage to virtually every country, race, and culture in the world. In fact, America is one of the world's most diverse countries, and this diversity is reflected in the workplace. Because of this, project managers can expect to lead diverse teams. The term *diversity* as applied in the current context refers to human differences of all kinds. Dealing with these differences in ways that make the team stronger is the responsibility of the project manager.

There are many ways in which people can be different, but the ways that always seem to command the most attention are race, culture, national heritage, gender, religion, politics, education, worldview, and personality. Because people in construction project teams can be different from each other in all of these ways, it is important for project managers to learn how to lead diverse teams and to teach team members, through words and by their example, how to work well in a diverse environment.

DIVERSITY DEFINED

Try this experiment. Ask several friends or fellow students what comes to mind when they hear the term *diversity*. Do not be surprised to find that people tend to associate the term with racial differences. This perception is partially correct, but only partially. Diversity includes racial differences but it is a much broader concept than just racial differences. As it applies in the context of project management, diversity encompasses all of the ways in which people can be different.

Following is a partial list of ways that people can be different: race, mental ability, physical ability, physical appearance, age, marital status, geographic status, religion, denominations within religions, ethnicity, nationality, worldview, education level, values, political beliefs, interests, personality, cultural background, height, weight, career status, white collar, blue collar, and personal preferences (food, clothing, music, hobbies, etc.) as shown in Figure 14.1. This is a long list but remember it is only a partial list. There are even more ways in which people can be different.

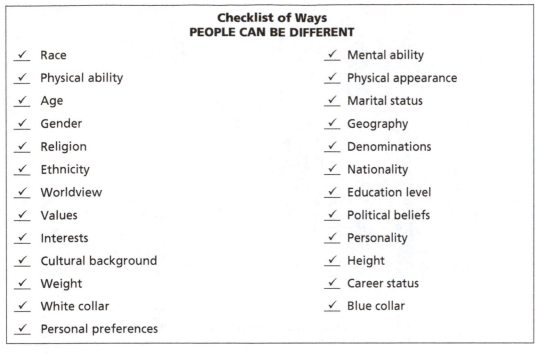

FIGURE 14.1   Members of project teams can be different in many ways.

Obviously, people can be different in a lot of ways. In fact, members of the same family—even identical twins—can be vastly different when it comes to their physical abilities, personal interests, religion, education, personality, work skills, appearance, attitudes, ambition, political beliefs, and worldviews. If there can be this many differences between family members, imagine all the ways in which people in construction project teams can be different. It is a certainty that project managers are going to have to lead team members who are different from them in how they talk, dress, eat, interact, socialize, believe, think, and approach their work. Being able to effectively lead diverse teams is an essential people skill for project managers.

DIVERSITY-RELATED CONCEPTS

In conversations about diversity, certain terms will surface over and over again. Project managers should be familiar with these terms themselves and help their team members become familiar with them. Important diversity-related concepts that project managers should understand include the following (Figure 14.2):

- Prejudice
- Stereotyping/labeling
- Discrimination
- Tolerance

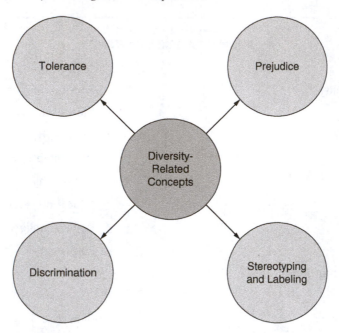

FIGURE 14.2 Project managers should understand the diversity-related concepts.

Prejudice

Prejudice is a predisposition to adopt negative perceptions about groups of people. For example, one might be prejudiced against conservatives, liberals, northerners, southerners, easterners, westerners, men, women, whites, blacks, accountants, engineers, Democrats, Republicans, youth, elderly, or any other identifiable group of people. People who are prejudiced have learned to harbor negative perceptions toward a distinct group or groups of people. These negative perceptions can result in ill feelings toward members of the group in question. For those who are prejudiced against a given group of people, all members of that group are subject to the same negative perceptions. For example, an individual who is prejudiced against teenagers might claim that all teenagers have slouchy posture and bad attitudes and that they waste time hanging out at shopping malls. Although this description might be accurate for some teenagers, it is not accurate for all teenagers. This anomaly is true of any group that is the target of prejudiced attitudes.

People who are prejudiced tend to divide the world into two distinct groups: "us" and "them." Those who appear to share their values, cultural mores, worldviews, and other pertinent characteristics are part of the "us" group. Everyone else is lumped into the group known as "them." Prejudice can manifest itself in project teams in a number of different ways—all negative. People in project teams may act out their prejudices through discrimination, stereotyping, and labeling. Others act out their prejudices in less overt, but equally negative ways. For example, an individual who is prejudiced against a given group might find it difficult to admit that anyone in that group ever had a good idea, did a good job, was worthy of promotion, or could sufficiently satisfy any performance measure. With prejudiced people, members of the "us" group receive the benefit of the doubt in all situations while members of the "them" group receive only the doubt.

Prejudice can blind project managers and members of their teams to the fact that people who they are prejudiced against are capable of excellent performance. As a result, prejudiced project managers might not want certain people to serve on their teams or they might not invest the time and effort in developing these team members that they invest in others. Not only is this wrong from the perspective of equity, it is unwise from the perspective of effective project management. Prejudice can blind a project manager to the talent and potential of others.

A team member's performance on a project is a function of individual talent, attitude, motivation, and other job-related factors, not diversity factors such as race, sex, politics, cultural heritage, or worldview. Consequently, when project managers allow prejudice to influence their actions and decisions, they limit their ability to get the best possible performance from their teams. In addition, they often pass on their prejudice to team members, corrupting their views concerning diversity.

Stereotyping/Labeling

Stereotyping is a negative by-product of prejudice. It is the act of generalizing certain characteristics to all members of a given identifiable group. For example, you may have heard so-called *blond jokes*. What blond jokes have in common is the assertion that all blonds are dumb. In order to "get" a blond joke, one must accept this assertion. Accepting this assertion is engaging in stereotyping. It was stereotyping that for years denied even the best African-American college quarterbacks opportunities to lead football teams at the professional level. For decades, outstanding college quarterbacks who happened to be black were automatically converted to receivers and defensive backs when drafted by professional football teams.

The stereotype behind this unfortunate practice was that blacks were not smart enough to lead teams at the professional level. Now that African-American quarterbacks are common in professional football, people who are too young to have experienced this stereotype have trouble believing it ever existed. But, of course, it did. The same kind of stereotype once applied to football coaches at the college and professional levels. Now that some of the most successful coaches who ever walked the sidelines are black, the stereotype has been discarded and forgotten. However, what should not be forgotten is how wrong the stereotype was because it shows how illogical and damaging stereotyping can be.

Stereotyping results in lose-lose-lose situations in project management. Victims of stereotyping lose because they are denied opportunities to prove themselves and to contribute to the success of the team. The team loses because it is denied the talent of the stereotyped individual. The individual who engages in stereotyping loses because he is responsible for the first two losing propositions. This, in turn, renders him less effective at his job be it CEO, project manager, jobsite superintendent, crew chief, or worker.

Labeling is an extension of stereotyping in that it involves attributing a certain characteristic to a distinct group of people and then using that characteristic to label all people in that group. For example, people who are talented in the fields of engineering, science, math, or computers are often labeled as "nerds" or "geeks." These labels are supposed to conjure up the image of a socially inept person who can solve quadratic equations but cannot make informal social conversation. While there may be some academically gifted individuals who are socially challenged, to lump all scholars into a uniform group in this way is not just illogical, it is narrow minded.

Prejudiced behavior such as stereotyping and labeling harm both the victims and the perpetrators. Just because some people in an identifiable group exhibit certain characteristics hardly means that all people in that group share those characteristics. After all, some people in almost any group will exhibit certain identifiable characteristics. The mistake made by prejudiced people is attributing those identifiable characteristics to all members of the group in question.

To understand how illogical the practices of stereotyping and labeling can be, take the prejudicial practice of referring to all southerners as rednecks. This is a stereotype that is easily disproven. Some of America's most brilliant scholars, poets, musicians, artists, scientists, and political leaders have been men and women from the South. In fact, several southerners have been president of the United States, including George Washington, who is known as "the father of our country" and Thomas Jefferson, the author of America's Declaration of Independence. An objective observation by anyone who is not prejudiced will reveal the fallacy of stereotyping and labeling.

Project managers who allow stereotyping and labeling to influence their decisions do their victims, their teams, and themselves a great disservice. Try to imagine a baseball coach leaving his best hitter on the bench because of stereotypical thinking. In essence, this is what project managers do when they allow prejudiced practices such as stereotyping and labeling to influence their decisions. Project managers who excel make their decisions on the basis of such factors as productivity, performance, and quality. In other words, their decisions are based on who can best help them get their projects completed on time, within budget, and according to specifications. This focus on performance rather than prejudice is essential to achieving excellence as a project manager.

Discrimination

Project managers must constantly discriminate. They discriminate between best and worst practices, high and low-performing team members, efficient and wasteful processes, good work and unacceptable work from subcontractors, and other factors that affect performance. Someone with discriminating taste is a person who recognizes good food. Someone with a discriminating mind is a person who is able to separate fact from opinion. Obviously, there are types of discrimination that are appropriate. However, discrimination—as it applies in the context of diversity—is not one of them.

Diversity-related discrimination is a negative concept. In this context, discrimination means putting prejudice into action. It means allowing diversity-related factors such as race, gender, culture, and worldview to influence one's actions and decisions, a practice that is illegal, unethical, and counterproductive. Diversity-related discrimination can quickly and effectively undermine a team's performance and the morale of its team members. Wherever it exists, discrimination is an obstacle to the achievement of excellence because it undermines individual performance, team performance, and team morale.

DISCRIMINATION AND TEAMWORK.   Project teams are like football teams. In football, every time the ball is snapped eleven people on the offense have their individual assignments that must be carried out properly if the ball is to be advanced. In addition to carrying out their individual assignments, each team member is responsible for cooperating with and supporting other team members.

Take, for example, when the quarterback throws a pass. Before the ball is thrown, the offensive line is responsible for protecting the quarterback from the rush of the opponent's defensive line and blitzing linebackers. In order to do this well, the offensive linemen must cooperate with each other in mutually supportive ways. Then, if the ball is caught, these same linemen immediately switch their support to the receiver and cooperate in trying to help him score by providing downfield blocks. Consider what would happen if the offensive linemen decided they would not block for the quarterback because he is an easterner or would not hustle downfield to block for the receiver because he is a westerner. Teamwork would be undermined and an attitude of us-against-them would set in. The same thing will happen in construction project teams where diversity-related discrimination is present.

DISCRIMINATION AND TEAM PERFORMANCE.   In order for a team to excel, the project manager must lead in ways that encourage, support, and facilitate peak performance. Few things will undermine a project manager's efforts to promote peak performance faster than diversity-related discrimination. This is because team members who feel discriminated against will quickly lose their motivation. Further, the energy they need to put into their work will be drained away by their frustration over being discriminated against.

Discrimination affects its victims on a deep and personal level. Because of this team members cannot give their best efforts when they are being discriminated against by others. The more energy team members invest in fighting discrimination, the less they will have to invest in achieving peak performance. Members of teams can spend their time trying to excel or they can spend it fighting discrimination, but they cannot do both. Few people have the time or energy to do both.

DISCRIMINATION AND MORALE.   Of the various factors that can affect the performance of teams, few are more powerful than morale. Morale refers to the spirit of team members—how they feel about themselves, the team, the project manager, and even the project itself. Team members with high morale exhibit such characteristics as trust, loyalty, and pride in the team. High morale is synonymous with team unity. On the other hand, team members with low morale are not likely to excel. In fact, the opposite is more likely to happen. Typically, low morale eventually results in poor performance.

Obviously maintaining the highest possible morale should be a goal of project managers. Morale is a function of various factors, one of the most important being the perceptions of team members concerning how they are treated by the project manager and their team mates. Members of project teams want to be appreciated, respected, and recognized for their contributions to the team. Team members with low morale typically feel as if the project manager does not care about them and does not appreciate, respect, or properly recognize them. It should come as no surprise then that discrimination results in low morale.

Discrimination against certain team members by the project manager is certainly evidence that he does not care about them, but discrimination goes well beyond just not caring. In fact, it is an oppressive practice that involves denying opportunities to some team members while providing them to others without even considering talent, motivation, commitment, or any other performance-related factors. Such a practice is not just illegal and unethical, it is counterproductive. Team members who believe they are being discriminated against are likely to suffer from low morale, and low morale is the enemy of team performance. Worse yet, low morale is contagious. Like chickenpox it can spread through contact. Team members suffering from low morale will soon spread their *disease* to other team members. The inevitable result of low morale is poor performance.

CONSTRUCTION PROJECT MANAGEMENT SCENARIO 14.1

I think you are prejudiced

Mack Murphy is known as a project manager who runs a tight ship. He always develops a detailed schedule and monitors closely to make sure that the project stays on schedule. Murphy monitors his team's budget down to the penny. He is also very particular when it comes to choosing team members for his project teams. Murphy has a well-deserved reputation for getting projects done on time, within budget, and according to specifications. But, there is a disturbing side to Mack Murphy that has become apparent only in recent months as his company's ratio of female construction professionals has increased. Murphy refuses to have women on his project staff teams. When asked about the issue, Murphy always responds, "I just pick the best people for my projects. It doesn't matter to me whether they are men or women. I just want people who will give me their best." This explanation worked for a while, but when he refused to select a high-performing women for his team—a women who is well-thought-of by other project managers in the company—eyebrows were raised. This time when asked about the issue, his standard answer did not ring true. Murphy's supervisors said, "Mack, I think you are prejudiced."

Discussion Questions

In this scenario, Mack Murphy—an excellent project manager—does not seem to want to have women on his project teams. Have you ever worked with someone who appeared to be prejudiced against people in some identifiable group? If so, discuss the situation. Did it cause problems? Have you ever been the victim of prejudice yourself? If so, discuss any aspect of the situation you are comfortable talking about? If you were Mack Murphy's supervisor, what would you tell him about the potential negative effects of prejudice?

Tolerance

The diversity-related concepts explained so far—prejudice, stereotyping, labeling, and discrimination—are all negative. The concept explained in this section—tolerance—is positive. At least for as far as it goes. Tolerance is a willingness to attempt to interact positively with people in spite of differences (e.g., race, gender, national origin, education level, political orientation, point of view). People who are tolerant do not automatically reject another person, opinion, or perspective on the basis of diversity-related factors. As a result, tolerant people are more likely to make decisions on the basis of performance than on diversity-related factors.

Tolerance is a step in the right direction in that it moves people away from prejudice, stereotyping, labeling, and discrimination. But, there are two potential problems with tolerance that project managers should understand and avoid: (1) in project teams striving for excellence simply tolerating others is not enough and (2) well-intentioned people can misinterpret the concept and end up tolerating behaviors that are unacceptable. Project managers should understand these two potential problems and take appropriate steps to prevent their occurrence.

TOLERANCE IS NOT ENOUGH.   It is often said that diversity is an asset to organizations. In fact, some organizations adopt slogans such as "…our diversity is our strength." A more accurate rendering of this philosophical ideal would be that diversity can be an asset. It is a concept with enormous potential for good, but potential seldom becomes reality without a concerted effort to make it so. The hard truth about diversity is that it becomes an asset only when it is handled well. Project managers who do a poor job of handling diversity will not just miss out on its potential benefits they will introduce diversity-based conflict into their teams. When it comes to making diversity an asset, tolerance is a step in the right direction but it is not enough.

In actual practice, tolerance too often amounts to people just putting up with each other, for the sake of harmony. But just achieving a forced harmony is not going to construction project teams more productive, and putting up with others is not the same thing as embracing them. Too often the unstated message behind tolerance is: *I will put up with you but only because I have to*. In practice there is often a begrudging aspect to tolerance. Begrudgingly putting up with others is certainly better than overtly or covertly acting on prejudice, but it is not enough to help a team achieve excellence. To enjoy the full benefits of diversity, project managers and their teams must go beyond just tolerating diversity it to embracing it. This is an important point for project managers to understand and help their team members understand.

Embracing diversity is tolerance taken to the highest level. People who embrace diversity do not just put up with others who are different. Rather, they seek them out and ask for their opinions, perspectives, and input. People who embrace diversity understand that human differences—race, gender, background, experiences, culture, education—can produce differences in perspectives, points of view, and opinion. These differences can, in turn, produce fresh ideas and different ways of looking at problems. Consequently, people who embrace diversity do so not just because it is the right thing to do but because it is also the smart thing to do.

People who are uniform in their perspectives, opinions, and points of view will also be uniform in their thinking. Consequently, they will tend to see problems in the same way. Give them a problem and they are all likely to see the same cause and same solution. But people with diverse perspectives are more likely to achieve a 360 degree view of the problem. They will see causes others miss and, as a result, be in positions to suggest solutions that others might overlook. Uniformity tends to narrow the possibilities when trying to solve problems or find ways to improve performance. Diversity expands the possibilities.

The opinions of project managers and their team members are informed by their individual backgrounds and experiences. This is why different people can look at the same problem and see different causes and solutions. One person, no matter how talented, can see a problem from just one perspective—his or her own. But a diverse group of people will view the problem from a variety of perspectives. This diversity of perspectives can lead to better decisions, better ideas, and better solutions. These, in turn, lead to better performance. Project managers can learn to embrace diversity and help their team members do the same by applying the following strategies:

- Talking with team members about diversity and making sure they understand its potential for improving the team's performance
- Making diversity one of the criteria when selecting the members of project teams
- Seeking out a broad base of opinions before making decisions
- Encouraging team members to play devil's advocate during team discussions to encourage differing points of view

- Encouraging team members to break out of their comfort zones and seek out team mates who are different from them
- Recognizing and rewarding team members strictly on the basis of performance rather than diversity-related factors
- Setting an example of embracing diversity

TOLERATING THE UNACCEPTABLE IS A MISTAKE.   Tolerance is a practice that can enhance a team's performance, especially when it is taken to the level of embracing diversity. When promoting tolerance among team members, project managers should keep its potential impact on performance foremost in their minds. This is important because when it comes to the practical application of tolerance, some project managers do not understand the concept. It is not uncommon for project managers to be so anxious to prove they are tolerant that they tolerate behavior that is unacceptable. One of the things tolerant project managers should not tolerate is behavior that undermines performance. Not only does this misapply the concept, but it can also quickly undermine its credibility.

The race, gender, background, culture, or education level of counterproductive or even nonproductive team members should never be used as an excuse for ignoring poor performance. Unproductive work practices and counterproductive behavior in a team are wrong, and should not be tolerated. Tolerance does not mean giving team members who do not contribute a free pass because of diversity-related factors. Tolerance practiced properly means embracing people who are different and seeking out their perspectives for the purpose of finding better solutions to problems and better ways to improve performance.

PREJUDICE IS LEARNED BEHAVIOR THAT CAN BE UNLEARNED

Prejudice in project teams is both unethical and counterproductive. In spite of this, there are still people in the field of construction, as well as all other fields, who struggle with embracing or even tolerating diversity. People who are uncomfortable working in a diverse environment are not necessarily bad people, but they have learned some bad habits. This concept of learned behavior is important for project managers to understand because it applies to them personally as well as to their team members.

People are not born prejudiced, but some learn to be. To test this theory, consider the example of young children. Left to themselves, little children will happily play with each other without concern for race, gender, or other differences. They are just happy to have playmates. It is only as they grow older that they learn, from a variety of different sources, to become suspicious of people who are different from them. These suspicions eventually produce negative attitudes that can, in turn, harden into prejudice. For this reasons, it is important for project managers and their team members to understand that prejudice is a learned behavior that can, like all learned behaviors, be unlearned. If people can learn negative behaviors they can learn positive behaviors. Once they learn positive behaviors, they can substitute them for the negative and, in the process, overcome them. This relearning process applies to prejudice in the same way it applies to any form of negative behavior.

OVERCOMING PREJUDICE AND EMBRACING DIVERSITY

As has already been explained, prejudice, stereotyping, labeling, and discrimination are learned behaviors. They are not genetic. This is good news because, as has already been stated, what can be learned can be unlearned, just ask a former smoker. The comparison

> ### Strategies for
> ### EMBRACING DIVERSITY
>
> - Focus on character not race, gender, culture, or other differences
>
> - Look for common ground with others
>
> - Focus on what really matters
>
> - Relate to people as individuals

FIGURE 14.3 Project managers should learn to embrace diversity and make it an asset.

with smoking is appropriate because giving up long-held prejudices can be as hard as giving up smoking or any other bad habit. It requires commitment, persistence, and a willingness to change. Further, it requires replacing a bad habit with something better.

Effective strategies for unlearning prejudice and embracing diversity include: (1) focus on the character of people rather than race, gender, culture, or other differences; (2) look for common ground with others; (3) focus on what really matters; and (4) relate to people as individuals (see Figure 14.3). These strategies can be used by project managers for overcoming their own prejudices and for helping team members overcome theirs.

Of course, stating these four strategies is easier than applying them. Remember, breaking bad habits and unlearning negative behaviors requires persistence, patience, and concerted effort. There will be ups and downs. There will be times when people appear to take one step forward and two back. When trying to break bad habits, taking two steps forward and one back is normal—it is to be expected. However, once started down the right path, project managers should keep going until they have replaced learned prejudice with a willingness to embrace diversity.

Focus on Character Rather Than Diversity-Based Differences

The various ways people can be different—race, gender, culture, age, background, politics, nationality, education—are the wrong things to consider when forming opinions of people. A better approach is to focus on the traits that really make people who they are, those known as *character traits*. Character traits that are especially important for project managers and their team members include the following (Figure 14.4):

- Honesty and integrity
- Selflessness
- Dependability and trustworthiness
- Initiative
- Tolerance/sensitivity
- Perseverance

Race, gender, culture, and the other ways people can be outwardly different are not character traits. These factors are either determined by birth or strongly influenced by it. They do not make people who they are. Character traits, on the other hand, are grounded in choice. People cannot choose their race, nationality, or culture of origin but they can choose to be honest, selfless, dependable, and so on. There are people of every race, both genders,

---

**CHECKLIST OF
IMPORTANT CHARACTER TRAITS**

- Honesty and integrity

- Selflessness

- Dependability and trustworthiness

- Initiative

- Tolerance/sensitivity

- Perseverance

FIGURE 14.4 Project managers should learn to focus on character rather than differences.

and all cultures who are honest, dependable, and selfless, and, of course, the obverse is also true. It is character, not race, gender, or culture that makes people who they are.

Look for Common Ground with Others

People who are willing to look beyond the superficial differences of others will typically find that they have more in common with them than differences. This is the case, even when the people in question appear on the surface to have little or nothing in common. Even people from different countries who speak different languages, are of different races, and have different cultural backgrounds have more similarities than differences. People tend to share the same desires, hopes, fears, and needs regardless of their apparent differences. The key to finding common ground with other people is to look past the surface-level differences and get to know them well enough to see the similarities.

In a construction project team composed of people of different races, genders, cultures, politics, and levels of education, there will still be plenty of common ground on which to build positive working relationships. This common ground will be both personal and professional. On a personal level, the members of the team will share the same concerns and worries about their families, homes, the economy, and other issues that affect their daily lives. In addition, they will have outside interests and hobbies as well as likes and dislikes in terms of music, food, and entertainment. All of these things are fertile ground for finding commonalities. On a professional level, the members of the team will be concerned about wages, salaries, promotions, working conditions, recognitions, rewards, and, of course, doing the best possible job on the project that brought them together.

Focus on What Really Matters

Project managers who feel the weight of responsibility and the pressure of accountability in leading teams will quickly come to realize what is important and what is not about the members of their teams. What project managers striving to complete projects on time, within budget, and according to specifications need above all else from their team members is peak performance. Once this fact is understood, the door to a whole new perspective can open up for even the most prejudiced project manager. This new perspective is gained when an individual is willing to acknowledge that what really matters are not diversity factors such as race, gender, age, and culture, but performance factors such as talent, experience, motivation, attitude, and teamwork.

This is precisely the perspective needed by project managers who work for construction firms that operate in a competitive environment. People who make project teams perform better should be the ones selected to serve on those teams regardless of such factors as race, gender, age, culture, politics, or any of the other diversity factors that make people different. In a competitive environment, project managers who make decisions on the basis of diversity factors that are unrelated to performance will lead their teams not to excellence but to mediocrity.

Relate to People as Individuals

One of the things that can make overcoming prejudice difficult is that prejudiced people can always find someone of another race, gender, age, or culture who fits their stereotypes; someone they can use to validate their prejudice. There are usually *some* people in any group who will actually display the characteristics that prejudiced people attribute to the whole group. This is one of the ways that people develop their prejudices—by attributing characteristics or behaviors they see in a few members of a group to everyone in that group.

An inescapable fact is that whenever people attribute any characteristic to everyone in a given group, they are automatically wrong. People, no matter how they might be grouped, are individuals. As was explained earlier in this chapter, even identical twins can be vastly different in a variety of ways. Consequently, the only dependable, fair, and equitable way to relate to people is as individuals rather than as members of racial, gender, age, cultural, or political groups. This is also the smart way to relate to people for project managers who are responsible for the performance of construction project teams.

In construction firms there are people of different races, ages, genders, and cultures who are positive, talented, motivated, team players who perform at peak levels. Conversely, there are also those who do not measure up. In either case, their work-related behavior is a product of their character, not their race, age, gender, or cultural background. Project managers who make an effort to relate to others as individuals will find it difficult to maintain the prejudices they have been taught by society. This, in turn, will make them better project managers.

CONSTRUCTION PROJECT MANAGEMENT SCENARIO 14.2

You need to overcome the prejudice you've learned

Before going away to college, Elizabeth Martin had led an insulated life. Growing up, the people in her neighborhood, the students in her schools, and the friends she spent the most time with were just like her. Martin's parents and the parents of her friends were all successful white-collar professionals who were able to give their children the material advantages of wealth. They talked alike, dressed alike, drove the same kinds of cars, liked the same kinds of food, were members of the same clubs, and voted for the same political candidates. There was little or no diversity in Elizabeth Martin's life from birth through high school. However, that changed radically the minute she stepped foot on her college campus.

Rather than enroll at the same elite private institution most of the young people from her neighborhood attended, Martin chose to attend her grandfather's alma mater—a public university located in a major city. As a result, overnight Martin went from a life of uniformity to a life of diversity. The starkness of the change in her life was shocking. Marin felt like she had been dropped onto an alien planet. She had never interacted with people of other races, cultures, worldviews, and perspectives. The change was overwhelming.

At first, Martin had difficulty adjusting. In fact, she became so distraught that her grades suffered and she contemplated dropping out and going home. After seeking the help of one of the universities' staff counselors, Martin decided to make some adjustments and persevere. As she walked out of his office, the counselor's words practically rung in her ears: "You need to overcome the prejudice you've learned."

Discussion Questions

In this scenario, Elizabeth Martin grew up in a corner of the world characterized by uniformity. Consequently, when she went away to college the diversity she encountered was so alien to her as to be overwhelming. To her credit, Martin decided to stay in college and work on overcoming her learned prejudice. Have you ever known someone who was uncomfortable in a diverse setting? If so, discuss the situation and its outcome. If Elizabeth Martin was your friend and she asked for advice about overcoming the prejudice she learned growing up, what would you tell her?

SUMMARY

America is one of the most diverse countries in the world and this diversity is reflected in the workplace. Consequently, project managers can expect to lead diverse teams. The term diversity as applied in the current context refers to all of the ways that people who serve on project teams can be different. Dealing with these differences in ways that make the team stronger is the responsibility of the project manager. Diversity is a concept that encompasses all of the ways that people can be different. Important diversity-related concepts include prejudice, stereotyping, labeling, discrimination, and tolerance.

Prejudice is a predisposition to adopt negative perceptions about groups of people. People who are prejudiced tend to divide the world into two groups: us and them. People in groups often act out their prejudice in negative and even harmful ways. Stereotyping is a by-product of prejudice. It is the act of generalizing certain characteristics of an individual to all members of a given group. Labeling is an extension of stereotyping in that it involves attributing a certain characteristic to a distinct group of people and then using that characteristic to label all people in that group.

Diversity-related discrimination is a negative concept. It means allowing diversity-related factors such as race, culture gender, and worldview to influence one's actions and decisions. Discrimination can quickly undermine the effectiveness of a project team and the credibility of project managers. Tolerance is a willingness to interact positively with people in spite of differences based on such factors as race, gender, culture, worldview, political beliefs, and so on. Tolerance is a step in the right direction, but as a concept it is not without problems. There are two potential problems with tolerance that project managers should understand and avoid: (1) in project teams striving for excellence simply tolerating others is not enough and (2) well-intentioned people can misinterpret the concept and end up tolerating behaviors that are unacceptable. Just tolerating differences in people is not enough to mold them into an effective team. Project managers and their team members need to embrace their differences. On the other hand, well-meaning project managers should never use tolerance as an excuse for ignoring behavior that is counterproductive.

Prejudice is learned behavior that can be unlearned. People are not born prejudiced—they learn to be as they age. Behavior that is learned can be unlearned and replaced with more appropriate behaviors. The key to overcoming learned prejudice and then embracing diversity in people include the following: (1) focus on character rather than differences, (2) look for common ground with others, (3) focus on what really matters rather than differences, and (4) relate to people as individuals.

KEY TERMS AND CONCEPTS

Diversity
Prejudice
Stereotyping
Labeling
Discrimination
Tolerance
Discrimination and teamwork
Discrimination and team performance

Discrimination and morale
Embracing diversity
Tolerating the unacceptable
Learned behavior
Embracing diversity
Diversity-based differences
Character
Common ground

REVIEW QUESTIONS

1. Define the term "diversity" as it relates to project management.
2. Why is it important for project managers to be able to lead diverse teams?
3. Define the term "prejudice."
4. Explain the concept of stereotyping. Give an example.
5. What is meant by labeling? Give an example.
6. Define the term "tolerance."
7. Explain the potential problems associated with tolerance.
8. Explain how discrimination can undermine teamwork.
9. Describe the effect discrimination can have on morale in project teams.
10. Explain what is meant by the phrase "prejudice is learned behavior…"
11. What is meant by embracing diversity? Explain how one goes about embracing diversity.

APPLICATION ACTIVITIES

The following activities may be completed by individual students or by students working in groups:

1. Have you ever experienced or observed an act of prejudice, stereotyping, labeling, or discrimination? Write down or discuss the situation. What happened? Were you the victim or just an observer? If you were the victim, how did the act of prejudice make you feel?
2. Have you ever experienced or observed tolerance being taken too far (i.e., someone tolerating behavior that was counterproductive)? Write down or discuss the situation. What happened? Why do you think the individual who used tolerance as an excuse for overlooking counterproductive behavior did so?
3. Do the research necessary to create a list of stereotypes that existed at one time but either no longer exist or are dying out. Discuss your list with other students.
4. Identify a project manager in a construction firm who is willing to cooperate in completing this project and discuss the issue of diversity with this person. Determine if he or she has ever experienced diversity-related problems that affected the morale or performance of the team. If so, how was the situation handled?

CHAPTER 15

Managing Adversity in Construction Projects

Difficult times are not uncommon in the construction business. Like other businesses, construction goes through peak periods and low periods depending on the state of the economy and other factors. This is one of the reasons that this book has stressed the need for project managers to complete their projects on time, within budget, and according to specifications. The ability to do so is what makes a construction firm competitive.

Being competitive is critical for construction firms because in good times and bad, the more competitive a construction firm is the better it will do. In the event of a prolonged recession, being competitive can mean the difference between survival and bankruptcy for a firm. But even in good economic times construction firms can face adversity from other sources. This is the nature of business. For example, a construction firm with an excellent environmental record might inadvertently spill a toxic substance at a jobsite. The resulting fines, cleanup costs, litigation, and bad publicity can put the company through a period of intense adversity. A firm with an excellent safety record might suddenly experience a tragic accident in which several of its workers are critically injured. The resulting litigation, medical costs, decline in employee morale, and damage to its reputation can put the construction firm through an extended period of adversity. Consequently, project managers in construction need to be prepared to lead their teams during times of adversity.

Construction projects like baseball games in that they consist of more than just one inning or tennis matches in that they consist of more than one set. In a baseball game, each batter comes to the plate several times and even the most talented player strikes out occasionally. In addition, players occasionally have to hit the dirt when the opposing pitcher brushes them back with a fastball thrown high and inside. But the best players do not let these temporary instances of adversity keep them down or make them quit. Instead, they pick themselves up, brush themselves off, and step right back up to the plate. They persevere no matter how difficult the game becomes and no matter how many times they strike out. Refusing to quit and never giving up are essential when leading teams through periods of adversity.

What applies to the construction firm applies to its project teams. There are going to be problems over the course of a construction project. Projects do not always run smoothly or as originally planned. Project managers and their team members will face adversity. Consequently, a willingness to persevere in times of adversity is critical to success as a project manager. It is one more factor that separates the best project managers from the mediocre.

One of the reasons champions become champions is because they refuse to give up and they never quit. On the other hand, even champions are human. Like all people, during times of adversity, they can become discouraged. But what separates champions from the mediocre is that they refuse to give in to discouragement. The characteristics of the most successful project managers are listed as follows: (1) they view setbacks as opportunities for comebacks, (2) they learn from mistakes and do better next time, (3) they refuse to accept failure as a permanent condition, and (4) when knocked down, they get back up and try again.

In a competitive environment, construction firms need project managers who are willing to persevere in spite of obstacles, setbacks, difficulties, unexpected problems, crises, or any other kind of adversity. Project managers, in turn, need team members who are willing to do the same. Having project managers who set a consistent example of persevering during times of adversity is essential to construction firms and their project teams. Following the project manager's example is how team members learn to persevere. Members of project teams who witness their project manager maintaining a positive can-do attitude during difficult times will have the right example to follow.

Every construction firm experiences adversity. This, in turn, means that every project team will experience adversity. Consequently, having project managers who can stay positive and persevere in spite of the difficulties is essential to getting projects completed on time, within budget, and according to specifications. Contracts will be revised, important team members will resign, deadlines will change, hardware will break down, bad weather will intercede, corporate mergers and buyouts will occur, shortages of materials will happen unexpectedly, work stoppages and strikes will be called, and the list goes on and on.

There is never a scarcity of adversity in construction firms that operate in a competitive environment, and adversity that affects the firm eventually affects its project teams. For this reason, construction firms need project managers who view adversity as a situation that comes with a gift in its hand. That gift is the opportunity to: (1) grow stronger by persevering and weathering the storm and (2) get better by learning from the experience and being better prepared next time. Project managers who view adversity in this way and pass their good example on to their team members will be valuable assets to their construction firms.

DO NOT GIVE UP AND NEVER QUIT

Excellence in project teams is not just a matter of talent. There are plenty of talented people who never achieve excellence because when the work becomes difficult—when faced with adversity—they become overwhelmed, disheartened, and just give up. This is unfortunate because the best results are achieved by those who are willing to keep trying—those who refuse to give up and quit. Rarely are construction projects easy and rarely do they go exactly according to plan. There is almost always some degree of adversity associated with completing construction projects.

A good example of the benefits of perseverance is America's Olympic women's softball team. The team had to work long and hard just to get the sport accepted as an Olympic event. Turned down numerous times, team leaders became frustrated, discouraged, and even angry, but they persevered and refused to quit. As a result, they were eventually successful and softball became an Olympic event. Just getting the sport accepted by the International Olympics Committee was a hard-won victory. But America's team did not stop there. Having won the battle for Olympic recognition the softball team went on to win three consecutive gold medals in its sport.

Examples abound of perseverance paying off. Think of the boxer who loses the first nine rounds of the fight, but perseveres and wins by a knockout in the 10th round. Think of all the times baseball games have been won when by a walk-off home run in the bottom of the ninth inning. Think of football games won by a field goal as the clock winds down in overtime. Think of great basketball games won at the buzzer by a desperation shot tossed up at the last possible second. In all of these examples, victory went to the team that persevered, that refused to give up and quit. Sometimes success is just one step beyond the last step you think you can take.

These examples come from the world of sports because that is where the best-known and most dramatic instances of victory through perseverance can be found. But the never-quit-never-give-up attitude applies just as directly to construction teams as it does to sports teams. Consequently, it is just as important for project managers and their team members to learn to persevere as it is for athletes. There will be times when changes will throw the project team's work into disarray, when so much work will pile up that team members begin to doubt they can ever get it done, and when long hours will result in fatigue and frustration. It is during times such as these that the willingness to persevere distinguishes project teams that excel from those that never get beyond mediocrity.

Perseverance Strategies

Perseverance means hanging in there a little longer when the natural inclination is to give up and quit. Persevering is more of a mental than a physical exercise. In fact, perseverance can be thought of as the mental equivalent of physical stamina. Achieving excellence in teams requires that team members do a lot of things well, often difficult things. The road to excellence is never easy, nor is it final. Once achieved, excellence must be achieved again and again on each successive project. Market forces, competition, supplier problems, the economy, close schedules, tight budgets, demanding specifications, technological developments, and a variety of other factors make achieving excellence a never-ending challenge for project managers and their team members.

Consequently, organizations need project managers who can bear up under the never-ending pressure of leading project teams and keep going in spite of anxiety, frustration, fear, uncertainty, and fatigue, both mental and physical. The following strategies will help project managers and their team members when faced with the challenge of persevering during times of adversity:

- When facing an intractable problem and it seems that everything has been tried without results, remember the lesson of the great inventor, Thomas Alva Edison. In trying to invent such useful products as the storage battery and a durable filament for the light bulb, he failed repeatedly. It is said that it took him almost 25,000 attempts to finally succeed with just these two inventions. But when others might have quit, Edison refused to give up. Instead, he persevered, and finally succeeded. The world can be thankful he did.

- When a team member fails at something important and he is discouraged, remind him that every time he tries something and fails, he is better prepared to succeed the next time. His failed attempt does not become a failure unless he quits. A failed attempt is just another opportunity to try again better prepared the next time.
- When a team member is unsure of her ability to complete a given assignment or meet a given challenge and when the consequences of failure are frightening, encourage her to focus on what will happen when she succeeds rather than what might happen if she fails. Tell her to stay focused on the ball that is in play, not on the scoreboard. Then play to win rather than to avoid losing.

CONSTRUCTION PROJECT MANAGEMENT SCENARIO 15.1

My team members give up too easily

Juan Padrone had to struggle to work his way out of poverty. Even as a child he worked to help his mother support their family after his father died in a tragic accident. Then he worked himself through college, secured a good job, and worked hard to climb the career ladder. Consequently, Padrone knows about persevering during hard times. As a result, he has high expectations of his team members when the work on a project becomes difficult. Padrone expects his team members to persevere when they face adversity. Unfortunately, his team members lack his experience at dealing with difficult times.

Padrone's team is going through tough times right now and he is not happy with how his team members are responding. Team members seem to have developed a defeatist attitude toward their work. They come to work, but just seem to be going through the motions without really trying. While discussing this situation with a colleague, Padrone shrugged in frustration and said: "My team members give up too easily. Do you have any suggestions for how I can get them to toughen up mentally?"

Discussion Questions

In this scenario, the project manager is an individual who knows all about dealing with hard times. His whole life has been characterized by the need to persevere against adversity. Consequently, he does not understand people who are unwilling to persevere. Have you ever been in a situation in which the people involved wanted to quit in the face of adversity? If so, explain the circumstances and what eventually happened. If you were Juan Padrone's colleague in this scenario, what advice would you give him?

FACE ADVERSITY AND OVERCOME IT

There is a tendency to think that success comes easily for those who make a name for themselves in a given profession. In reality, this is hardly ever the case. Further, the concept of the overnight success is typically a myth. Few people who succeed do so overnight. Most so-called overnight successes are people who struggled long and hard to finally succeed. Successful people are rarely strangers to adversity. An excellent example of facing adversity and overcoming it is Franklin Delano Roosevelt, President of the United States, leading up to and during World War II.

Roosevelt was elected President when the United States was being squeezed in the economic vise of the Great Depression. Unemployment was at its highest level in America's history, the nation's banking system had crashed, small businesses were closing daily, people were losing their homes because they could not pay their mortgages, the Midwestern farming states had become a vast dust bowl no longer suited for farming, and many people woke up every morning wondering what, if anything, they would have to eat that day.

Into this bleak picture stepped Franklin Delano Roosevelt, the former governor of New York and a man of great optimism. Soon after taking office he began to use radio broadcasts he called "fireside chats" to reassure Americans that the economy would pick up again and their lives would get better. He spoke to the country in such calm and optimistic terms that Americans began to gain a sense of hope. Then, before any of the President's economic recovery programs had time to produce significant results, the Japanese attacked Pearl Harbor and the United States found itself embroiled in World War II.

Faced with an even bigger crisis than the Great Depression, President Roosevelt again calmed the anxiety of Americans. In a nationally broadcast address he set a confident and hopeful tone when he said: "All we have to fear is fear itself." Roosevelt used the same calm optimism to face the adversity of World War II that he had used to face the adversity of the Great Depression. The President's optimism during times of unprecedented adversity is a story unto itself, but what is even more instructive about his example is that he held the country together while suffering from a severe case of polio. President Roosevelt—a man who portrayed himself as a robust, vibrant leader—could not even walk. But he hid this fact so well that many Americans were not even aware of it until after his death.

In spite of having to struggle daily against the increasingly debilitating and painful effects of a crippling disease, Franklin Delano Roosevelt remained calm and optimistic in the face of tremendous adversity. He could have simplified his life and eased his daily pain by agreeing to use in public the wheel chair he used in private. But the attitudes of Americans toward people in wheelchairs were less enlightened in those days, not to mention the attitudes of America's enemies. Roosevelt knew that the times and circumstances demanded an image of strength. Consequently, he endured incredible pain and inconvenience to provide that example.

Not only did the president fight courageously every day to win the war and revitalize the nation's economy, but he did it while supporting himself, when in public, with heavy metal braces worn under his trousers that bit painfully into his frail, useless legs. The leg braces and the façade of healthful vigor he felt it necessary to maintain only added to the adversity this courageous man faced every day. But in spite of the pain and inconvenience, Roosevelt maintained the necessary façade from the moment he was elected President of the United States until the day his heart finally gave out during a rest-and-recuperation visit to his "Little White House" in Warm Springs, Georgia.

One might wonder why a wealthy President from Hyde Park, New York, would have a vacation retreat in the rural, poverty-stricken town of Warm Springs, Georgia. The reason reveals even more about the incredible courage and perseverance of this unique individual. Using his own money, Roosevelt had founded a treatment center at Warm Springs for polio victims. The naturally occurring warm springs that bubbled up in this tiny, out-of-the-way Georgia town had a stimulating effect on the withered limbs of polio victims, especially children. Although serving as President of the United States during some of America's darkest hours left him no time to undergo the physical therapy offered at his own treatment center,

Roosevelt wanted to make sure that future generations of polio victims had a chance to receive the treatments. Thanks to Roosevelt, they did.

Project managers facing periods of adversity would do well to remember the example of President Franklin Delano Roosevelt and set a similar example for their team members. They should stay positive and optimistic, be calm and reassuring, and focus on solutions rather than problems. No matter how difficult the problems they must face become, project managers should provide a consistent, calm, and resolute example of perseverance that sends the message, "We can get through this." Of course, this kind of example is important at all times, but it is even more important during times of adversity. During times of adversity, project managers are just like emergency room doctors. They have to maintain their calm and their focus no matter what is going on around them because others will take their cues from them.

Challenging Construction Project

Great Pyramid of Giza

Try to image managing a construction project that will take 20 years to complete, will be the tallest building in the world for more than 3,800 years, will encompass a volume of approximately 2,500,000 cubic meters, and will be made of more than two million limestone blocks, some of which will weigh more than 80 tons. Try to imagine using 5.5 million tons of limestone, 8,000 tons of granite, and 500,000 tons of mortar on the project. Now try to imagine doing all of this without trains or trucks to transport the materials, earthmoving equipment to excavate the site, backhoes to dig the foundation, or cranes to lift and set the granite blocks. This was the challenge faced by Hemiunu, vizier to Khufu, the Egyptian pharaoh whose tomb the building was supposed to be. The building, of course, is the Great Pyramid of Giza, the oldest of the Seven Wonders of the Ancient World. In the language of project management, Hemiunu was the project manager and Khufu was the owner.

It is necessary to imagine how this incredible feat of construction was accomplished because even after many years of study, experts still do not agree on what construction techniques were used to build the Great Pyramid. Some Egyptologists belief the massive limestone blocks were mined a quarries and transported to the sight. Others believe the blocks were made of crushed limestone formed and poured on site in the manner of poured-in-place concrete. Regardless of how the limestone blocks—weighing 25 to 80 tons—were procured they still had to be hoisted and put in place which was quite a feat of construction ingenuity for crews that had no powered equipment.

Construction experts are still mystified by the Great Pyramid. Some of the more frequently asked questions about its construction that still have not been answered conclusively include: 1) If the largest of the limestone blocks used in constructing the Great Pyramid were mined in a quarry in Aswan more than 500 miles away, how were the enormous blocks transported to the sight? and 2) Once the limestone blocks were on sight, how were they hoisted into place and set as part of the structure? The Great Pyramid would be a challenging construction even if built today. Imagine ingenuity and determination required to build it more than 4,000 years ago.

Source: Based on John Romer, The Great Pyramid: Ancient Egypt Revisited (Cambridge, Massachusetts: Cambridge University Press, 2007).

ADVERSITY IS A NORMAL PART OF WORK AND LIFE

If achieving excellence in project teams were easy, all teams and all project managers would complete all projects on time, within budget, and according to specifications. Unfortunately, achieving excellence is seldom easy. Rather, it is difficult to achieve and even more difficult to maintain over time. There are and always will be plenty of obstacles on the road to excellence—inhibitors that undermine performance. It is when dealing with these inevitable obstacles that a willingness to persevere will separate effective teams from mediocre teams. This is why it is so important that, during hard times, project managers set a calm and reassuring example for their team members that says "Adversity is normal—just keep going."

In a competitive environment, adversity is often the rule rather than the exception. Consequently, the project managers and team members most likely to thrive in such an environment are those who can maintain their focus, positive attitude, and solution-oriented outlook in spite of the difficulties. People who approach adversity as if it is a one-time event to be survived rather than as the normal state of things to be dealt with on a regular basis will eventually feel overwhelmed and give up. This is important because in a competitive environment, adversity can be the normal state of things. Even in the best of times there will be periods of adversity. This means that project managers and their team members must learn to deal with adversity in a positive way as a normal part of doing their jobs.

Some strategies that will help project managers and their team members persevere in approaching adversity as a normal part of their work are as follows:

- Accept that adversity is to be expected.
- Refuse to get caught up in the here and now of the situation—look down the road past the difficulties.
- Focus on solutions rather than the problems.
- Develop a course of action for getting past the difficulties faced and implement it.
- Prepare physically and mentally for the next crisis and accept that there will be one.
- Stay positive and take adversity in stride.

Not only should project managers internalize these rules of thumb, but they should also teach them to their team members through both conversation and example.

ACCEPT THAT LIFE CAN BE UNFAIR

One of the factors that can make it difficult for people to deal with adversity in a positive manner is the seeming unfairness of it. People facing adversity often wonder, "Why is this happening to me? It's not fair." While it might be true that the circumstances in question are unfair, getting caught up in the unfairness of adversity is a mistake. Focusing on the unfairness of a situation can cause people to become discouraged and give up. This is a point that project managers should understand and help their team members understand.

During times of adversity, it is important for project managers to help their team members understand that they are not alone. The maxim about misery loving company certainly applies in project teams facing adversity. A bad situation can seem less unfair when it is known that others are facing similar circumstances. If the project team is going through hard times, all team members are probably affected. Perhaps in different ways, but they are all affected nonetheless. Making this point to team members can help them see that they are not alone in dealing with adversity.

Watch people in the aftermath of a natural disaster such as an earthquake, tornado, or hurricane. Amazingly, people who have lost everything can be seen working to help others restore a semblance of normalcy to their wrecked lives. Knowing that they are not alone in their trials and that there are others going through the same difficulties bolsters people and gives them hope. But when people feel isolated in their misery, the unfairness of the situation can give rise to frustration, hopelessness, and despair. People who feel this way are inclined to give up and quit.

This is why it is important for project managers to let their team members know that work and life are not always fair, and they should not expect them to be. Further, no matter how hard well-intended people try to make them so work and life are not likely to ever be completely fair. Project managers should endeavor to be fair and equitable with their team members, colleagues, and customers. Construction firms should adopt policies, procedures, and practices that ensure as much fairness as possible for their personnel. But project managers should never lose sight of the fact that bad things do happen to good people, to good teams, and to good construction firms. This fact should also be understood by team members. When the concept of unfairness is understood and accepted, team members will be better prepared to deal with the adversity they will inevitably face.

HELP TEAM MEMBERS WHO ARE FACING ADVERSITY

When facing adversity, it is easy to become self-absorbed. After all, the fear, frustration fatigue, and uncertainty associated with adversity are felt on a deeply personal level. On the other hand, people who respond in a self-absorbed way to adversity are not likely to get through it. Experience shows over and over that one of the most effective ways to deal with adversity is to help someone else who is facing difficulty. This is a strategy that project managers should adopt themselves and teach to their team members.

In any construction firm, if one project team is facing hard times there are probably others going through similar difficulties. Reaching out to other project managers who are facing budget, schedule, or personnel problems may be the best way there is for project managers to avoid becoming self-absorbed and giving in to gloom and defeatism. The same is true for their team members.

This concept of helping oneself by helping others should be thoroughly understood by project managers who are trying to encourage perseverance among their team members. When a project team goes through hard times, getting its personnel to focus on helping each other is an excellent way to lessen everybody's burden and to teach team members how to deal with adversity. Looming deadlines, demanding, tight budgets, poor-performing subcontractors, late deliveries, labor problems, and other adversity-inducing events are a normal part of life for project managers and their teams. Consequently, helping team members develop the willingness and ability to persevere during hard times is an essential ingredient in the formula for excellence in project teams. Project managers who wish to develop peak-performing team members should teach them, especially by example, how to persevere in times of adversity.

DEALING WITH MICROMANAGERS WHO CREATE ADVERSITY

Construction project managers must be prepared to deal with owners, architects, and higher-level managers who are micromanagers. Micromanagers can produce more than just frustration for project managers. If not handled well, they can create adversity. A micromanager is a stakeholder in a superior position who constantly looks over the shoulder of the project manager offering unsolicited input, often getting in the way and clouding the issue of who is in charge. Micromanagers often insert themselves in the everyday details of the project manager's job to an extent that is inappropriate and even counterproductive. They constantly look over the project managers' shoulders and question every detail of what they do and second guess all of their decisions.

Not only are micromanagers annoying, but they can also create problems by undermining the project manager's credibility with team members. When a stakeholder in the construction project micromanages a project manager, team members begin to wonder "who is in charge here and who do we report to?" If not handled properly, a micromanager can render a project manager irrelevant, create stress in the team, generate divided loyalties, and damage morale to the point that the team's work suffers.

The first step in dealing with micromanagers is to understand why they micromanage. There are a variety of reasons, and not all of them apply to every micromanager. However, when project managers find themselves dealing with micromanagers, the factors explained below will help in establishing an understanding of them. Once it is understood why an individual micromanages, it is easier to know how to deal with that person.

What follows are common reasons why people who micromanage feel compelled to do so. People are compelled to micromanage because of the following reasons: (1) they think no one can do the job correctly but them, (2) they struggle with letting go of the type of work they used to do (This typically applies to higher-level managers within the construction firm, but not necessarily. The author once worked on a project with a micromanaging owner who had worked as a construction project manager.), (3) they do not understand the concept of delegation, and (4) they do not have confidence in the project manager. All of these factors apply to some micromanagers, but more commonly just one or two of these factors apply. The key is for project managers to be observant enough to determine which factor or factors apply to the stakeholder who is micromanaging them.

Strategies for Overcoming Micromanagement

In general terms, project managers who must work with a micromanager should be patient, refuse to take the intrusiveness personally, and understand that the micromanager is an individual who is dealing with issues that might be deep-seated and personal. It can take time to cure a micromanager of his propensities. Project managers who work for a micromanager—who thinks only he or she can do the job correctly—should patiently demonstrate that they can do the job. They should keep the individual informed of the team's progress on a regular basis and show that the project is moving forward on time, within budget, and according to specifications. When a problem arises, project managers should explain the situation to the micromanager and describe what they plan to do about it.

Project managers must work with a micromanager who struggles with letting go of the kind of work he or she once did and should: (1) make a point of involving him or her in ways that will satisfy his or her need to stay in touch with the work (e.g., periodic meetings, requests for advice, frequent e-mail updates) and (2) involve him or her in the problem-solving process when unexpected difficulties arise. When this kind of micromanager realizes that he or she is going to be involved and engaged in appropriate ways, the need that is driving him or her to micromanage will be satisfied or at least will be partially so. With persistence and patience the need that drives an individual to micromanage can be satisfied sufficiently to at least give the project managers room to do the job.

Project managers who work for superiors in the construction firm who micromanage because they do not understand the concept of delegation have an upward-mentoring challenge. Mentoring is usually viewed as more senior, more experienced people helping less experienced people develop the knowledge, skills, and attitudes they need to succeed in their jobs. A less known kind of mentoring is upward mentoring: when a subordinate mentors a superior. Upward mentoring is a concept that should be labeled "handle with care." Project managers who approach a superior and say "It is obvious you don't understand the concept of delegation so I am going to teach you" are not likely to get a good reception. Consequently, a more subtle approach must be used.

An effective way to approach this issue is to begin giving the micromanager the following message: "I appreciate your help, but know how busy you are. You can delegate this task to me and I will keep you informed of progress." Once the micromanager is able to see that the project manager is getting the work done, and keeping him informed, he will begin to form a better appreciation of the concept of delegation. However, patience is in order because it can take micromanagers' time to break old habits.

When a project manager senses that a superior is micromanaging because of a lack of confidence, the challenge is straightforward and obvious: demonstrate competence. Concentrate on getting the work of the project done on time, within budget, and according to specifications. In other words, project managers in this situation should simply prove to the micromanager that they can do the job well. This will happen as a matter of course over the duration of the project. It might take completing an entire project or even more than one project before the micromanager becomes a believer, but with patience and persistence it will happen.

CONSTRUCTION PROJECT MANAGEMENT SCENARIO 15.2

My team is not handling adversity very well

Janice Carter had been through tough times before at ABC Construction, Inc. Because the economy had been in a recession for an extended period of time, ABC is accepting almost any contract it can get. The company is getting enough work to keep it going, but taking on any and all projects is creating problems. Budgets and schedules are tight and vacant positions are not being filled. Carter has recently lost two members of her project staff team. Those who remain are expected to do their work in addition to that of the team members who have resigned. Carter has never been expected to do so much with so little and in so little time.

Carter is holding up well in view of the circumstances, but her team members aren't. Her team members are beginning to show signs of stress, frustration, and fatigue. Carter has tried several strategies for bolstering the morale in her team, but without much success. She is beginning to worry. When asked by a colleague how she was holding up, Carter responded "I'm doing fine but my team is not handling adversity very well. I need some help. What can I do to give them some hope until things turn around?"

Discussion Questions

In this scenario, Janice Carter is handling some tough times well but the members of her project team are not. Have you ever worked with people who did not handle adversity well? If so, describe the circumstances and how people responded to them. If you were Janice Carter's mentor, what advice would you give her concerning bolstering the morale of her team members?

SUMMARY

There is never a scarcity of adversity in organizations that are trying to compete in a competitive environment, and adversity that affects the construction firm soon affects its project teams. For this reason, construction firms need project managers who understand that adversity comes with a gift in its hand. That gift is the opportunity to grow stronger by persevering and weathering the storm and get better by learning from the experience and being better prepared next time.

Excellence in project teams is not just a matter of talent. There are plenty of talented people who never achieve excellence because when the work becomes difficult—when they are faced with adversity—they become overwhelmed and just give up. This is unfortunate because the best results are achieved by those who are willing to keep trying—those who refuse to give up and quit.

Perseverance means hanging in there a little longer when the natural inclination is to give up and quit. Persevering is more of a mental than a physical challenge. In fact, perseverance can be thought of as the mental equivalent of physical stamina. Strategies for persevering include: (1) thinking of others who have persevered and emulating their examples, (2) remembering that every time one fails at something she is better prepared to succeed the next time, and (3) focusing on the benefits of success rather than the consequence of failure.

Those who approach adversity as a one-time event will be unprepared when the next problem arises, as it surely will. Strategies for facing adversity include the following: (1) understand that adversity is a normal part of work and life, (2) refuse to get caught up in the here and now of the situation—look down the road past the difficulties, (3) focus on solutions rather than problems, (4) develop a course of action for getting past the difficulties and implement it, (5) prepare mentally and physically for the next round of adversity, and (6) stay positive and take adversity in stride.

It is important for project managers and their team members to understand and accept that life is not always fair. Getting hung up on the unfairness of life is a mistake. One of the best ways to respond to adversity is to find someone who is going through hard times too and help that person. Helping someone else who is facing difficulties will keep one from becoming self-focused and giving in to self-pity.

Micromanaging superiors in the organization can cause difficulties for project managers. This is a sufficiently common problem that project managers need to understand why some people feel compelled to micromanage as well as how to deal with micromanagers. People who micromanage do so because they: (1) think no one can do the job right but them, (2) struggle with letting go of work they used to do, (3) do not understand the concept of delegation, and (4) have no confidence in the project manager. Project managers who report to a micromanager should be patient and refuse to give in to frustration. They should attempt to determine which of these reasons apply in their case and then do what is necessary to overcome that reason.

KEY TERMS AND CONCEPTS

Adversity

Do not give up

Never quit

Perseverance strategies

Face adversity

Adversity is normal

Focus on solutions

Stay positive

Take adversity in stride

Life can be unfair

Help team members

Micromanagers

REVIEW QUESTIONS

1. What is one of the most important factors that make champions into champions?
2. Why do construction firms need project managers who are willing to persevere?
3. Explain why talent alone will not produce excellence in a project team.
4. What is meant by the term "perseverance"? Give an example.
5. Explain three strategies for persevering in times of adversity.
6. What lesson does the life of President Franklin Delano Roosevelt teach about dealing with difficulties?
7. Explain what is meant by the phrase "adversity is a normal part of life."
8. List six strategies that will help project managers and their team members persevere when facing hard times.
9. Why is it important for project managers and their team members to understand that life is not always fair?
10. How can helping someone else who is facing difficulty help one to overcome adversity?
11. Explain the various reasons why some people become micromanagers.
12. How should project managers deal with superiors and stakeholders who are micromanagers?

APPLICATION ACTIVITIES

The following activities may be completed by individual students or by students working in groups:

1. Assume that you are a new project manager. On your first project, the team has run into several difficulties. Some of these difficulties are listed as follows: (1) contract revisions have shortened an already tight schedule by two months and (2) an important member of the project team has been injured in an automobile accident and will be out of work for six months. The team is not responding well to the adversity. Frustration is setting in, tempers are getting short, and complaints are mounting. Explain how you will go about leading the team to persevere through the hard times.
2. Assume you are an experienced project manager, but for the first time you have to report to a micromanager who does not seem to be able to let go of the kind of work he used to do. His micromanaging is causing problems in your team. Team members are beginning to question who is in charge. Explain how you will go about reclaiming control of your team and project.

INDEX

Note: Page numbers followed by "*f*" indicate figures; those followed by "*t*" indicate tables